Current Progress in Cancer Cell Signaling

Current Progress in Cancer Cell Signaling

Editor: Molly Cullison

AMERICAN
MEDICAL PUBLISHERS
www.americanmedicalpublishers.com

AMERICAN
MEDICAL PUBLISHERS
www.americanmedicalpublishers.com

Cataloging-in-Publication Data

Current progress in cancer cell signaling / edited by Molly Cullison.
    p. cm.
Includes bibliographical references and index.
ISBN 978-1-63927-646-2
1. Cancer cells. 2. Cellular signal transduction. 3. Cancer cells--Regulation.
4. Cancer--Genetic aspects. 5. Cancer--Treatment. I. Cullison, Molly.
RC269.7 .C87 2023
616.994 07--dc23

American Medical Publishers,
41 Flatbush Avenue,
1st Floor, New York,
NY 11217, USA

ISBN 978-1-63927-646-2 (Hardback)

Contents

Preface

The world is advancing at a fast pace like never before. Therefore, the need is to keep up with the latest developments. This book was an idea that came to fruition when the specialists in the area realized the need to coordinate together and document essential themes in the subject. That's when I was requested to be the editor. Editing this book has been an honour as it brings together diverse authors researching on different streams of the field. The book collates essential materials contributed by veterans in the area which can be utilized by students and researchers alike.

Cancer is referred to as a type of disease characterized by the growth of aberrant cells that divide irrepressibly and have the capacity to invade and destroy normal body tissue. It has the potential to spread throughout the body. Cancer is caused by epigenetic and genetic changes which enable cells to proliferate and evade mechanisms that would usually control their migration and survival. These numerous changes map to signaling pathways that regulate cell growth and division, cell motility, cell fate and cell death. They can be interpreted in the context of larger signaling network distortions that fuel the progression of cancer, including inflammation, modifications to the tumor microenvironment and angiogenesis. The majority of cancers start in epithelial cells and manifest as carcinomas in organs like the breast, pancreas, lung, liver and skin. This book contains some path-breaking studies on cancer cell signaling. Those in search of information to further their knowledge will be greatly assisted by it.

Each chapter is a sole-standing publication that reflects each author's interpretation. Thus, the book displays a multi-facetted picture of our current understanding of application, resources and aspects of the field. I would like to thank the contributors of this book and my family for their endless support.

Editor

Crosstalk between SHH and FGFR Signaling Pathways Controls Tissue Invasion in Medulloblastoma

Anuja Neve [1,†], Jessica Migliavacca [1,†], Charles Capdeville [1,†], Marc Thomas Schönholzer [1], Alexandre Gries [1], Min Ma [2], Karthiga Santhana Kumar [1], Michael Grotzer [1] and Martin Baumgartner [1,*]🄳

[1] Department of Oncology, University Children's Hospital Zürich, CH-8032 Zürich, Switzerland; anujaneve@googlemail.com (A.N.); Jessica.Migliavacca@kispi.uzh.ch (J.M.); Charles.Capdeville@kispi.uzh.ch (C.C.); marcthomas.schoenholzer@bluewin.ch (M.T.S.); Alexandre.Gries@kispi.uzh.ch (A.G.); Karthiga.Kumar@kispi.uzh.ch (K.S.K.); Michael.Grotzer@kispi.uzh.ch (M.G.)
[2] Faculty of Biology and Medicine, University of Lausanne, Biochemistry, CH-1066 Epalinges, Switzerland; Min.Ma@unil.ch
*   Correspondence: Martin.Baumgartner@kispi.uzh.ch
†   These authors contributed equally to this work.

Abstract: In the Sonic Hedgehog (SHH) subgroup of medulloblastoma (MB), tumor initiation and progression are in part driven by smoothened (SMO) and fibroblast growth factor (FGF)-receptor (FGFR) signaling, respectively. We investigated the impact of the SMO-FGFR crosstalk on tumor growth and invasiveness in MB. We found that FGFR signaling represses GLI1 expression downstream of activated SMO in the SHH MB line DAOY and induces *MKI67*, *HES1*, and *BMI1* in DAOY and in the group 3 MB line HD-MBO3. FGFR repression of GLI1 does not affect proliferation or viability, whereas inhibition of FGFR is necessary to release SMO-driven invasiveness. Conversely, SMO activation represses FGFR-driven sustained activation of nuclear ERK. Parallel activation of FGFR and SMO in ex vivo tumor cell-cerebellum slice co-cultures reduced invasion of tumor cells without affecting proliferation. In contrast, treatment of the cells with the SMO antagonist Sonidegib (LDE225) blocked invasion and proliferation in cerebellar slices. Thus, sustained, low-level SMO activation is necessary for proliferation and tissue invasion, whereas acute, pronounced activation of SMO can repress FGFR-driven invasiveness. This suggests that the tumor cell response is dependent on the relative local abundance of the two factors and indicates a paradigm of microenvironmental control of invasion in SHH MB through mutual control of SHH and FGFR signaling.

Keywords: medulloblastoma; FGFR; Sonic Hedgehog signaling; organotypic culture; cell invasion; signal crosstalk; MAP kinase signaling

1. Introduction

The crosstalk between signaling pathways in eukaryotic cells controls cellular functions involved in proliferation, differentiation, viability, and motile behavior. Although the cause of cancer is usually a genetic or an epigenetic alteration, the aggressiveness of the disease is ultimately determined by the interplay of the cell with its environment, resulting in abundant signal crosstalk in the cancer cells [1]. Medulloblastoma (MB) is the most common malignant pediatric brain tumor. It is divided into four subgroups with a total of twelve molecularly distinct subtypes [2]. In Sonic Hedgehog (SHH) MB, the crosstalk between basic fibroblast growth factor (bFGF/FGF2) activated fibroblast growth factor (FGF)-receptor (FGFR) and tumor growth factor receptor beta (TGF-β) signaling determines the migratory and the invasive capabilities of the tumor cells [3]. In granule cell precursors (GCPs), the putative cells of origin of SHH MB, bFGF abolishes the potent proliferative response of SHH and accelerates differentiation of GCPs [4]. Moreover, bFGF pre-treatment of SHH MB tumor cells derived from a Ptch+/− mouse or direct injection of high concentrations of bFGF into Ptch+/− MB tumors that were orthotopically implanted in recipient mice prevented tumor growth or caused regression, respectively [5]. This raises the question of a dichotomous pro-invasive/anti-proliferative function of bFGF in MB tumors with activated SHH signaling that may impact on the growth and the dissemination of the tumor cells.

Hedgehog (Hh) signaling is an evolutionary conserved signaling pathway. It regulates a multitude of cellular processes during development, embryonic patterning, organ morphogenesis, and growth control by regulating cell proliferation, differentiation, and migration [6]. The core components of Hh signaling are Patched1 (PTCH1), a twelve transmembrane receptor that binds Hh ligands including SHH, and smoothened (SMO), a seven-transmembrane receptor relieved from PTCH1 repression after Hh ligand binding to PTCH1. This activation of SMO causes the processing of the glioma-associated oncogene transcription factors (GLI [7]) GLI2/3 from gene repressing to inducing activity and results in the expression of the constitutive activator GLI1 [8]. In addition to the repression of SMO by PTCH1, Hh signaling is regulated by other mechanisms, including the suppressor of fused (SUFU)-mediated cytoplasmic retention and degradation of GLI and the phosphorylation of GLI1, 2, 3 by protein kinases or their acetylation, ubiquitination, or sumoylation [6]. In SHH-driven MB, the induction of GLI is repressed by the phosphorylation of GLI1 by MEKK2 and MEKK3 downstream of FGFR signaling [9]. In contrast, an earlier study in fibroblasts found that bFGF activation of FGFR could induce GLI1 transcriptional activity, and it was suggested that mitogen-activated protein (MAP) kinase and SHH signaling synergize to drive proliferation [10].

In the developing central nervous system (CNS), Hh signaling exerts pleiotropic activity and drives proliferation, specification, and axonal targeting. In the adult CNS, Hh signaling modulates self-renewal and specification of neural stem cells [11]. bFGF also tunes diverse functions in the developing and the adult brain [12,13]. It is secreted by the choroid plexus, the microglia, and the astrocytes and is concentrated on laminin-containing fractone structures in the brain [14]. bFGF accumulates in MB tumor tissue in a pattern reminiscent of tumor-infiltrating cells, and it is a potent promoter of cell dissemination by signaling via an FRS2-dependent cascade [3]. The conflicting results on the role of bFGF in MB growth and invasiveness and its regulatory impact on Hh signaling highlight the unclear functional significance of the FGFR-SHH crosstalk in MB. Moreover, whether SMO activation alters FGFR function has not been addressed. In this study, we therefore test whether combined activation of FGFR and SMO affects proliferation and invasiveness in MB cells.

2. Results

2.1. bFGF-Induced FGFR Activation Represses SAG-Induced GLI1 Expression

SHH signaling in MB can be repressed by bFGF-induced FGFR activation [9]. To explore the FGFR-SMO crosstalk in the context of cell migration control, we used the SHH MB cell line DAOY and determined whether SMO activation could also affect FGFR signaling. In these cells and also in a mouse model of gr3 MB, FGFR activity promotes cell invasion and tumor progression [3]. In DAOY cells, parallel stimulation with both secreted SHH and bFGF represses GLI1 transcriptional activity through direct phosphorylation of GLI1 by MEKK2 and MEKK3 [9]. We first tested whether SMO agonist (SAG) also leads to SHH pathway activation in DAOY MB cells and whether GLI1 expression is altered in response to parallel FGFR activation by bFGF at a concentration that promotes invasiveness. We found that SAG treatment of DAOY cells induced *GLI 1* mRNA (Figure 1A) and GLI protein expression (Figure 1B). Co-stimulation of the cells with bFGF reduced SAG-induced *GLI1* transcription and protein expression (Figure 1A,B). Treatment of the cells with the pan-FGFR inhibitor BGJ398 rescued *GLI1* expression (Figure 1C) and GLI protein levels (Figure 1D) in the presence of SAG-bFGF co-stimulation. GLI1 was not detectable in the gr3 line HD-MBO3. In combination with SAG, BGJ398 treatment also caused a dramatic increase in *GLI1* expression, whereas BGJ398 treatment alone only moderately increased *GLI1* expression (Figure 1C) but not GLI1 protein levels. This indicates that the induction of GLI1 by BGJ398 treatment both at mRNA and protein levels is effective only when FGFRs and SMO are activated.

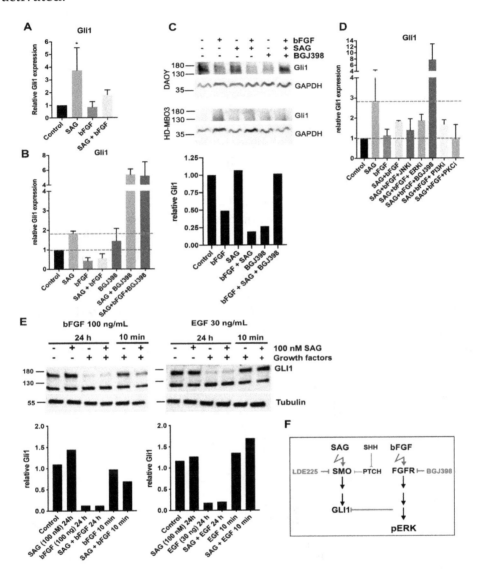

Figure 1. Growth factor signaling represses GLI1 expression. (**A**) qrt-PCR analysis of *GLI1* expression in DAOY cells stimulated with smoothened (SMO) agonist (SAG) (100 nM), basic fibroblast growth factor (bFGF) (100 ng/mL) or in combination for 24 h ($n = 3$, mean and SD, $*$ $p < 0.05$). (**B**) qrt-PCR analysis of BGJ398 (1 μM) effects on SAG and bFGF-induced *GLI1* expression ($n = 2$, mean and SD). (**C**) Immunoblot (IB) analysis of GLI1 expression in response to treatment as in C. No GLI1 expression at protein levels was detected in the gr 3 medulloblastoma (MB) line HD-MBO3. Relative integrated pixel densities of GLI1 bands in DAOY cells are shown below (normalized to Glycerinaldehyd-3-phosphat-Dehydrogenase (GAPDH). (**D**) qrt-PCR analysis of kinase inhibitors against c-jun N-termina kinase (JNK), extracellular-signal regulated kinase (ERK), phosphatidylinositol 3'kinase (PI3K), and protein kinase C (PKCs) (all at 1 μM) effects on SAG plus bFGF-induced *GLI1* expression ($n = 2$, mean and SD). (**E**) Upper: IB analysis of SAG-induced GLI1 expression after 24 h or 10 min stimulation with bFGF (100 ng/mL) or epidermal growth factor (EGF) (30 ng/mL). Right: Integrated densities of GLI1 bands relative to tubulin. (**F**) Schema depicting the observed impact of fibroblast growth factor (FGF)-receptor (FGFR) signaling on GLI1 expression.

Kinase inhibitors of extracellular-signal regulated kinase (ERK), phosphatidylinositol 3′kinase (PI3-K), or protein kinase C (PKC) did not rescue *GLI1* expression (Figure 1D). Thus, none of these putative effectors of FGFR alone are involved in GLI1 repression. Interestingly, epidermal growth factor (EGF) stimulation for 24 h also repressed basal and SAG-induced GLI1 (Figure 1E). Thus, receptor tyrosine kinase (RTK)-dependent repression of GLI1 is not specific for bFGF. These findings show that the activation of SMO promotes *GLI1* transcription and leads to GLI1 expression in DAOY cells. Parallel activation of FGFR signaling represses GLI1 expression both at the transcriptional and the protein level (Figure 1F). Furthermore, pharmacological repression of FRGR with BGJ398 in the presence of active SMO causes a very pronounced induction of *GLI*.

2.2. SMO Activation by SAG Does Not Repress bFGF-Induced Collagen Invasion

Since bFGF is a strong promoter of invasion in DAOY cells [15], we tested whether SAG treatment would affect bFGF-induced collagen I invasion. We performed spheroid invasion assays in the absence of stimulation and with SAG or bFGF alone or with a combination of both. SAG treatment did not promote collagen I invasion (Figure 2A), whereas bFGF stimulation caused a robust increase in the average distance of invasion. Co-stimulation with SAG did not significantly alter bFGF induced invasion, suggesting that there is no inhibitory crosstalk between SHH and FGFR signaling with respect to invasion control. To exclude the possibility that FGFR-mediated repression of SHH signaling (Figure 1C) impedes SAG-induced collagen I invasion, we treated bFGF-stimulated cells with BGJ398 and SAG in parallel. BGJ398 treatment completely abrogated bFGF-induced invasion (Figure 2B). Co-treatment with bFGF, SAG, and BGJ398 caused a significant increase in collagen I invasion compared to bFGF plus BGJ398 treatment alone. Although BGJ398-only treatment caused some increase in invasion, its impact on invasion was only significant on the distance of invasion (Figure 1C) as well as on the cumulated distance of invasion (Figure 1D) when SAG was also present. This indicates that SMO activation can moderately increase the invasion capabilities of the cells in vitro when FGFR signaling is repressed.

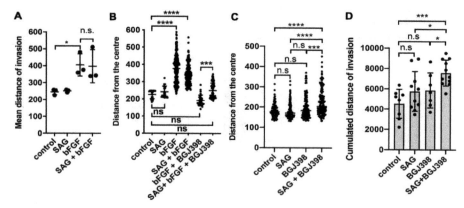

Figure 2. SMO activation does not repress bFGF-induced collagen invasion. (**A**) Mean distance of invasion quantified using Spheroid Invasion Assay (SIA) was compared between treatments as indicated. Each dot represents mean of independent experiment (* $p = 0.0152$, n.s. = not significant unpaired *T*-test). (**B**) Distance of invasion after treatments as indicated. Each dot represents a cell. Data from multiple spheroids are combined (*** $p < 0.001$, **** $p < 0.0001$, one-way ANOVA with Bonferroni's multiple comparisons test). (**C**) Analysis of BGJ398 impact on distance of invasion compared to BGJ398 plus SAG (*** $p < 0.001$, **** $p < 0.0001$, one-way ANOVA with Bonferroni's multiple comparisons test). (**D**) As C but total invasion was calculated from the cumulated invasion distances of all cells. Each dot represents the cumulated invasion distance of one spheroid. Mean and SD are shown (* $p < 0.05$, *** $p < 0.001$, n.s. = not significant, unpaired *T*-test).

2.3. SMO Activity Is Necessary for Proliferation without Affecting bFGF-Dependent Gene Expression

bFGF and FGFR1-4 signal via fibroblast growth factor receptor substrate 2 (FRS2)-dependent Ras/mitogen-activated kinase (RAS/MAPK) and phosphatidylinositol 3′kinase (PI3K)/ protein kinase

B (PKB) and FRS2-independent phospholipase C-gamma (PLC-γ), janus kinase-signal transducer and activator of transcription (JAK-STAT) pathways that contribute to proliferation and viability [16]. To test whether SMO activation by SAG affects FGFR signaling towards proliferation and viability, we counted the cells grown in low (1%) serum without or with bFGF or SAG or a combination of both (Figure 3A). Neither bFGF nor SAG caused a marked change in cell number, whereas the inhibition of FGFRs with BGJ398 or of SMO with LDE225 significantly reduced proliferation. To compare treatment effects in cell lines of SHH (DAOY) and group 3 (gr 3, HD-MBO3 [17]) MB, we used CellTrace Violet dye intensity as a readout. In-cell dye fluorescence was increasingly diluted with each additional generation number and thus servd as a measurand for proliferation (Supplementary Figure S1A). Dye dilution was comparable under control conditions in DAOY and HD-MBO3, and blockade of proliferation with mitomycin C caused a comparable dye retention per cell in both cell lines (Supplementary Figure S1B). LDE225 treatment completely repressed proliferation in a subset of cells between 24 and 72 h (right peak in the histogram in Figure 3B, X/Y plot of time resolved mean fluorescence in Figure 3C). In DAOY cells, BGJ398 also reduced proliferation, although this effect became evident only at 72 h. In HD-MBO3 cells, both LDE225 and BGJ398 caused a similar reduction in cell proliferation compared to control, whereas bFGF caused a moderate increase in proliferation. In contrast to DAOY cells, there was no indication of a bi-phasic fluorescence intensity distribution in HD-MBO3 after LDE225 treatment. Thus, a subset of DAOY cells is highly susceptible to SMO inhibition, and HD-MBO3 cell proliferation depends in part on active FGFR signaling.

We next determined whether FGFR and SHH pathway modulation alters the expression of *GLI1*, *HES1*, *MKI67*, and *BMI1*, a small subset of genes involved in proliferation and differentiation control. We found that SAG treatment caused a significant increase in *GLI1* expression in DAOY cells without affecting the expression of the other genes. bFGF repressed SAG-induced *GLI1* and caused significant increases in *HES1*, *MKI67*, and *BMI1* in both cell lines (Figure 3D,E). Expressions of *GLI1* and *HES1* were also significantly increased in primary SHH MB compared to the other three subgroups (Supplementary Figure S2A,B). None of the three bFGF-induced genes were repressed by parallel SAG stimulation. Consistent with the proliferation data, BGJ398 treatment reduced *MKI67* expression in HD-MBO3 cells, an effect that was not rescued by bFGF or SAG treatment (Figure 3E). These data confirm the repressive activity of FGFR activation on *GLI1* expression and reveal *HES1* as a novel target gene of FGFR signaling.

2.4. SMO Activation Represses Nuclear ERK Activity after bFGF Stimulation

FGFR activation causes the induction of the MAP kinase signaling pathway and an increase in the phosphorylation of ERK on residues Thr202/Tyr204 (Figures 1B and 4A). To test whether parallel SMO activation could influence the extent of ERK activation in response to growth factor treatment, we compared the level of phosphorylated ERK (pERK). We observed no reduction of pERK in cells co-stimulated for ten minutes with bFGF and SAG compared to cells stimulated with bFGF only (Figure 4A). We also observed no differences in phospho-ERK when SAG stimulation was combined with hepatocyte growth factor (HGF) or EGF. We additionally explored the levels of phosphorylated ERK in three different cellular fractions [cytosolic: (C), membranes: (M) and nucleus/cytoskeleton: (N)], encompassing the totality of all isolated proteins after 10 and 90 min of stimulation with bFGF, SAG, or the combination of bFGF plus SAG. We repeated the experiment three times and observed some variations in pERK distribution. A representative blot (90 min stimulation) is shown in Figure 4B and the corresponding quantification of pERK in Figure 4C. However, statistical analysis revealed no significant difference when bFGF-alone treatment was compared to bFGF plus SAG. We furthermore compared the cumulated pERK levels from all fractions, confirming that bFGF-induced pERK is not influenced by parallel SMO activation (Figure 4D). As we could not confirm the expected increase in nuclear pERK after bFGF stimulation by immunoblot (IB), we used a synthetic kinase activation relocation sensor (SKARS) for ERK [18,19], which is designed to measure the nuclear translocation of activated ERK in real-time. We found that bFGF (Figure 4E, left) caused a dose-dependent and persistent increase in sensor translocation. Co-stimulation of the cells with 100 ng/mL bFGF plus 100 nM SAG markedly reduced bFGF-induced sensor translocation compared to bFGF stimulation alone (Figure 4E, right). This suggested that one immediate, transcription-independent effect of SHH pathway activation is the repression of bFGF-induced nuclear ERK activity (Figure 4F).

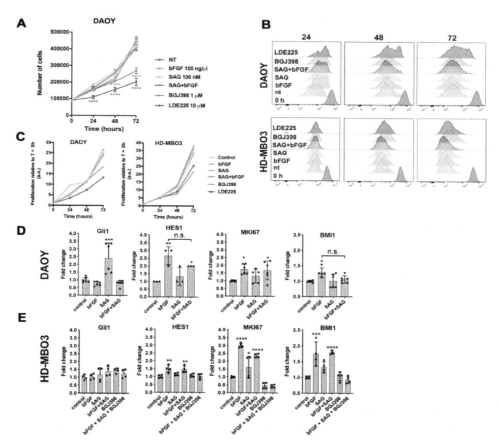

Figure 3. SMO activity is necessary for proliferation without affecting bFGF-dependent gene expression. (**A**) DAOY cells were counted every 24 h over a 72 h period to determine treatment effects on proliferation. Data represent mean values and SEM of three independent experiments. (**B**) CellTrace Violet dye dilution analysis in DAOY and HD-MBO3 cells by flow cytometry every 24 h over a 72 h period with treatments as in A. The fluorescence intensity of the dye measured immediately after labeling the cells at the time of seeding is T0. (**C**) Relative mean fluorescence intensity (MFI) plots of CellTrace Violet dye measurements at indicated times normalized to MFI at T = 0 h (red histogram in B). (**D,E**) Gene expression analysis by qrt-PCR after 48 h of the indicated treatments in DAOY (**D**) and HD-MBO3 (**E**) cells (mean and SD, * $p < 0.05$, ** $p < 0.01$, *** $p < 0.001$, **** $p < 0.0001$, one-way ANOVA with Tukey's multiple comparisons test).

2.5. Combined, Exogenous Activation of SHH and FGFR Signaling Prevents Tissue Invasion Ex vivo

To evaluate the impact of FGFR and SHH signaling on tumor cell growth and tissue invasion, we performed organotypic cerebellum slice-tumor cell co-culture [18] experiments in the absence and the presence of exogenously added bFGF, SAG, or with a combination of both. DAOY cells were implanted as spheroids and treated for five days after implantation with the different growth factors and drugs. Under control conditions, the tumor cells expanded from a spheroid into the surrounding brain tissue, which resulted in a tumor cell mass of irregular shape and low circularity (Figures 5A and 6A, Supplementary Figure S3A). Treatment of the co-culture with LDE225 prevented the invasion of the surrounding tissue (Figures 5A and 6B), resulting in a spherical, well-circumscribed tumor cell mass with significantly increased circularity (Figure 5A, Supplementary Figure S3A). Conversely, activation of SHH signaling with SAG increased invasion, caused reduced circularity (Figure 5A), and single cells invading deep in the tissue slice were observed (Supplementary Figure S3B). Co-stimulation of SAG-treated cells with bFGF did not further increase invasion; it rather caused a phenotype similar to the LDE225-treated condition characterized by reduced invasion and increased circularity of the tumor cell mass (Figures 5A and 6C). We also determined the impact of the treatments on proliferation by performing EdU staining followed by a click-iT reaction on the co-cultured cerebellum slices for fluorescent labeling of newly synthesized DNA. Nuclear fluorescence is indicative of active DNA synthesis, which was considerably decreased in cells treated with LDE225 (Figure 5A, Supplementary Figure S3A). EdU incorporation relative to area was also decreased in cells treated with SAG or bFGF alone, whereas the combination of SAG plus bFGF rescued proliferation. We also observed increased abundance and intensity of glial fibrillary acidic protein (GFAP)-positive cell protrusions and infiltrates at the site of the tumor cell mass in SAG and SAG plus bFGF-treated slices; this accumulation of GFAP-positive cells was not observed in the samples treated with BGJ398 or LDE225 (Supplementary Figure S3A). BGJ398 and LDE225 treatment also caused morphological alterations in Purkinje cells reminiscent of axon retraction (Supplementary Figure S3A).

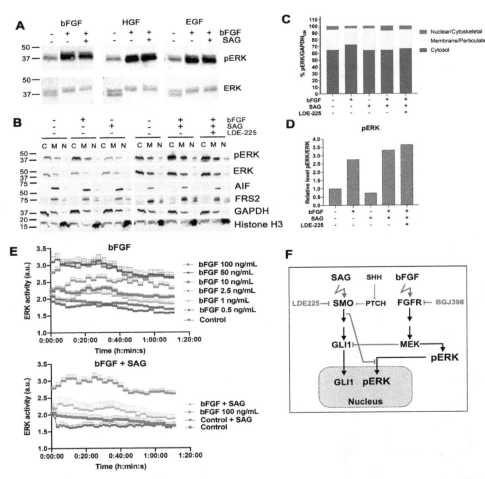

Figure 4. SMO activation represses nuclear ERK activity after bFGF stimulation. (**A**) IB analysis of pERKT202Y204 and ERK after 10 min stimulation with bFGF (100 ng/mL) or hepatocyte growth factor (HGF) (20 ng/mL) or EGF (30 ng/mL). (**B**) IB analysis of proteins indicated in cytosolic (C), membrane/particulate (M), and nuclear/cytoskeleton (N) fractions. (**C**) Integrated densities of pERK bands relative to the control (UN: untreated) GAPDH in the different cellular fractions. (**D**) Relative pERK integrated chemiluminescence intensities of all three fractions in C relative to ERK. (**D**) Cumulated integrated fluorescence intensities of all three fractions in C relative to control. (**E**) Time course of ratio nuclear/cytosolic ERK/SKARS (ERK activity) with indicated treatments. Mean and SEM are shown. (**F**) Schema depicting the proposed impact of SMO activation on nuclear translocation of activated ERK in the SHH-FGFR crosstalk.

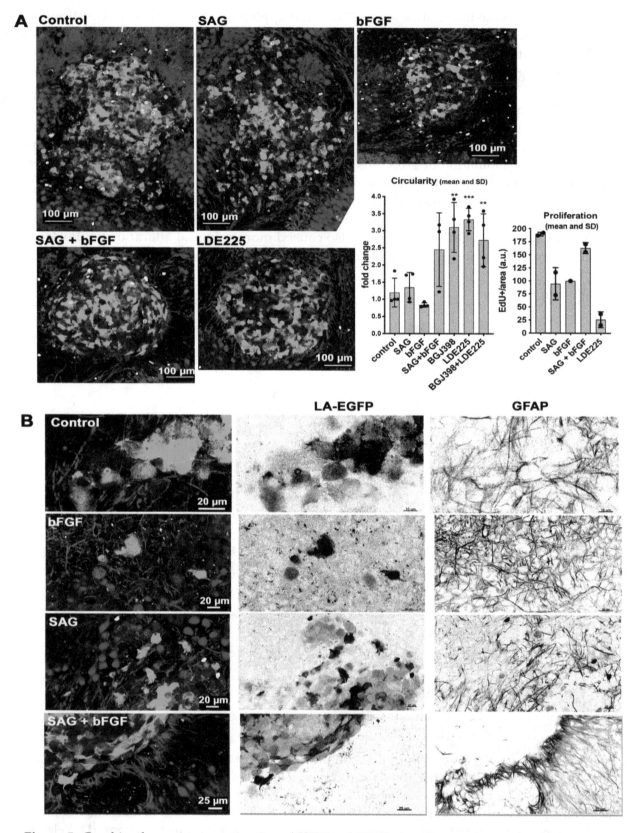

Figure 5. Combined, exogenous activation of SHH and FGFR signaling prevents tissue invasion ex vivo. (**A**) Confocal microscopy analysis of organotypic cerebellum slice cultures (OCSCs) five days after spheroid implantation. DAOY tumor cells express LA-EGFP. Anti-GFAP is in red, anti-Calbindin (Purkinje cells) in blue. Right quantification of circularity ($n = 4$, ** $p < 0.01$, *** $p < 0.001$, one-way ANOVA with Bonferroni's multiple comparisons test) and proliferation ($n = 2$). (**B**) Higher-resolution confocal microscopy images of LA-EGFP-positive tumor cells at the spheroid-tissue margins.

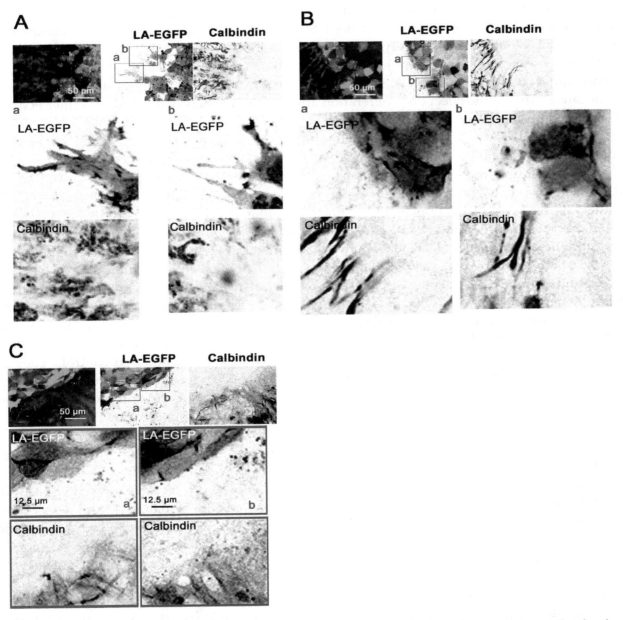

Figure 6. SMO inhibition or bFGF plus SAG co-stimulation restrict invasive phenotype. Confocal microscopy analysis of control (**A**) LDE225 (**B**) and bFGF plus SAG (**C**) co-stimulated OCSCs. Magnifications in (a) and (b) are 4× images of the boxed areas in the upper row. Inverted greyscale was used for better visualization of stains.

These data indicate that the level of SMO activity determines the invasive behavior of the tumor cells in the cerebellar slice tissue. Inactivation of SMO prevents, whereas its exogenous activation by SAG increases tissue invasion. Co-activation of both SMO and FGFRs results in an intermediate phenotype with reduced invasion (compared to control or SAG or bFGF stimulation) but comparable proliferative activity.

3. Discussion

In this study, we investigated the crosstalk between the FGFR and the SHH signaling pathways in cell-based models of MB in the context of cell invasion control. We found that SHH pathway activation with the smoothened agonist SAG triggered sustained expression of the GLI1 transcriptional activator and contributed to tissue invasion ex vivo. Repression of SMO activity by LDE225 only moderately impaired cell growth in vitro but abrogated proliferation and tissue invasion ex vivo.

FGFR activation caused collagen and tissue invasion, promoted transcription of *HES1*, and resulted in nuclear translocation of activated ERK. Parallel activation of FGFR and SMO blocked GLI1 expression, markedly reduced nuclear ERK activity, and repressed tissue invasion ex vivo. Neither in vitro nor ex vivo did we observe altered proliferation in response to parallel pathway activation, suggesting that the main impact of the FGFR-SMO crosstalk is on invasion control. In gr3 MB cells, SMO activation did not cause a detectable effect on proliferation and invasion, whereas blockade of either FGFRs or SMO reduced proliferation.

The impact of FGFR activity on MB growth and progression is controversial. FGFR activation was found to block proliferation, to induce differentiation, and to ultimately cause cell death in a cell line derived from a surgical specimen [19]. This activity was later found to be dependent on FGF2 (bFGF) and FGF9 and was only effective in cell lines derived from classic but not desmoplastic tumors [20]. In granule cell precursors (GCPs) isolated from mice, the putative cells of origin of MB, SHH promoted and bFGF repressed proliferation and also ablated GLI1 target gene expression induced by SHH [4]. The antiproliferative effect of bFGF on GCPs was confirmed throughout postnatal development of the GCPs, and both pre-implantation exposure to and injection of bFGF into MB tumors of Ptch+/− mice blocked tumor growth [5]. In contrast to the anti-proliferative effect found in these murine SHH MB models, Zomerman et al. found that bFGF stimulation significantly increased proliferation in human MB cell lines and noted high levels of bFGF release in all lines tested [21]. Our data on the gr3 line HD-MBO3 support this observation, as blockade of FGFR slows proliferation, and bFGF stimulation causes transcription of *MKI67* and *HES1*, two genes involved in proliferation control. They also corroborate our previous findings, where genetic blockade of FGFR signaling impaired tissue invasion ex vivo and pharmacological inhibition of FGFR signaling reduced growth and progression in a gr3 MB in vivo model [3]. Thus, the consequence of FGFR activation on tumor cell growth and progression may depend on the organism's developmental state and corresponding microenvironmental cues. In support of this is our observation herein, where the activation of SMO in the context of repressed FGFR leads to very robust *GLI1* expression. Conversely, the activation of FGFR causes robust nuclear ERK activation and represses SHH signaling, GLI1 expression, and pro-invasive functions of activated SMO, which may explain the tight spatio-temporal control of FGFR signaling during development, where both SMO and FGFRs play essential roles. GLI1 repression may also explain why engraftment and growth of Ptch+/− derived tumors were blocked by bFGF, as this likely caused repression of GLI1 activity in a susceptible phase of tumor growth [5]. Whether this was a consequence of repressed invasion, which we observed in cerebellar slices when both pathways were activated in parallel, remains to be elucidated. Our findings of GLI1 repression by FGFR signaling are in line with [9] and additionally indicate the lack of implication of c-jun N-terminal kinase (JNK), phosphatidylinositol 3′kinase (PI3-K), and protein kinase C (PKC) in FGFR-induced repression of GLI1. This is in contrast to the repression of GLI1 target genes by bFGF in GCPs, where JNK inhibitors were also found to be effective [4].

The fact that SMO activation in a condition with complete FGFR inhibition confers moderate invasive properties independent of FGFR indicates that SMO activity toward cell locomotion is constitutively repressed by FGFR signaling. In the absence of activated FGFRs, SMO signaling may thus contribute to tumor cell dissemination and accelerate expansion and spreading of the tumor independent of its growth-promoting functions. Repression of the growth-promoting functions by LDE225 in DAOY cells revealed susceptible and resistant cells in this laboratory line. Proliferation of the gr3 MB line HD-MBO3 is moderately susceptible to LDE225 and requires FGFR signaling for maximal proliferation in vitro. This is consistent with our observation of BGJ398 effect in vivo, where both invasion and tumor growth were blocked with this FGFR inhibitory compound [3].

The small selection of genes analyzed does not provide a clear indication that SMO activation negatively affects FGFR-induced transcription, as the increased expressions of *HES1*, *MK67*, and *MBI1* in bFGF-stimulated cells were not repressed by SAG treatment. Increased *MKI67* transcription in HD-MBO3 cells corroborates the observation that FGFR activation in gr3 MB promotes proliferation and that therapeutic targeting FGFR signaling with small molecule compound inhibitors in this subgroup could provide a clinical benefit.

Cerebellum tissue invasion of DAOY MB cells is exquisitely sensitive to LDE225. Parallel activation of FGFR and SHH signaling partially phenocopies SMO inhibition by LDE225 and reduces invasion. LDE225 treatment can repress phosphorylation of focal adhesion kinase (FAK) and paxillin [22,23], two molecules critically implicated in integrin-dependent motility and invasiveness in numerous cancers, including MB [24]. SHH signaling promotes epithelial to mesenchymal transition (EMT) and lymph node invasion in bladder cancer [25,26] and hypoxia-induced up-regulation of cancer stem cell genes and EMT in cholangiocarcinoma [27]. Wang et al. [28] furthermore proposed that EMT in esophageal adenocarcinoma is mediated through increased GLI1 expression and is associated with PI3-K AKT signaling. Whether SMO-dependent tissue invasion is mediated through GLI1 in MB is unclear. Our data indicating that FGFR blockade by BGJ398, which triggers *GLI1* expression, is necessary for SAG-induced collagen I invasion, support a mechanism depending on GLI1. However, more direct functions independent of transcriptional control may also need to be considered, in particular as we observed direct repression of nuclear ERK by SAG, which cannot be explained by GLI1 transcriptional activity. The only partial phenotype triggered by FGFR-SMO co-activation indicates transcriptional control of invasion rather than direct repression of—for example—FAK phosphorylation. Despite the remarkable efficacy of LDE225, it is unlikely to be effective in all SHH-driven MB, particularly not in those dependent on oncogenic alterations in the SHH pathway downstream of SMO. Co-treatment with bFGF and SAG does not cause the near-complete blockade of EdU incorporation observed with LDE225, suggesting that the mechanisms promoting invasion and proliferation are not identical. In vitro, blockade of proliferation by LDE225 treatment was only partial with a subset of cells being resistant. The ex vivo analysis did not completely recapitulate this finding, as we observed only a few proliferating cells. However, since the tumor cell mass in LDE225-treated organotypic cerebellum slice cultures (OCSCs) was relatively large at endpoint, it is also possible that either proliferation-arrest occurs in all cells after a lag phase or the low-fetal bovine serum (FBS) medium in the in vitro culture supplied pro-proliferative signals.

An intriguing observation of our study is that SMO activation causes sustained repression of nuclear accumulation of bFGF-activated ERK. ERK has numerous targets in the nucleus [29], and the level of ERK activation in the nucleus was shown to balance proliferation (if nuclear ERK was high) versus differentiation (if nuclear ERK was low) in stem and progenitor cells [30]. Nuclear translocation of ERK is initiated by phosphorylation of its regulatory Tyr and Thr residues followed by exposure of the nuclear localization signal (NLS), subsequent phosphorylation by CKII, nuclear import through importin 7, and the regulation of up to 125 substrates (reviewed in [29]). Nuclear ERK substrates include transcription factors including c-MYC [31] and chromatin modifying enzymes, which makes it difficult to predict the functional consequence of repressed nuclear ERK activity in FGFR-SMO co-activated MB cells. Correlative gene expression analysis between GLI1 and c-MYC expression indicates negative correlation across all subgroups of MB (r-value = −0.236, $p = 4.03^{-11}$, source: R2 genomics analysis and visualization platform), suggesting that SMO activation could repress c-MYC in MB. In contrast, GLI1 expression correlates positively with MYCN expression across all subgroups of MB (r-value = 0.619, $p = 8.93^{-82}$, source: R2 genomics analysis and visualization platform). The mechanistic basis and the functional consequences of the repression of FGFR-induced nuclear ERK activation by activated SMO merit further in-depth investigation, as this regulatory network affects key genes such as c-MYC and MYCN, both relevant for MB tumor initiation and progression of different MB subtypes [2].

Activation of MAP/ERK signaling was previously shown through SMO agonist Purmorphamine in fibroblast-like synoviocytes and was found to promote proliferation and motility [32]. We observed no increase in pERK by IB after SMO activation, indicating that ERK regulation downstream of SMO is cell-type dependent, and that the modest increase in SAG-dependent invasion in cells with repressed FGFR is not mediated through ERK. We observed a discrepancy between nuclear ERK activation measured by the ERK relocation sensor and ERK phosphorylation assessed by IB. Although we can only speculate about the underlying cause, a general impact of SAG on the regulation of nuclear import and export can be excluded, as the NLS-mediated nuclear translocation of mCherry-nuc was not affected. However, to mechanistically understand and eventually therapeutically exploit the control of nuclear ERK activation by SHH signaling, further investigation of the regulation of CKII activation and importin7 function by SHH signaling will be necessary.

4. Material and Methods

4.1. Reagents

Kinase inhibitors BGJ398 (FGFRs, S2183), Go6983 (PKCs, S2911), LY2157299 (PI3-K, S2230), and SP600125 (JNK, S1460), as well as SMO inhibitor LDE225 (S2151) were purchased from Selleckchem, Houston, TX, USA. bFGF (100-18B), HGF (100-39), EGF (100-47), and SAG (9128694) were purchased from PeproTech EC LtD (London, UK).

4.2. Cells and Cell Culture

DAOY human MB cells were purchased from the American Type Culture Collection (ATCC, Rockville, MD, USA). HD-MBO3 [17] was generously provided by Till Milde (DKFZ, Germany). DAOY cells were cultured as described in [33]. DAOY Lifeact-enhanced green fluorescent protein LA-EGFP cells were produced by lentiviral transduction of DAOY cells with pLenti-LA-EGFP. Cell line authentication and cross-contamination testing were performed by Multiplexion GmbH (Heidelberg, Germany) by single nucleotide polymorphism (SNP) profiling.

4.3. Mouse Maintenance

Mouse protocols for organotypic brain slice culture were approved by the Veterinary Office of the Canton Zürich (Approval ZH134/17). Wild type C57BL/6JRj pregnant females were purchased from Janvier Labs (Le Genest-Saint-Isle, France) and were kept in the animal facilities of the University of Zürich Laboratory Animal Center.

4.4. qrt-PCR Analysis of Gene Expression

In this study, 2×10^5 DAOY wild type or 2.5×10^5 HDMBO3 wild type cells were seeded per well in 6 well plates in complete growth medium and incubated overnight at 37 °C. Medium was replaced with serum-free medium. After overnight incubation at 37 °C, cells were treated with bFGF (100 ng/mL), SAG (100 nM), BGJ398 (1 μM), or in combination for 48 h with a treatment change after 24 h. For quantitative real-time PCR (qRT-PCR) analysis of target genes, total RNA was isolated using Qiagen RNeasy Mini Kit (74106, Qiagen, Hilden, Germany). Then, 1 μg of total RNA was used as a template for reverse transcription, which was initiated by random hexamer primers. The cDNA synthesis was carried out using High Capacity cDNA Reverse Transcription Kit (4368813, Applied Biosystems, Foster City, CA, USA). qRT-PCR was performed using PowerUp Syber Green (A25776, Thermo Scientific, Waltham, MA USA) under conditions optimized for the ABI7900HT instrument. The ΔΔCT method was used to calculate the relative gene expression of each gene of interest.

4.5. Spheroid Invasion Assay (SIA)

In this experiment, 2000 cells/100 μL per well were seeded in cell-repellent 96 well microplates (650790, Greiner Bio-one). The cells were incubated at 37 °C overnight to form spheroids. Then, 70 μL of the medium were removed from each well, and there remained a medium with spheroid overlaid with a solution containing 2.5% bovine collagen 1. Following the polymerization of collagen, fresh medium was added to the cells and treated with growth factors and/or with inhibitors. The cells were allowed to invade the collagen matrix for 24 h, after which they were fixed with 4% PFA and stained with Hoechst. Images were acquired on an Axio Observer 2 mot plus fluorescence microscope using a 5× objective (Zeiss, Munich, Germany). Cell invasion is determined as the average of the distance invaded by the cells from the center of the spheroid, which was determined using automated cell dissemination counter (aCDc) with our cell dissemination counter software aSDIcs [15].

4.6. Immunoblot (IB)

To assess treatment effects on target proteins, 2.5×10^5 DAOY wild type or 2.5×10^5 HDMBO3 wild type cells were seeded per well in 6 well plates in complete growth medium and incubated overnight at 37 °C. Medium was replaced with serum-free medium. After overnight incubation at 37 °C, cells were treated with bFGF (100 ng/mL), SAG (100 nM), BGJ398 (1 μM), or in combination for 24–48 h. Treatments were changed every 24 h. Cells were lysed using RIPA buffer and processed for immunoblot (IB) with antibodies against GLI1 (1:1000, Cell Signaling Technologies, Danvers, MA, USA), ERK1/2 (1:1000, Cell Signaling Technologies), and phospho-ERK1/2 (1:1000, Cell Signaling Technologies, Danvers, MA, USA). Loading was normalized using GAPDH (1:1000, Cell Signaling Technologies) or Tubulin (1:1000, Sigma) detected on the same membrane. HRP-linked secondary antibodies (1:5000, Cell Signaling Technologies) were used to detect the primary antibodies. Chemiluminescence detection was performed using ChemiDoc Touch Gel and Western Blot imaging system (BioRad, Hercules, CA, USA) and FujiFilm LAS 3000 (Bucher Biotech, Basel Switzerland) Integrated density of immuno-reactive bands was quantified using ImageJ open source image processing program (https://imagej.net/).

4.7. Cell Fractionation

Cell fractionation assay was performed to determine phosphorylated ERK level in subcellular compartment. The 3×10^6 DAOY cells were seeded in 10 cm dishes 24 h before drug treatment, and cells had fetal bovine serum deprivation overnight (12 h). The next day, DAOY cells were pre-treated or not with LDE-225 (10 μM) and completed after 2 h with bFGF (100 ng/mL) and/or with SAG (100 nM) for 10 or 90 min. Following the recommendations of the provider, we used the FractionPREP™ Cell fractionation kit (K270-50, BioVision, Milpitas, CA, USA) to obtain the cytosolic C, the membrane/particulates M, and the nuclear/cytoskeletal N subcellular protein fractions. During the fractionation, washing steps were done twice between each buffer with cold EDTA-EGTA (1 mM) (Fluka Biochemika, Buchs, Switzerland) containing cocktails of protease and phosphatase inhibitors (Roche). We also added phosphatase inhibitor cocktail in all kit buffers. To perform immunoblots, we added in the final protein fractions, Laemmli buffer (BioRad) and DTT (50 mM) (Sigma). Samples were sonicated and denatured before deposit of 3% of final volume of each fraction on SDS-PAGE gels. We processed for IB with anti-phosphorylated ERK (#9101, Cell Signaling, 1:1000) and anti-total ERK (#9102, Cell Signaling Technologies, 1:1000). To check the correct cellular fractionation, we used anti-FRS2 (Sc-8318, Santa Cruz Biotechnology, Santa Cruz, USA, 1:200) and anti-AIF (Apoptosis Inducible Factor, D39D2, #5318, Cell Signaling Technologies, 1:1000) for the membrane/particulate M fraction, anti-Histone H3 (#4499, Cell Signaling Technologies, 1:2000) and anti-HDAC2 (3F3, #5113, Cell Signaling Technologies, 1:1000) for the nuclear/xytoskeletal N fraction, and anti-GAPDH (#2118, Cell Signaling Technologies, 1:2000) for the xytosolic C fraction. Loading was normalized using GAPDH detected on the same membrane. Relative pERK and ERK protein levels were determined compared to untreated cells. HRP-linked secondary antibodies (Cell Signaling, 1:10.000) were used to detect the primary antibodies. Chemiluminescence detection was performed using ChemiDoc Touch Gel and Western Blot imaging system (BioRad, Hercules, CA, USA) and FujiFilm LAS 3000 (Bucher Biotech AG, Basel, Switzerland. Integrated density of immuno-reactive bands was quantified using ImageJ.

4.8. Ex vivo Organotypic Cerebellum Slice Culture (OCSC)

Wild type C57BL/6JRj mice pups were sacrificed at postnatal day (PND) 8–10 by decapitation. Cerebella were dissected and placed in cold Geys balanced salt solution containing kynurenic acid (GBSSK) and then embedded in 2% low melting point agarose gel. Solidified agarose blocks were glued onto the vibratome (VT 1200S, Leica, Wetzlar, Germany) disc with Roti Coll1 glue (0258.1 Carl Roth, Karlsruhe, Germany), mounted in the vibratome chamber filled with cold GBSSK, and 350 μm thick sections were cut. Slices were transferred to petri dishes filled with cold GBSSK. Millipore inserts (PICM 03050, Merck Millipore, Burlington, VT, USA) were placed in six well plates filled with 1 mL cold slice culture medium (SCM) onto which the slices were then transferred using a Rotilabo-embryo spoon (TL85.1, Carl Roth GmbH, Karlsruhe, Germany). A maximum of three slices were placed per insert, and excess medium was removed. Slices were monitored for any signs of apoptosis, and media was changed daily for the first week and once in two days thereafter. Tumor spheroids were formed with DAOY LA-EGFP cells. The co-culture was treated with bFGF (100 ng/mL), SAG (100 nM), LDE225 (10 μM), or BGJ398 (10 μM). Spheroids were incubated for 5 days. Following the treatment, the co-cultures were fixed as described in [18]. After PFA fixation, the slices were incubated in standard cell culture trypsin EDTA and incubated at 37 °C in a humidified incubator for 23 min. The slices were blocked in phosphate buffered saline (PBS) containing 3% fetal calf serum, 3% bovine serum albumin (BSA), and 0.3% triton × 100 for 1 h at room temperature (RT). Primary antibodies were diluted in the blocking solution and incubated overnight on a shaker at 4 °C. Following 3 washes at RT using 5% BSA in PBS, secondary antibodies were incubated for 3 h at RT. The inserts were flat mounted in glycergel mounting medium (C0563, Dako, Jena, Germany). The slice-spheroid co-cultures were stained for GFAP and calbindin, and three-color image acquisition was performed on an SP8 Leica confocal microscope (Leica Microsystems, Mannheim, Germany).

4.9. ERK-SKARS

Next, 5000 cells/100 μL were seeded in 96-well plates (Greiner μClear, Greiner Bio One GmbH, Kremsmünster, Austria) in full growth medium. The medium was replaced with serum-free medium after 6 hours and was incubated overnight at 37 °C. After starvation, bFGF and/or SAG were added simultaneously using a multichannel pipette (Gilson, Middleton, WI, USA). Serum-free medium was added to control conditions. The image acquisition was started 1 min after treatment with a Nikon Ti2 widefield microscope using a 20× objective (Nikon Instruments Inc., Melville, NY, USA). The chamber for the plate was constantly held at 37 °C in humidified air (95% air, 5% CO_2). The cells were imaged for 25–35 ms using the excitation wave lengths of 531 nm (red channel for mCherry-nuc-9) and 482 nm (green channel for SKARS). The intervals between the acquisitions were set to 60 s. The images were analyzed using CellProfiler (version 2.2.0, free open source software, https://cellprofiler.org/).

4.10. Cell Proliferation Analysis

To assess cell proliferation, cells were stained with CellTrace Violet cell proliferation kit (C34571, ThermoFisher Scientific) according to the manufacturer's protocol. Labeled cells were seeded at the concentration of 6×10^4 cells in a 6-well plate with complete growth medium. The medium was then replaced after 6 hours with media containing 1% FBS. After overnight incubation at 37 °C, cells were treated with bFGF (100 ng/mL), SAG (100 nM), BGJ398 (1 μM), LDE225 (10 μM), or in combination for 24, 48, or 72 h. Media containing treatment agents was changed every 24 h. Fluorescence dilution of the dye was measured at the indicated time points by flow cytometry using an LSRFortessa (BD Bioscience, San Jose, CA, USA) and analyzed using FlowJo Software v10.0 (BD Bioscience, San Jose, CA, USA). Treatment with 10 μM of Mitomycin C (M4287, Sigma Aldrich, St Louis, MO, USA) for 16 h was used as a negative control for cell proliferation. For the proliferation curve, 8×10^4 DAOY cells were seeded in a 6-well plate with complete growth medium. The medium was then replaced after 6 h with media containing 1% FBS. After overnight incubation at 37 °C, cells were treated with bFGF (100 ng/mL), SAG (100 nM), BGJ398 (1 μM), LDE225 (10 μM), or in combination. Cells were trypsinized and counted every 24 h over a 72 h period using the Trypan blue exclusion method (Thermo Fisher, Waltham, MA, USA).

4.11. Confocal Microscopy

Confocal microscopy of cerebellar slice cultures (Leica Microsystems, Mannheim, Germany) was performed as described in [18].

4.12. Gene Expression Analysis

Gene expression data were obtained from the R2 genomics and visualization platform (http://hgserver1.amc.nl/cgi-bin/r2/main.cgi, accessed 12 August 2019) using the Tumor Medulloblastoma-Cavalli-763 dataset [2].

5. Conclusions

FGFR activation in MB cells promotes an invasive phenotype in vitro in a 3D collagen invasion model and ex vivo in tissue slices, which is accompanied by increased expression of *HES1, MKI67,* and *BMI1.* Concomitant FGFR activation represses SMO-induced expression of Gl1, a key effector of SHH pathway activation. Although such parallel activation of SHH signaling does not impede FGFR-driven expression of *HES1, MKI67,* and *BMI1,* it ablates bFGF-induced nuclear translocation and accumulation of activated ERK. The functional consequence of parallel activation of FGFR and SHH signaling ex vivo in tissue slices is the repression of the invasive phenotype induced by FGFR or SHH signaling alone. We conclude from this and from previously published data that, on the one hand, SMO activation promotes proliferation and contributes to tissue invasion through transcriptional control of a set of genes that mediate EMT. On the other hand, FGFR activation promotes invasiveness by directly targeting actin modulators, leading to a transcription-independent pro-invasive phenotype and to proliferation through MAP kinase pathway activation. Concomitant signaling through both pathways ablates GLI1 induction and prevents accumulation of activated ERK in the nucleus, together resulting in a non-invasive cellular state. The susceptibility of the cells to concomitant exposure may depend on the differentiation stage of the cell and the intrinsic status of activation of the SHH pathway with granule cell precursor and early stage MB tumor cells that are highly dependent on SHH signaling for proliferation being more susceptible to FGFR-mediated perturbations.

Author Contributions: Conceptualization, A.N. and M.B.; methodology, A.N., K.S.K., M.M.; investigation, A.N., J.M., C.C., M.T.S., A.G., K.S.K.; writing—original draft preparation, M.B.; writing—review and editing, A.N., J.M., C.C., M.T.S., A.G., K.S.K., M.M., M.G., M.B.; funding acquisition, M.G. and M.B.

Acknowledgments: The authors thank Serge Pelet for generously sharing the ERK-SKARS construct. Imaging was performed with equipment maintained by the Center for Microscopy and Image Analysis, University of Zurich.

References

1.   Sever, R.; Brugge, J.S. Signal transduction in cancer. *Cold Spring Harb. Perspect. Med.* **2015**, *5*, a00098. [CrossRef] [PubMed]

2.   Cavalli, F.M.G.; Remke, M.; Rampasek, L.; Peacock, J.; Shih, D.J.H.; Luu, B.; Garzia, L.; Torchia, J.; Nör, C.; Morrissy, A.S.; et al. Intertumoral Heterogeneity within Medulloblastoma Subgroups. *Cancer Cell* **2017**, *31*, 737–754.e6. [CrossRef] [PubMed]

3.   Kumar, K.S.; Neve, A.; Stucklin, A.S.G.; Kuzan-Fischer, C.M.; Rushing, E.J.; Taylor, M.D.; Tripolitsioti, D.; Behrmann, L.; Kirschenbaum, D.; Grotzer, M.A.; et al. TGF-β Determines the Pro-migratory Potential of bFGF Signaling in Medulloblastoma. *Cell Rep.* **2018**, *23*, 3798–3812.e8. [CrossRef] [PubMed]

4.   Fogarty, M.P.; Emmenegger, B.A.; Grasfeder, L.L.; Oliver, T.G.; Wechsler-Reya, R.J. Fibroblast growth factor blocks Sonic hedgehog signaling in neuronal precursors and tumor cells. *Proc. Natl. Acad. Sci. USA* **2007**, *104*, 2973–2978. [CrossRef] [PubMed]

5. Emmenegger, B.A.; Hwang, E.I.; Moore, C.; Markant, S.L.; Brun, S.N.; Dutton, J.W.; Read, T.-A.; Fogarty, M.P.; Singh, A.R.; Durden, D.L.; et al. Distinct roles for fibroblast growth factor signaling in cerebellar development and medulloblastoma. *Oncogene* **2012**, *32*, 4181–4188. [CrossRef] [PubMed]

6. Pak, E.; Segal, R.A. Hedgehog Signal Transduction: Key Players, Oncogenic Drivers, and Cancer Therapy. *Dev. Cell* **2016**, *38*, 333–344. [CrossRef]

7. Niewiadomski, P.; Niedziółka, S.M.; Markiewicz, Ł.; Uśpieński, T.; Baran, B.; Chojnowska, K. GLI Proteins: Regulation in Development and Cancer. *Cells* **2019**, *8*, 147. [CrossRef]

8. Garcia, A.D.R.; Han, Y.-G.; Triplett, J.W.; Farmer, W.T.; Harwell, C.C.; Ihrie, R.A. The Elegance of Sonic Hedgehog: Emerging Novel Functions for a Classic Morphogen. *J. Neurosci.* **2018**, *38*, 9338–9345. [CrossRef]

9. Lu, J.; Liu, L.; Zheng, M.; Li, X.; Wu, A.; Wu, Q.; Liao, C.; Zou, J.; Song, H. MEKK2 and MEKK3 suppress Hedgehog pathway-dependent medulloblastoma by inhibiting GLI1 function. *Oncogene* **2018**, *37*, 3864–3878. [CrossRef]

10. Riobo, N.A.; Haines, G.M.; Emerson, C.P. Protein kinase C-delta and mitogen-activated protein/extracellular signal-regulated kinase-1 control GLI activation in hedgehog signaling. *Cancer Res.* **2006**, *66*, 839–845. [CrossRef]

11. Alvarez-Buylla, A.; Ihrie, R.A. Sonic hedgehog signaling in the postnatal brain. *Semin. Cell Dev. Biol.* **2014**, *33*, 105–111. [CrossRef] [PubMed]

12. Gómez-Pinilla, F.; Lee, J.W.; Cotman, C.W. Distribution of basic fibroblast growth factor in the developing rat brain. *Neuroscience* **1994**, *61*, 911–923. [CrossRef]

13. Reuss, B.; Dono, R.; Unsicker, K. Functions of fibroblast growth factor (FGF)-2 and FGF-5 in astroglial differentiation and blood-brain barrier permeability: Evidence from mouse mutants. *J. Neurosci.* **2003**, *23*, 6404–6412. [CrossRef] [PubMed]

14. Bjornsson, C.S.; Apostolopoulou, M.; Tian, Y.; Temple, S. It Takes a Village: Constructing the Neurogenic Niche. *Dev. Cell* **2015**, *32*, 435–446. [CrossRef]

15. Kumar, K.S.; Pillong, M.; Kunze, J.; Burghardt, I.; Weller, M.; Grotzer, M.A.; Schneider, G.; Baumgartner, M. Computer-assisted quantification of motile and invasive capabilities of cancer cells. *Sci Rep.* **2015**, *5*, 15338. [CrossRef]

16. Turner, N.; Grose, R. Fibroblast growth factor signalling: From development to cancer. *Nat. Rev. Cancer* **2010**, *10*, 116. [CrossRef]

17. Milde, T.; Lodrini, M.; Savelyeva, L.; Korshunov, A.; Kool, M.; Brueckner, L.M.; Antunes, A.S.L.M.; Oehme, I.; Pekrun, A.; Pfister, S.M.; et al. HD-MB03 is a novel Group 3 medulloblastoma model demonstrating sensitivity to histone deacetylase inhibitor treatment. *J. Neurooncol.* **2012**, *110*, 335–348. [CrossRef]

18. Neve, A.; Kumar, K.S.; Tripolitsioti, D.; Grotzer, M.A.; Baumgartner, M. Investigation of brain tissue infiltration by medulloblastoma cells in an ex vivo model. *Sci. Rep.* **2017**, *7*, 5297. [CrossRef]

19. Kenigsberg, R.L.; Hong, Y.; Yao, H.; Lemieux, N.; Michaud, J.; Tautu, C.; Théorêt, Y. Effects of basic fibroblast growth factor on the differentiation, growth, and viability of a new human medulloblastoma cell line (UM-MB1). *Am. J. Pathol.* **1997**, *151*, 867–881.

20. Duplan, S.M.; Théorêt, Y.; Kenigsberg, R.L. Antitumor activity of fibroblast growth factors (FGFs) for medulloblastoma may correlate with FGF receptor expression and tumor variant. *Clin. Cancer Res.* **2002**, *8*, 246–257.

21. Zomerman, W.W.; Plasschaert, S.L.A.; Diks, S.H.; Lourens, H.-J.; Meeuwsen-de Boer, T.; Hoving, E.W.; den Dunnen, W.F.A.; de Bont, E.S.J.M. Exogenous HGF Bypasses the Effects of ErbB Inhibition on Tumor Cell Viability in Medulloblastoma Cell Lines. *PLoS ONE* **2015**, *10*, e0141381. [CrossRef]

22. Zhang, H.; Chen, Z.; Neelapu, S.S.; Romaguera, J.; McCarty, N. Hedgehog inhibitors selectively target cell migration and adhesion of mantle cell lymphoma in bone marrow microenvironment. *Oncotarget* **2016**, *7*, 14350–14365. [CrossRef]

23. D'Amato, C.; Rosa, R.; Marciano, R.; D'Amato, V.; Formisano, L.; Nappi, L.; Raimondo, L.; Di Mauro, C.; Servetto, A.; Fulciniti, F.; et al. Inhibition of Hedgehog signalling by NVP-LDE225 (Erismodegib) interferes with growth and invasion of human renal cell carcinoma cells. *Br. J. Cancer* **2014**, *111*, 1168–1179. [CrossRef]

24. Tripolitsioti, D.; Kumar, K.S.; Neve, A.; Migliavacca, J.; Capdeville, C.; Rushing, E.J.; Ma, M.; Kijima, N.; Sharma, A.; Pruschy, M.; et al. MAP4K4 controlled integrin β1 activation and c-Met endocytosis are associated with invasive behavior of medulloblastoma cells. *Oncotarget* **2018**, *9*, 23220–23236. [CrossRef]

25. Islam, S.S.; Mokhtari, R.B.; Noman, A.S.; Uddin, M.; Rahman, M.Z.; Azadi, M.A.; Zlotta, A.; van der Kwast, T.; Yeger, H.; Farhat, W.A. Sonic hedgehog (SHH) signaling promotes tumorigenicity and stemness via activation of epithelial-to-mesenchymal transition (EMT) in bladder cancer. *Mol. Carcinog.* **2016**, *55*, 537–551. [CrossRef]

26. Nedjadi, T.; Salem, N.; Khayyat, D.; Al-Sayyad, A.; Al-Ammari, A.; Al-Maghrabi, J. Sonic Hedgehog Expression is Associated with Lymph Node Invasion in Urothelial Bladder Cancer. *Pathol. Oncol. Res.* **2019**, *25*, 1067–1073. [CrossRef]

27. Bhuria, V.; Xing, J.; Scholta, T.; Bui, K.C.; Nguyen, M.L.T.; Malek, N.P.; Bozko, P.; Plentz, R.R. Hypoxia induced Sonic Hedgehog signaling regulates cancer stemness, epithelial-to-mesenchymal transition and invasion in cholangiocarcinoma. *Exp. Cell Res.* **2019**, *385*, 111671. [CrossRef]

28. Wang, L.; Jin, J.Q.; Zhou, Y.; Tian, Z.; Jablons, D.M.; He, B. GLI is activated and promotes epithelial-mesenchymal transition in human esophageal adenocarcinoma. *Oncotarget* **2018**, *9*, 853–865. [CrossRef]

29. Maik-Rachline, G.; Hacohen-Lev-Ran, A.; Seger, R. Nuclear ERK: Mechanism of Translocation, Substrates, and Role in Cancer. *IJMS* **2019**, *20*, 1194. [CrossRef]

30. Michailovici, I.; Harrington, H.A.; Azogui, H.H.; Yahalom-Ronen, Y.; Plotnikov, A.; Ching, S.; Stumpf, M.P.H.; Klein, O.D.; Seger, R.; Tzahor, E. Nuclear to cytoplasmic shuttling of ERK promotes differentiation of muscle stem/progenitor cells. *Development* **2014**, *141*, 2611–2620. [CrossRef]

31. Plotnikov, A.; Flores, K.; Maik-Rachline, G.; Zehorai, E.; Kapri-Pardes, E.; Berti, D.A.; Hanoch, T.; Besser, M.J.; Seger, R. The nuclear translocation of ERK1/2 as an anticancer target. *Nat. Commun.* **2015**, *6*, 6685. [CrossRef] [PubMed]

32. Liu, F.; Feng, X.X.; Zhu, S.L.; Huang, H.Y.; Chen, Y.D.; Pan, Y.F.; June, R.R.; Zheng, S.G.; Huang, J.L. Sonic Hedgehog Signaling Pathway Mediates Proliferation and Migration of Fibroblast-Like Synoviocytes in Rheumatoid Arthritis via MAPK/ERK Signaling Pathway. *Front. Immunol.* **2018**, *9*, 18001. [CrossRef] [PubMed]

33. Fiaschetti, G.; Schroeder, C.; Castelletti, D.; Arcaro, A.; Westermann, F.; Baumgartner, M.; Shalaby, T.; Grotzer, M.A. NOTCH ligands JAG1 and JAG2 as critical pro-survival factors in childhood medulloblastoma. *Acta Neuropathol. Commun.* **2014**, *2*, 39. [CrossRef] [PubMed]

2

Long-Pentraxin 3 Affects Primary Cilium in Zebrafish Embryo and Cancer Cells via the FGF System

Jessica Guerra [1,†], Paola Chiodelli [1,†], Chiara Tobia [1], Claudia Gerri [1,2]🆔 and Marco Presta [1,3,*]🆔

1   Department of Molecular and Translational Medicine, University of Brescia, 25123 Brescia, Italy;
     j.guerra@unibs.it (J.G.); paola.chiodelli@unibs.it (P.C.); chiara.tobia@unibs.it (C.T.);
     claudia.gerri@crick.ac.uk (C.G.)
2   Francis Crick Institute, London NW1 1AT, UK
3   Italian Consortium for Biotechnology (CIB), 25123 Brescia, Italy
*   Correspondence: marco.presta@unibs.it
†   The two authors contributed equally to this work.

Abstract: Primary cilium drives the left-right asymmetry process during embryonic development. Moreover, its dysregulation contributes to cancer progression by affecting various signaling pathways. The fibroblast growth factor (FGF)/FGF receptor (FGFR) system modulates primary cilium length and plays a pivotal role in embryogenesis and tumor growth. Here, we investigated the impact of the natural FGF trap long-pentraxin 3 (*PTX3*) on the determination of primary cilium extension in zebrafish embryo and cancer cells. The results demonstrate that down modulation of the *PTX3* orthologue *ptx3b* causes the shortening of primary cilium in zebrafish embryo in a FGF-dependent manner, leading to defects in the left-right asymmetry determination. Conversely, *PTX3* upregulation causes the elongation of primary cilium in FGF-dependent cancer cells. Previous observations have identified the PTX3-derived small molecule NSC12 as an orally available FGF trap with anticancer effects on FGF-dependent tumors. In keeping with the non-redundant role of the FGF/FGR system in primary cilium length determination, NSC12 induces the elongation of primary cilium in FGF-dependent tumor cells, thus acting as a ciliogenic anticancer molecule in vitro and in vivo. Together, these findings demonstrate the ability of the natural FGF trap PTX3 to exert a modulatory effect on primary cilium in embryonic development and cancer. Moreover, they set the basis for the design of novel ciliogenic drugs with potential implications for the therapy of FGF-dependent tumors.

Keywords: FGF; long-pentraxin 3; primary cilium; cancer; zebrafish

---

1. Introduction

Primary cilia are antenna-like organelles protruding from most mammalian cells [1] to act as chemosensors and mechanosensors for external stimuli [2]. The primary cilium is composed by an axoneme made by microtubule triplets anchored at the cell by a basal body [3]. The basal body derives from the mother centriole of the centrosome [4], which is fundamental to cell division. The cilium is resorbed prior to mitosis to release the centrioles. Then, ciliogenesis occurs shortly after cytokinesis has been completed [3].

Recently, the zebrafish embryo has been used as a model for the study of primary cilia functions in development and diseases [5,6]. During embryonic development, primary cilia in the embryonic node are involved in the left-right asymmetry process [7] and genetic defects in primary cilia are associated with a variety of pathological conditions that are grouped under the name "ciliopathies" (reviewed in [8–10]). Due to their role in the modulation of various signaling pathways, including Hedgehog (Hh) and Wnt, dysregulation of primary cilia plays an important role also in cancer progression [11–13].

Thus, "ciliotherapy" approaches have been proposed for cancer therapy in which ciliogenic drugs hamper tumor cell proliferation in part through induction of the primary cilium [14].

Long-pentraxin 3 (*PTX3*) is a soluble patter recognition receptor involved in the innate immunity arm [15]. PTX3 plays a role in wound healing/tissue remodeling, cardiovascular diseases, fertility, and infectious diseases [16]. In common with other pentraxins, the *C*-terminal portion of PTX3 includes the pentraxin-signature [17] whereas its unique *N*-terminal extension is responsible for binding to different members of the fibroblast growth factor (FGF) superfamily (including FGF2, FGF6, FGF8b, FGF10, and FGF17), thus preventing their interaction with all members of the FGFR family (FGFR1–4) [18–20]. Thus, PTX3 inhibits FGF-dependent responses, such as endothelial cell proliferation in vitro and angiogenesis in vivo [19,21,22], and exerts an oncosuppressive effect on FGF-dependent tumors, including multiple myeloma, melanoma, fibrosarcoma, lung, prostate, and bladder cancers [23–29]. In a therapeutic perspective, these findings led to the identification of the PTX3-derived small molecule NSC12 as the first orally available FGF trap able to inhibit the activity of all the members of the canonical FGF family by preventing their binding to FGFR1, 2, 3, and 4 [24]. This confers to NSC12 the capacity to inhibit the tumorigenic, angiogenic, and metastatic activity of tumors in which ligand-dependent activation of the FGF/FGF receptor (FGFR) system represents a driving force [24].

The FGF/FGFR system modulates primary cilium length in human and murine fibroblasts and chondrocytes [30]. In addition, it controls primary cilium extension in zebrafish embryos [31]. However, despite its ability to act as a natural FGF trap, no data are available about a possible involvement of PTX3 in primary cilium length determination under physiological and pathological conditions, including cancer. In the present research, we investigated the effect of the modulation of *PTX3* expression on primary cilium extension in zebrafish embryo and cancer cells. Our data demonstrate that down modulation of the *PTX3* orthologue *ptx3b* causes the shortening of primary cilia in zebrafish embryo in a FGF-dependent manner, leading to defects in the left-right asymmetry determination. Conversely, human *PTX3* (*hPTX3*) upregulation causes the elongation of primary cilia in different FGF-dependent cancer cell lines, including TRAMP-C2 prostate cancer cells that originate from the transgenic adenocarcinoma of the mouse prostate (TRAMP) model [32]. Accordingly, the PTX3-derived FGF trap NSC12 acts in vitro and in vivo as a ciliogenic anticancer molecule on FGF-dependent tumor cells. Together, our findings demonstrate for the first time the ability of PTX3 to exert a modulatory effect on primary cilium, shedding a new light on the manifold biological functions of this soluble pattern recognition receptor in embryonic development and cancer. In addition, they set the basis for the design of novel PTX3-derived ciliogenic drugs able to affect a different aspect of the biology of FGF-dependent tumors.

2. Results

2.1. In Silico Analysis of hPTX3 Co-Orthologs in Zebrafish

According to the Gene and HomoloGene databases at NCBI [PMID: 25398906], two putative co-orthologs of *hPTX3*, named *ptx3a* and *ptx3b*, are present in the zebrafish genome. They are located on the chromosomes 18 and 2, respectively, and are organized in three exons and two introns as their human counterpart.

CLUSTAL Omega alignment (https://www.ebi.ac.uk/Tools/msa/clustalo/) of the FASTA protein sequences of *hPTX3* (NP_002843.2), zPtx3a (XP_021329017.1), and zPtx3b (XP_694358.3) showed that *hPTX3* shares 39.74% amino acid sequence identity with Ptx3a and 41.13% identity with Ptx3b (Figure S1). Moreover, the canonical pentraxin signature and the conserved cysteine residues Cys-210 and Cys-271 are present in both zebrafish co-orthologs. Based on the Synteny Database program (http://syntenydb.uoregon.edu/synteny_db/), both zebrafish genes share a syntenic cluster of genes with *hPTX3*. In detail, when considering a window site of 50 genes, *ptx3a* shows three conserved genes (*selt1b, veph1, ccnL1*) (Figure S2A) whereas *ptx3b* shows ten conserved genes (*ccnl1b, golim4b, pccD10, slitrk3, samd7, sec62, gpr160, skil, phc3, prkcl*) (Figure S2B). Together, in silico data indicate that *ptx3a* and *ptx3b* are two bona-fide co-orthologs of *hPTX3*. In this research, we focused our attention on *ptx3b* due to its higher amino acid identity and conserved synteny with *hPTX3*.

2.2. Temporal and Spatial Expression of ptx3b during Zebrafish Development

The expression of *ptx3b* was analyzed at different stages of zebrafish embryo development by RT-PCR and whole-mount in situ hybridization (WISH). As shown in Figure 1A, *ptx3b* expression, detectable in the ovary, is absent at the four-cell stage, increases during epiboly, and remains constant from the five-somite stage (ss) to the 72 h post-fertilization (hpf) stage. During somitogenesis, the expression of *ptx3b* is restricted to the pronephric duct primordia where it was observed up to 48 hpf (Figure 1D–J). In addition, *ptx3b* is expressed in a transient manner at 26 hpf also in the notochord (Figure 1F,G), as highlighted by the analysis of paraffin-embedded transverse cross sections of the embryo trunk (Figure 1H), to be lost at 48 hpf (Figure 1I,J).

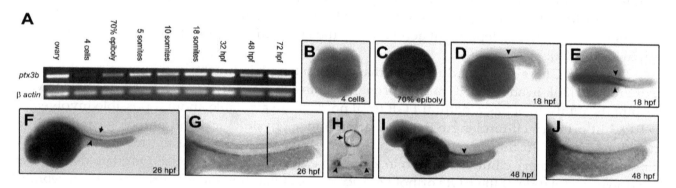

Figure 1. Zebrafish down modulation of the long-pentraxin 3 orthologue (*ptx3b*) expression. (**A**) RT-PCR analysis of *ptx3b* expression in the ovary and at the indicated developmental stages; (**B–J**) whole-mount in situ hybridization (WISH) analysis of *ptx3b* expression in zebrafish embryo at the indicated developmental stages; (**B,E**) dorsal view; (**C,D,F,G,I,J**) lateral view; (**G,J**) high magnification of the trunk region of embryos in (**F**) and (**I**); respectively; (**H**) transverse cross section of the trunk region of a 26 h post-fertilization (pf) embryo at the level of the black bar in (**G**); arrowheads in (**D–F,H,I**) indicate the pronephric ducts; arrows in (**F**) and (**G**) indicate the notochord.

2.3. ptx3b Knockdown Causes Defects in the Determination of Left-Right Asymmetry in Zebrafish

In order to assess the role of *ptx3b* on zebrafish embryo development, we used an antisense morpholino (MO) knockdown approach. To this purpose, a splicing MO (*ptx3b* MO) was designed to target the exon 2/intron 2 border of the *ptx3b* transcript. A five-mismatch nucleotide MO, unable to bind the *ptx3b* mRNA (ctrl MO), was used as control. As shown in Figure 2A, RT-PCR analysis performed at 28 hpf confirmed the targeting efficacy of the *ptx3b* MO that caused the skipping of exon 2 in the *ptx3b* transcript of embryo morphants when compared to controls. Based on these results, the dose of 0.6 pmol/embryo was considered as the optimal dose of *ptx3b* MO to be used for further studies.

When compared to controls, 40% of *ptx3b* morphants showed no morphologic alterations and 50% of them exhibited only moderate defects (Figure 2B,C). Given the hypothesis that the FGF trapping activity of *PTX3* may result in primary cilium alterations and consequent defects in the left-right asymmetry process [33], WISH analysis was performed to investigate the expression of laterality genes and the positioning of visceral organs in control and *ptx3b* morphants. As shown in Figure 2D,E, the majority of 16-ss and 22 hpf *ptx3b* morphants showed bilateral, right-sided, or absent expression of *spaw* and *pitx2* transcripts, two Nodal-related genes normally expressed at the left side of the embryo [34].

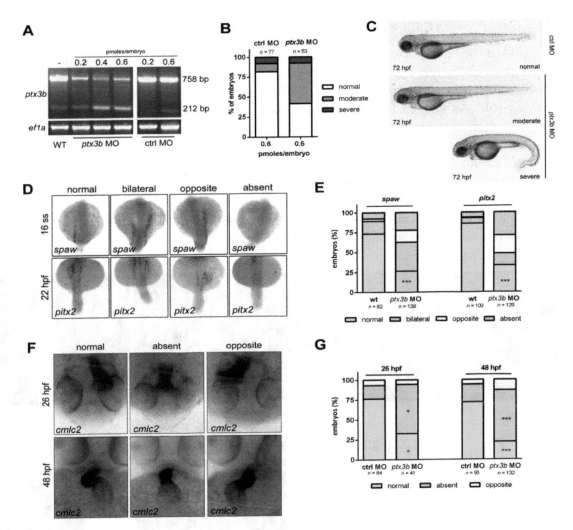

Figure 2. *ptx3b* downregulation causes left-right asymmetry defects in zebrafish embryo. (**A**) RT-PCR analysis showing the effect of different doses of ctrl or *ptx3b* morpholino (MO) on *ptx3b* expression in 28 hpf injected embryos. The efficacy of *ptx3b* MO is demonstrated by the presence of a specific 212 bp band in *ptx3b* MO-injected embryos, which confirms the occurrence of exon skipping. *ef1a* serves as control; (**B**) percentage of embryos showing normal, moderate, or severe phenotype at 72 hpf after the injection of 0.6 pmoles of ctrl MO or *ptx3b* MO, respectively; (**C**) representative bright field whole mount pictures of the phenotypes observed in ctrl and *ptx3b* morphants; (**D**) WISH representative pictures of the alterations of the expression of the laterality genes *spaw* and *pitx2* observed in *ptx3b* morphants at 16 ss and 22 hpf, respectively; (**E**) percentage of embryos with normal or altered expression of *spaw* and *pitx2* in WT (untreated) and *ptx3b* MO-injected embryos. Data are the mean of 4 and 3 independent experiment, respectively; (**F**) WISH representative pictures of *cmlc2* expression during normal and altered cardiac jogging and looping at 26 and 48 hpf, respectively; (**G**) percentage of embryos with normal or altered cardiac jogging and looping in ctrl MO and *ptx3b* MO-injected embryos. Data are the mean of 2 and 3 independent experiments, respectively. * $p < 0.05$, *** $p < 0.001$, ANOVA. *n*, total number of analyzed embryos.

Next, we examined the positioning of heart, liver, and pancreas in control and *ptx3b* MO-injected embryos by WISH analysis using the tissue-specific probes *cmlc2*, *prox1*, and *islet1*, respectively. As shown in Figure 2F,G, the *cmlc2*+ cardiac jogging and looping processes, which occur at 26 and 48 hpf, respectively, were absent in the majority of *ptx3b* morphants. Moreover, *ptx3b* down-modulation caused significant alterations in the positioning of *islet1*+ dorsal pancreatic bud and of *prox1*+ liver when assessed at 24 hpf and 48 hpf, respectively (Figure S3).

2.4. ptx3b Modulates Primary Cilium Length Determination via the FGF/FGFR System in Zebrafish

Primary cilia of the embryonic node, named Kupffer's vesicle (KV) in zebrafish [5], are involved in the determination of the left-right asymmetry during development [7,35]. On this basis, the length of KV primary cilia was measured in control and *ptx3b* morphants after acetylated α-tubulin immunostaining [6]. As shown in Figure 3, primary cilia of the KV are significantly shorter in *ptx3b* morphants when compared to control animals. At variance, no difference in the number of KV primary cilia was observed in *ptx3b* MO-injected embryos versus controls (mean ± S.E.M. equal to 21.5 ± 2.7 (n = 30) versus 22.7 ± 2.0 (n = 35) cilia per KV, respectively). To confirm the specificity of the *ptx3b* MO effects, one-cell stage embryos were co-injected with the *ptx3b* MO and an excess of *hPTX3* mRNA. As shown in Figure 3D,E, *hPTX3* mRNA was able to rescue the shortening of KV primary cilia caused by the *ptx3b* MO. Accordingly, rescued embryos showed a reduced percentage of cardiac looping defects when compared to *ptx3b* morphants (Figure 3F).

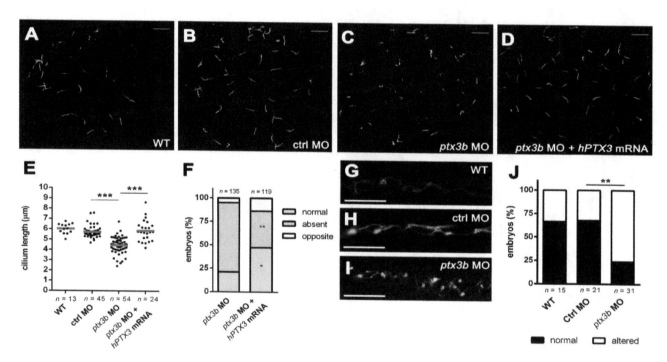

Figure 3. *ptx3b* knockdown alters primary cilium in zebrafish embryo. (**A–D**) Representative images of acetylated α-tubulin⁺ ciliary axonemes in the Kupffer's vesicle (KV) of 10 ss WT (untreated) embryos (**A**) and of embryos injected with ctrl MO (**B**); *ptx3b* MO (**C**); and *ptx3b* MO plus human *PTX3* (*hPTX3*) mRNA (**D**); scale bar: 10 μm; (**E**) primary cilium length (in μm) in the KVs of untreated embryos (WT) and of embryos injected with ctrl MO, *ptx3b* MO or *ptx3b* MO plus *hPTX3* mRNA (500 pg/embryo). Each dot represents the mean length of primary cilia measured in a single KV. *n*, total number of examined KVs. Data were obtained in three independent experiments, *** $p < 0.001$, Student's *t*-test; (**F**) percentage of embryos showing *cmcl²⁺* normal or altered cardiac looping at 48 hpf after injection with *ptx3b* MO or *ptx3b* MO plus *hPTX3* mRNA. Data are the mean of three independent experiments. * $p < 0.05$, ** $p < 0.01$, ANOVA. *n*, total number of examined embryos; (**G–I**) representative pictures of acetylated α-tubulin⁺ ciliary axonemes in pronephric ducts of WT, ctrl MO, and *ptx3b* MO-injected embryos at 48 hpf stage. Scale bar: 10 μm; (**J**) percentage of embryos with normal or altered primary cilia organization in pronephric ducts. ** $p < 0.01$, Z-test, two-tailed.

Notably, in keeping with the alterations observed in the KV, *ptx3b* morphants showed also structural defects of the primary cilia of the pronephric ducts [36] (Figure 3G–J).

Given the capacity of *PTX3* to bind different members of the FGF family [29], these data raise the hypothesis that a modulation of the activity of the FGF/FGFR system might be responsible for the shortening of primary cilia in *ptx3b* morphants. Indeed, *ptx3b* MO injection caused a significant increase of FGFR1 phosphorylation in zebrafish embryos that was fully abolished by the co-injection with an excess of *hPTX3* mRNA (Figure 4A,B). To substantiate this hypothesis, *ptx3b* MO-injected embryos were treated at the shield stage with the selective tyrosine kinase FGFR inhibitor BGJ398 (50 nM in fish water). As anticipated, BGJ398 treatment was able to rescue the length of KV primary cilia in *ptx3b* morphants (Figure 4C–F).

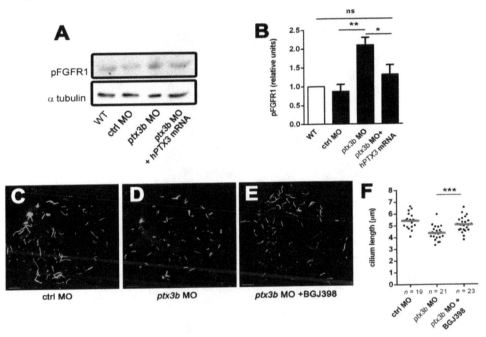

Figure 4. *ptx3b* regulates primary cilium length in zebrafish via the fibroblast growth factor (FGF)/FGF receptor (FGFR) system. (**A**) Western blot analysis of protein extracts from WT (untreated), ctrl MO, *ptx3b* MO, and *ptx3b* MO plus *hPTX3* mRNA-injected embryos probed with an anti-pFGFR1 antibody; uniform loading of the gel was confirmed using an anti-α tubulin antibody; (**B**) densitometric analysis of pFGFR1 levels normalized to α-tubulin expression. Data are the mean ± S.E.M. of 3 independent experiments, ** $p < 0.01$, * $p < 0.05$, Student's *t*-test; (**C–E**) representative images of the acetylated α-tubulin$^+$ KV ciliary axoneme of 10 ss embryos injected with ctrl MO or *ptx3b* MO and of *ptx3b* MO-injected embryos treated at shield stage with BGJ398 (50 nM). Scale bar: 10 μm; (**F**) primary cilium length (in μm) in the KVs of embryos treated as in (**C–E**). Each dot is the mean length of primary cilia measured in a single KV. *n*, total number of examined KVs. Data are from three independent experiments, *** $p < 0.001$, Student's *t*-test.

Together, these data show for the first time that *ptx3b* regulates the length of primary cilium axoneme during zebrafish embryo development by modulating the activity of the FGF/FGFR system, thus playing a non-redundant role in the left-right asymmetry process and visceral organ positioning in zebrafish.

2.5. PTX3 Regulates Primary Cilium Length in Cancer Cells via the FGF/FGFR System

Alterations of primary cilia may contribute to cancer progression [11,12]. Based on the results obtained in *ptx3b* zebrafish morphants and the well-known impact of *PTX3* on FGF-dependent tumors [23–27,37], we decided to investigate whether the modulation of *PTX3* expression may affect primary cilia in FGF-dependent cancer cells.

Stemming from the observation that primary cilium number and length decrease in a subset of pre-invasive human prostatic lesions [13], we measured the length of primary cilia in TRAMP-C2 prostate cancer cells and in their *hPTX3*-overexpressing counterpart (*hPTX3*-TRAMP-C2 cells), characterized by a reduced FGFR signaling and tumorigenic potential [37]. As anticipated, *hPTX3* overexpression resulted in a significant elongation of primary cilium in *hPTX3*-TRAMP-C2 cells when compared to TRAMP-C2 cells (Figure 5 and Figure S4). Similar results were obtained in *hPTX3* transfectants that originated from FGF-dependent human bladder cancer 5637 cells [27] and murine fibrosarcoma MC17-51 cells [26] (Figure 5). At variance, *hPTX3* overexpression did not affect the percentage of ciliated tumor cells in all the cell lines tested (Figure S5).

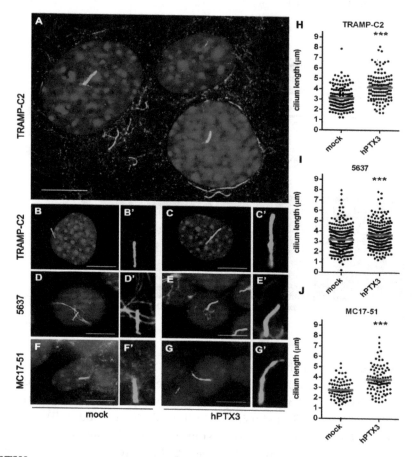

Figure 5. *PTX3* overexpression increases primary cilium length in different tumor cell lines. (**A**) Representative image of murine prostate TRAMP-C2 cells double immunostained with anti-acetylated α-tubulin (green) and anti-γ-tubulin (red) antibodies to visualize primary cilium axoneme and basal body, respectively. Nuclei were counterstained with DAPI (blue); (**B–G′**) representative images of primary cilium axoneme in mock (**B,B′**) and *hPTX3* overexpressing (**C,C′**); TRAMP-C2 cells; mock (**D,D′**) and *hPTX3* overexpressing (**E,E′**) human bladder carcinoma *hPTX3*-5637 cells; mock (**F,F′**) and *hPTX3* overexpressing (**G,G′**) murine fibrosarcoma MC17-51 cells. Scale bar: 10 μm; (**H–J**) the length (μm) of acetylated α-tubulin$^+$ cilia was measured in mock and *hPTX3* overexpressing cells using the ImageJ software. Black dots represent individual cilia; red bars show the mean value. Data are from three independent experiments, *** $p < 0.001$, Student's *t*-test.

PTX3 binds FGFs via its *N*-terminal domain [20]. To confirm the hypothesis that PTX3 may affect primary cilium length via the FGF/FGFR system in cancer cells, we measured the length of primary cilium in TRAMP-C2 cells stably transfected with the *N*-terminal *hPTX3* cDNA (*N*-term-*hPTX3*-TRAMP-C2 cells) or with the *C*-terminal *hPTX3* cDNA (*C*-term-*hPTX3*-TRAMP-C2 cells) [37]. As anticipated, overexpression of the FGF-binding *N*-terminal fragment of PTX3 resulted in an increase of the length of primary cilium when compared to mock-transfected cells, whereas overexpression of the PTX3 *C*-terminus was ineffective (Figure 6A). These results were confirmed by a rescue experiment in which treatment with exogenous recombinant FGF2 protein caused the shortening of the primary cilium in *N*-term-*hPTX3*-TRAMP-C2 cells with no effect on mock-TRAMP-C2 cells (Figure 6B).

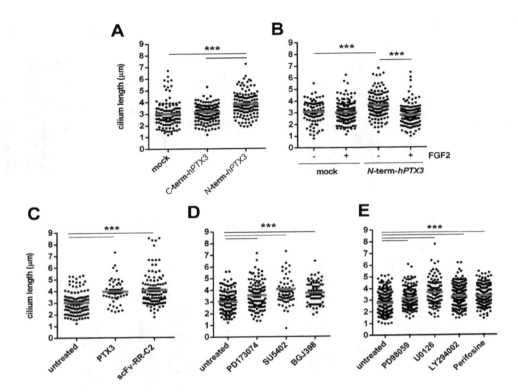

Figure 6. Inhibition of FGF signaling increases the length of primary cilium in TRAMP-C2 cells. (A–E) Cilia were visualized in serum-starved cells by acetylated α-tubulin immunostaining and their length was measured using the ImageJ software. Black dots represent individual cilia; red bars show the mean values. Data were obtained from three independent experiments, *** $p < 0.001$, Student's t-test; (A) primary cilium length in TRAMP-C2 cells overexpressing the C-terminal or the N-terminal fragment of human PTX3; (B) primary cilium length in mock_TRAMP-C2 and N-term-*hPTX3*-TRAMP-C2 cells treated for 48 h with 30 ng/mL FGF2; (C) primary cilium length in TRAMP-C2 cells treated for 48 h with recombinant *PTX3* protein (66 nM) or with anti-FGFR1 single-chain antibody fragment scFv-RR-C2 (300 nM); (D) primary cilium length in TRAMP-C2 cells treated for 48 h with the tyrosine kinase FGFR inhibitors PD173074 (100 nM), SU5402 (100 nM), or BGJ398 (100 nM); (E) primary cilium length in TRAMP-C2 treated for 48 h with the MAPK inhibitors PD98059 (10 μM) or U0126 (1.0 μM) or with the PI3K inhibitors LY294002 (10 μM) or perifosine (1.0 μM).

In keeping with these observations, treatment with recombinant PTX3 protein or with the neutralizing anti-FGFR1 single-chain antibody fragment scFv-RR-C2 [38] resulted in the elongation of primary cilium in TRAMP-C2 cells (Figure 6C). Accordingly, treatment with the tyrosine kinase FGFR inhibitors PD173074, SU5402 or BGJ398, or with different inhibitors of FGFR downstream signaling pathways, including the MAPK inhibitors PD98059 and U0126 and the PI3K inhibitors LY294002 and perifosine, were able to extend the length of primary cilium in TRAMP-C2 cells (Figure 6D,E). Together, the results indicate that the blockade of the FGF/FGFR system by extracellular or intracellular inhibitors exerts a significant impact on the primary cilium in TRAMP-C2 cells.

Activation of different tyrosine kinase receptors suppresses ciliogenesis in retinal epithelial cells by stabilizing the trichoplein-Aurora A complex following phosphorylation of the deubiquitinase USP8 [39]. Accordingly, the elongation of primary cilium observed in *hPTX3*-TRAMP-C2 cells was paralleled by a significant reduction of the intracellular levels of trichoplein (Figure 7A,B). Similar results were obtained in TRAMP-C2 cells treated with the tyrosine kinase FGFR inhibitors BGJ398 and SU5402 (Figure 7C,D).

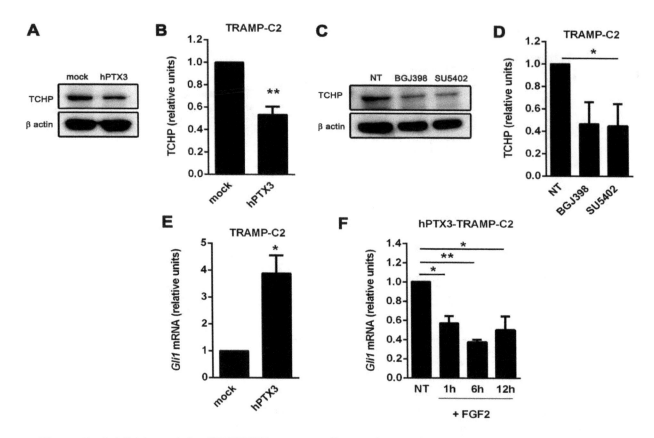

Figure 7. Inhibition of the FGF/FGFR system affects cilium-related signaling. (**A**) Serum-starved mock and *hPTX3*-TRAMP-C2 cell protein extracts were probed with an anti-trichoplein (TCHP) antibody. Uniform loading of the gel was assessed by probing the membrane with an anti-β actin antibody; (**B**) densitometric analysis of trichoplein levels normalized to β actin; (**C**) serum-starved mock-TRAMP-C2 cells were treated with the FGFR inhibitors BGJ398 (100 nM) or SU5402 (100 nM) for 48 h. After lysis, the extracts were probed with an anti-trichoplein antibody; (**D**) densitometric analysis of trichoplein levels normalized to β-actin; (**E**) *Gli1* expression in serum-starved mock and *hPTX3_*TRAMP-C2 cells; (**F**) serum-starved *hPTX3*-TRAMP-C2 cells were incubated with recombinant FGF2 (30 ng/mL) for 1, 6, or 12 h. Then, *Gli1* expression was evaluated by qRT-PCR and normalized to *Gaphd* mRNA levels. All data are the mean ± S.E.M. of 3–4 independent experiment, * $p < 0.05$, ** $p < 0.01$, Student's *t*-test.

Sustained activation of the FGF/FGFR system may affect primary cilium-dependent Hh signaling following the shortening of primary cilia [30]. On this basis, we investigated the impact of PTX3 on the activation of the Hh pathway in TRAMP-C2 cells using the downregulation of the expression of the Hh target transcriptional regulator GLI1 as a readout [40,41]. As shown in Figure 7E, *hPTX3* overexpression results in a significant increase of the levels of the *Gli1* transcript in *hPTX3*-TRAMP-C2 cells that was abolished by stimulation with exogenous FGF2 (Figure 7F).

2.6. The PTX3-Derived FGF Trap NSC12 Modulates Primary Cilium Length in Cancer Cells

Previous observations from our laboratory led to the discovery of NSC12 as the first PTX3-derived small molecule endowed with a potent anti-FGF activity [24,42]. Orally available, NSC12 inhibits the growth, angiogenic potential, and metastatic activity of various FGF-dependent tumors with potential implications in cancer therapy [24]. On this basis, we evaluated the effect of NSC12 on cell proliferation and primary cilium length in FGF-dependent TRAMP-C2, 5637, and MC17-51 cells. As shown in Figure 8A and Figure S6, NSC12 inhibits the proliferation of FGF-dependent tumor cells with an ID_{50} equal approximately to 1.0 μM. Accordingly, when administered at 1.0 μM concentration, NSC12 causes a significant elongation of the primary cilium in all the cell lines tested (Figure 8B).

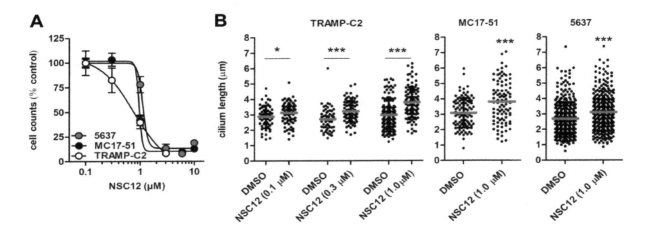

Figure 8. Ciliogenic activity of the FGF trap NSC12. Serum-starved TRAMP-C2, 5637, and MC17-51 cells were treated with the indicated concentrations of NSC12 or with the corresponding volume of vehicle (DMSO). After 48 h, cells were counted (**A**) and cilia lengths were evaluated with ImageJ software (**B**). Data are from two independent experiments, * $p < 0.05$, *** $p < 0.001$, Student's t-test.

Next, the capacity of NSC12 to exert a ciliogenic effect was evaluated in vivo on TRAMP-C2 tumor grafts. To this purpose, TRAMP-C2 cells were injected subcutaneously (s.c.) in syngeneic mice. At tumor take (30 days post-implantation), animals were treated i.p three times a week with NSC12 (5.0 mg/kg body wt) or vehicle. After 17 days, tumors were measured with calipers and harvested. Then, OCT embedded samples were immunostained with anti-acetylated α-tubulin antibody. In keeping with previous observations [24], NSC12 causes a significant inhibition of TRAMP-C2 tumor growth that was paralleled by a significant increase of the length of acetylated α-tubulin+ tumor cilium (Figure 9). At variance, NSC12 did not cause any change in α-tubulin immunoreactivity in TRAMP-C2 grafts (Figure S7).

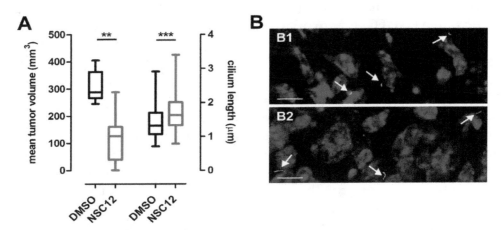

Figure 9. The FGF trap NSC12 exerts a ciliogenic activity on TRAMP-C2 tumors. (**A**) TRAMP-C2 cells were injected into the flank of C57BL/6 mice (10–15 animals/group). At tumor take, animals were treated i.p with NSC12 or vehicle. After 17 days, tumors were measured, harvested and frozen tissue sections were immunostained with an anti-acetylated α-tubulin antibody. The length of tumor cilia was measured in 40 microscopic fields from 3–5 tumors/group. Data are represented by box and whisker plots where the boxes extend from the 25th to the 75th percentiles, lines indicate the median values, and whiskers indicate the range of values. ** $p < 0.01$, *** $p < 0.001$, Student's t-test; (**B**) representative images of TRAMP-C2 tumors treated with vehicle (**a**) or NSC12 (**b**) and immunostained with anti-acetylated α-tubulin antibody (green) to visualize primary cilia (arrows). Nuclei were counterstained with DAPI (blue). Scale bar: 10 μm.

3. Discussion

The primary cilium is a microtubule-based structure that protrudes from the surface of most mammalian cells. It functions as a cellular antenna that captures environmental signals and serves as a hub for developmental and homeostatic signaling pathways [43].

PTX3 is a soluble patter recognition receptor and a key player of the innate immunity arm with non-redundant functions in pathogen recognition and inflammatory responses [15,16]. In addition, PTX3 is endowed with the capacity to exert an antitumor activity by acting as a natural FGF trap [29]. In this research, we provide experimental evidence that ptx3b, an ortholog of hPTX3, affects zebrafish development by regulating the length of the axoneme of the primary cilia of the KV, playing a non-redundant role in the left-right asymmetry process and visceral organ positioning during zebrafish embryogenesis. These effects appear to be mediated by the capacity of Ptx3b to modulate FGF/FGFR signaling in zebrafish embryo, its downregulation leading to increased FGFR1 phosphorylation. This was paralleled by shortening of the primary cilium axoneme in the KV, rescued by treatment with the selective tyrosine kinase FGFR inhibitor BGJ398. These data extend recent observations about the capacity of FGFR1 or FGF ligand inactivation to cause primary cilium shortening in zebrafish and Xenopus embryos [31]. Accordingly, FGF signaling regulates the length of primary cilia in various stem, embryonic, and differentiated mammalian cell types [30]. In addition, human skeletal dysplasia caused by activating FGFR3 mutations are characterized by alterations of the primary cilium [30,44].

Compelling experimental evidence indicates that alterations of primary cilium may play an important role also in tumors by affecting various intracellular signal transduction pathways (reviewed in [8,12]). Moreover, alterations of primary cilium affect different aspects of tumor biology, including cancer cell autophagy and apoptosis, response to hypoxia, epithelial–mesenchymal transition, and drug resistance [45]. In this frame, various types of cancer cells fail to express the primary cilium, including pancreatic, breast, prostate cancer, and melanoma cells [13,45–50]. Hence, restoration of the primary cilium in cancer cells may provide novel opportunities for therapeutic antineoplastic interventions. In this frame, drug repurposing "ciliotherapy" approaches have been proposed to hamper tumor growth by induction of the primary cilium [14].

Deregulation of the FGF/FGFR network occurs in tumors due to gene amplification, activating mutations and oncogenic fusions [51–53]. The multifaceted FGFR signaling network is engaged in the progression of different FGF-dependent tumors by acting on both tumor and stromal cell compartments, thus affecting oncogenesis through different mechanisms, including cell-signaling deregulation, angiogenesis, and resistance to cancer therapies [54]. On this basis, different classes of therapeutics have been developed, including non-selective and selective tyrosine kinase FGFR inhibitors, monoclonal antibodies, and FGF ligand traps [55,56]. Some of them are under investigation in clinical trials in different FGFR-related cancer settings [53].

In this frame, PTX3 has been shown to act as an oncosuppressor in different FGF-dependent tumors [24] by inhibiting tumor growth, neovascularization, and metastatic dissemination [29]. Here, we demonstrate that PTX3 has a ciliogenic effect on different cancer cell types. Experimental evidences indicate that this capacity is due to its FGF trap activity. (i) Overexpression in TRAMP-C2 cells of the FGF binding N-terminal fragment of PTX3 resulted in the elongation of primary cilium, whereas overexpression of the PTX3 C-terminus was ineffective; (ii) tyrosine kinase FGFR inhibitors and inhibitors of FGFR downstream signaling pathways cause primary cilium elongation in TRAMP-C2 cells. In parallel, tyrosine kinase FGFR inhibitors cause a significant reduction of the intracellular levels of trichoplein that, in complex with Aurora A, exerts suppressive effects on ciliogenesis [39]; (iii) hPTX3 overexpression leads to Gli1 upregulation, a marker of cilium-dependent Hh signaling [40], which is prevented by treatment with exogenous FGF2 protein.

NSC12 is a PTX3-derived small molecule able to bind all the extracellular members of the FGF family [24]. Numerous experimental evidences indicate that NSC12 may represent a prototype for the development of novel orally available therapeutic agents targeting FGF-dependent tumors [53]. Indeed, when compared to tyrosine kinase FGFR inhibitors, NSC12 is characterized by a reduced toxicity that may result in a more favorable therapeutic window [54]. Here we show that NSC12 inhibits the proliferation of FGF-dependent murine prostate cancer TRAMP-C2 cells, human bladder carcinoma 5637 cells, and murine fibrosarcoma MC17-51 cells (see also [26,27,37]) and that this inhibition is paralleled by the elongation of the primary cilium in all the cell lines tested. These data were confirmed in vivo, where NSC12 was able to inhibit tumor growth, and to induce a ciliogenic effect in TRAMP-C2 tumor grafts.

The present research focused mainly on the effect on primary cilium of FGF2 and FGFR1, the prototypic members of the FGF/FGFR system. Thus, further experiments will be required to assess the impact exerted on primary cilium by the modulation of different FGFs and FGFRs. Nevertheless, given the capacity of NSC12 to inhibit the activity of all the canonical FGFs and their interaction with all FGFRs [24], this small molecule FGF trap may represent the basis for the design of novel ciliogenic anticancer drugs able to affect different aspect of the biology of FGF-dependent tumors.

4. Materials and Methods

4.1. Chemicals

FGF2 was obtained from Peprotech (Rocky hill, NJ, USA), NSC12 was synthesized by M. Mor (University of Parma, Parma, Italy) as described [22], ScFv RR-C2 was isolated as described [42], rhPTX3 was kindly provided by B. Bottazzi, (Humanitas Institute, Milan, Italy). PD173074, SU5402, PD98059, and LY294002 were from Sigma-Aldrich (St. Louis, MO, USA), BGJ393 was from Selleckchem (Houston, TX, USA), UO126 was from MedChem Express, (Monmouth Junction, NJ, USA) and perifosine was from Aeterna Zentaris (Frankfurt, Germany).

4.2. Bioinformatic Analysis

Zebrafish genomic sequences were analyzed using the Gene and HomoloGene databases at the National Center for Biotechnology Information (NCBI, https://www.ncbi.nlm.nih.gov/) (PMID: 25398906). The protein sequences of the two putative orthologue Ptx3a (XP_021329017.1) and Ptx3b (XP_694358.3) were aligned to the human *PTX3* sequence found on the NCBI database (NP_002843.2) with the program CLUSTAL Omega (https://www.ebi.ac.uk/Tools/msa/clustalo/), while syntenic regions were analyzed with the Synteny Database program (http://syntenydb.uoregon.edu/synteny_db/).

4.3. Zebrafish Maintenance and Collection

The wild-type zebrafish AB line was maintained at 28 °C under standard conditions and embryos were staged as described [57]. To examine embryos older than 22 hpf, fish water was added with 0.2 mM 1-phenil-2-thiourea (PTU, Sigma-Aldrich). For the observation of the in vivo phenotypes, embryos were anesthetized using 0.16 mg/mL Tricaine (Sigma).

4.4. Morpholino Injection

ptx3b MO (5'-CTGAATCATGTACCTGAGGGCAGAT-3'; Gene Tools, Philomath, OR, USA) targeting the exon 2/intron 2 border of the *ptx3b* transcript was injected at the indicated concentrations in 1–4 cell stage embryos. A five-mismatch nucleotide ctrl MO (5'-CTCAATGATCTACCTCACGGCAGAT-3') was used as control. To confirm the targeting efficacy of the *ptx3b* MO, alternative splicing pattern analysis was performed on zebrafish embryos using appropriate primers (Table 1).

Table 1. Oligonucleotide primers used for RT-PCR and qPCR analysis.

Gene	Forward	Reverse
ptx3b	5'-TTGGCAGACTGAAGACATGG-3'	5'-GGGCAGATAGCGGTTTACGGT-3'
ptx3b skipping exon	5'-ACGTACCAGAATGTATGCG-3'	5'-CATGGAGGGTGTTACTTC-3'
β-actin	5'-CGAGCAGGAGATGGGAACC-3'	5'-CAACGGAAACGCTCATTGC-3'
ef-1a	5'-GGTACTTCTCAGGCTGACTGT-3'	5'-CAGACTTGACCTCAGTGGTTA-3'
Gli1	5'-TTCAAGGCCCAATACATGCT-3'	5'-GCGTCTTGAGGTTTTCAAGG-3'
Gapdh	5'-CATGGCCTTCCGTGTTCCTAC-3'	5'-TTGCTGTTGAAGTCGCAGGAG-3'

4.5. Whole-Mount In Situ Hybridization

Digoxigenin-labelled RNA probes were transcribed from linear cDNA constructs (Roche Applied Science, Penzberg, Germany). WISH was performed on embryos fixed in 4% PFA as described [58]. For sectioning, zebrafish embryos were post-fixed in 4% PFA after WISH, dehydrated in ethanol series, cleared in xilol, and paraffin-embedded overnight.

4.6. RT-PCR and qPCR Analysis

Total RNA was isolated from untreated and MO-injected embryos at different stages of development using TRIzol® Reagent (Invitrogen, Carlsbad, CA, USA) according to the manufacturer's instructions. Two micrograms of total RNA were retrotranscribed and 100 ng of cDNA were used for the evaluation of the alternative *ptx3b* splicing pattern by semi-quantitative RT-PCR analysis. The whole gels are shown in Figure S4.

For the evaluation of *Gli1* expression in cancer cells, total RNA was isolated with the same procedure. Then, 1/10th of the retrotranscribed cDNA was used for quantitative qPCR that was performed with the ViiA 7 Real-Time PCR System (Thermo Fisher Scientific, Waltham, MA, USA) using iTaq Syber Green Supermix (Biorad, Hercules, CA, USA) according to the manufacturer's instructions. Samples were analyzed in triplicate and normalized with respect to the levels of *Gapdh* expression using the appropriate primers (Table 1).

4.7. Cell Cultures

Murine prostate adenocarcinoma TRAMP-C2 cells (ATCC CRL-2731) and *hPTX3* transfectants [37] were maintained in DMEM supplemented with 10% FBS, penicillin/streptomycin (100 U/mL and 10 mg/mL, respectively), 10 mM HEPES buffer, 0.5 mM 2-mercaptoethanol, 2.0 mM glutamine, 5 mg/mL bovine insulin (Sigma), and 10 nM DHT. Human bladder carcinoma 5637 cells (ATCC HTB-9) and murine fibrosarcoma MC17-51 cells (MC-TGS17-51) and their corresponding *hPTX3* transfectants were grown as described [26,27]. All cells were maintained in a humidified 5% CO_2 incubator at 37 °C. Cells were maintained at low passage, returning to original frozen stocks every 3–4 months, and tested regularly for Mycoplasma negativity.

4.8. Cell Proliferation Assay

Cells were seeded on 48-well plates at 8×10^3 cells/cm². After 24 h, cells were treated with increasing concentrations of NSC12 in serum free DMEM (TRAMP-C2 and MC17-51 cells) or RPMI (5637 cells). After a further 48 h incubation, cells were trypsinized and viable cell counting was performed with the MACSQuant® Analyzer (Miltenyi Biotec, Bergisch Gladbach, Germany).

4.9. Tumor Grafts in Mice

Animal studies were approved by the local animal ethics committee (Organismo Preposto al Benessere degli Animali, Università degli Studi di Brescia, Brescia, Italy) and by the Italian Ministero della Salute. All the procedures and animal care were conformed to institutional guidelines that comply with national and international laws and policies (EEC Council Directive 86/609, OJL 358, 12 December 1987).

C57BL/6 (Charles River, Calco, Italy) male mice were maintained under standard housing conditions. TRAMP-C2 cells (5×10^6 in 200 μL of PBS) were injected s.c. into the dorsolateral flank of nine week-old animals. At tumor take (30 days post-implantation), animals were treated i.p three times a week with NSC12 (5.0 mg/kg body wt) or vehicle. Tumors were measured in two dimensions and volume was calculated according to the formula $V = (D \times d^2)/2$, where D and d are the major and minor perpendicular tumor diameters, respectively. After 17 days, tumors were harvested, embedded in OCT compound, and immediately frozen.

4.10. Immunoflurescence Analysis

Control and *ptx3b* MO injected embryos were fixed in 4% paraformaldehyde for 2 h at room temperature. Whole-mount immunofluorescence analysis of cilium axoneme was performed using a mouse anti-acetylated α-tubulin antibody (1:1000, Sigma) followed by Alexa Fluor 488 anti-mouse IgG (Molecular Probes, Eugene, OR, USA).

Tumor cells were seeded on glass coverslips and serum-starved for 48–72 h. Then, cells were fixed in 4% paraformaldehyde, permeabilized with 0.5% Triton X-100, saturated with 3% BSA in PBS, and incubated with an anti-acetylated α-tubulin antibody (1:1000, Sigma) and an anti-γ-tubulin antibody (1:1000, Sigma). Secondary antibodies anti-mouse Alexa Fluor 488 and anti-rabbit Alexa Fluor 594 (Molecular Probes) were used at 1:300 dilutions. Nuclei were counterstained with DAPI.

For tumor samples, 15 μm cryostat sections were air dried, fixed with 4% paraformaldehyde for 15 min, and permeabilized with 0.5% Triton X-100/PBS for 15 min at room temperature (RT). Sections were blocked for 2 h at RT with 1% BSA, 0.1% fish gelatin (Sigma-Aldrich), 0.1% Triton X-100 and 0.05% Tween20 (blocking solution), and incubated for a further 1 h at RT with a M.O.M blocking reagent (Vector Laboratories, Burlingame, CA, USA) to block endogenous mouse IgGs. Then, sections were incubated overnight at 4 °C with anti-acetylated α-tubulin antibody (1:1000 in blocking solution) and for one further hour at RT with anti-mouse Alexa Fluor 488 secondary antibody (1:250) in blocking solution. Nuclei were counterstained with DAPI.

Immunostained cells, embryos, and tissue sections were analyzed using a Zeiss Axiovert 200M epifluorescence microscope equipped with a Plan-Apochromat × 63/1.4 NA oil objective. Cilium length (in μm) was quantified manually using the "segmented line" tool of ImageJ (NIH).

4.11. Western Blot Analysis

Five zebrafish embryos were pooled for each experimental point and were sonicated in sample buffer (250 mM Tris-HCl pH 6.8, 8% SDS, 40% glycerol, 200 mM DTT, 0.02% bromophenol blue), while tumor cells were directly harvested in the same buffer. The protein concentration in the lysates was assessed using the Bradford protein assay (Bio-Rad Laboratories, Milano, Italy). Then, lysates were run on a SDS-PAGE gel, transferred onto a PVDF membrane, and immunocomplexes were visualized by chemiluminiscence (Bio-Rad). The following primary antibodies were used: rabbit anti-pFGFR1 (1:1000, Santa Cruz Biotechnology, Dallas, TX, USA), mouse anti-α-tubulin (1:1000, Sigma), rabbit anti-trichoplein (TCHP) (1:700, Bioss, Woburn, UK), mouse anti-β-actin (1:1000, Sigma). The whole blots showing all the bands with molecular weight markers are shown in Figure S5.

5. Conclusions

In summary, our study demonstrates the ability of the natural FGF trap *PTX3* to exert a modulatory effect on primary cilium in embryonic development and cancer. Moreover, they set the basis for the design of novel *PTX3*-derived ciliogenic drugs with potential implications for the therapy of FGF-dependent tumors.

Supplementary Materials
Figure S1: *ptx3a* and *ptx3b* are zebrafish co-orthologue genes of human *PTX3*, Figure S2: Zebrafish *ptx3a/ptx3b* and human *PTX3* synteny, Figure S3: *ptx3b* knockdown causes defects in liver and pancreas positioning in zebrafish embryo, Figure S4: Effect of *PTX3* overexpression on primary cilium length in TRAMP-C2 cells, Figure S5: *hPTX3* overexpression does not affect the percentage of ciliated tumor cells, Figure S6: Effect of NSC12 on the proliferation of TRAMP-C2 cells, Figure S7: Anti-α-tubulin immunoreactivity in TRAMP-C2 tumors, Figure S8: RT-PCR raw data, Figure S9: Western blot raw data.

Author Contributions: Conceptualization, J.G., P.C., C.T., and M.P.; methodology, J.G., C.T., and P.C.; investigation, J.G., C.T., P.C., and C.G.; writing—original draft preparation, J.G., C.T., and M.P.; writing—review and editing, J.G., C.T., and M.P.; supervision, M.P.; project administration, M.P.; funding acquisition, M.P. All authors have read and agreed to the published version of the manuscript.

Acknowledgments: Probes for *spaw* and *pitx2* transcripts were kindly provided by M. Beltrame (University of Milan, Italy). NSC12 was provided by M. Mor (University of Parma, Italy). Tumor tissues were provided by R. Ronca (University of Brescia, Italy). We wish to thank G. Borsani for his support in the bioinformatics analysis and G. Gariano (University of Brescia, Italy) for technical assistance.

References

1. Plotnikova, O.V.; Pugacheva, E.N.; Golemis, E.A. Primary cilia and the cell cycle. *Methods Cell. Biol.* **2009**, *94*, 137–160. [PubMed]

2. Oh, E.C.; Katsanis, N. Cilia in vertebrate development and disease. *Development* **2012**, *139*, 443–448. [CrossRef] [PubMed]

3. Gerdes, J.M.; Davis, E.E.; Katsanis, N. The vertebrate primary cilium in development, homeostasis, and disease. *Cell* **2009**, *137*, 32–45. [CrossRef] [PubMed]

4. Nigg, E.A.; Stearns, T. The centrosome cycle: Centriole biogenesis, duplication and inherent asymmetries. *Nat. Cell. Biol.* **2011**, *13*, 1154–1160. [CrossRef]

5. Essner, J.J.; Amack, J.D.; Nyholm, M.K.; Harris, E.B.; Yost, H.J. Kupffer's vesicle is a ciliated organ of asymmetry in the zebrafish embryo that initiates left-right development of the brain, heart and gut. *Development* **2005**, *132*, 1247–1260. [CrossRef]

6. Zaghloul, N.A.; Katsanis, N. Zebrafish assays of ciliopathies. *Methods Cell. Biol.* **2011**, *105*, 257–272.

7. Drummond, I.A. Cilia functions in development. *Curr. Opin. Cell. Biol.* **2012**, *24*, 24–30. [CrossRef]

8. Reiter, J.F.; Leroux, M.R. Genes and molecular pathways underpinning ciliopathies. *Nat. Rev. Mol. Cell. Biol.* **2017**, *18*, 533–547. [CrossRef]

9. Waters, A.M.; Beales, P.L. Ciliopathies: An expanding disease spectrum. *Pediatric Nephrol.* **2011**, *26*, 1039–1056. [CrossRef]

10. Yuan, S.; Sun, Z. Expanding horizons: Ciliary proteins reach beyond cilia. *Annu. Rev. Genet.* **2013**, *47*, 353–376. [CrossRef]

11. Eguether, T.; Hahne, M. Mixed signals from the cell's antennae: Primary cilia in cancer. *EMBO Rep.* **2018**, *19*. [CrossRef]

12. Peixoto, E.; Richard, S.; Pant, K.; Biswas, A.; Gradilone, S.A. The primary cilium: Its role as a tumor suppressor organelle. *Biochem. Pharmacol.* **2020**, *175*, 113906. [CrossRef]

13. Hassounah, N.B.; Nagle, R.; Saboda, K.; Roe, D.J.; Dalkin, B.L.; McDermott, K.M. Primary cilia are lost in preinvasive and invasive prostate cancer. *PLoS ONE* **2013**, *8*, e68521. [CrossRef]

14. Khan, N.A.; Willemarck, N.; Talebi, A.; Marchand, A.; Binda, M.M.; Dehairs, J.; Rueda-Rincon, N.; Daniels, V.W.; Bagadi, M.; Thimiri Govinda Raj, D.B.; et al. Identification of drugs that restore primary cilium expression in cancer cells. *Oncotarget* **2016**, *7*, 9975–9992. [CrossRef]

15. Bottazzi, B.; Garlanda, C.; Salvatori, G.; Jeannin, P.; Manfredi, A.; Mantovani, A. Pentraxins as a key component of innate immunity. *Curr. Opin. Immunol.* **2006**, *18*, 10–15. [CrossRef]

16. Garlanda, C.; Bottazzi, B.; Bastone, A.; Mantovani, A. Pentraxins at the crossroads between innate immunity, inflammation, matrix deposition, and female fertility. *Annu. Rev. Immunol.* **2005**, *23*, 337–366. [CrossRef]

17. Doni, A.; Garlanda, C.; Mantovani, A. Innate immunity, hemostasis and matrix remodeling: PTX3 as a link. *Semin. Immunol.* **2016**, *28*, 570–577. [CrossRef]

18. Presta, M.; Foglio, E.; Churruca Schuind, A.; Ronca, R. Long Pentraxin-3 Modulates the Angiogenic Activity of Fibroblast Growth Factor-2. *Front. Immunol.* **2018**, *9*, 2327. [CrossRef]

19. Leali, D.; Alessi, P.; Coltrini, D.; Rusnati, M.; Zetta, L.; Presta, M. Fibroblast growth factor-2 antagonist and antiangiogenic activity of long-pentraxin 3-derived synthetic peptides. *Curr. Pharm. Des.* **2009**, *15*, 3577–3589. [CrossRef]

20. Camozzi, M.; Rusnati, M.; Bugatti, A.; Bottazzi, B.; Mantovani, A.; Bastone, A.; Inforzato, A.; Vincenti, S.; Bracci, L.; Mastroianni, D.; et al. Identification of an antiangiogenic FGF2-binding site in the N terminus of the soluble pattern recognition receptor PTX3. *J. Biol. Chem.* **2006**, *281*, 22605–22613. [CrossRef]

21. Rusnati, M.; Camozzi, M.; Moroni, E.; Bottazzi, B.; Peri, G.; Indraccolo, S.; Amadori, A.; Mantovani, A.; Presta, M. Selective recognition of fibroblast growth factor-2 by the long pentraxin PTX3 inhibits angiogenesis. *Blood* **2004**, *104*, 92–99. [CrossRef] [PubMed]

22. Presta, M.; Camozzi, M.; Salvatori, G.; Rusnati, M. Role of the soluble pattern recognition receptor PTX3 in vascular biology. *J. Cell. Mol. Med.* **2007**, *11*, 723–738. [CrossRef] [PubMed]

23. Ronca, R.; Di Salle, E.; Giacomini, A.; Leali, D.; Alessi, P.; Coltrini, D.; Ravelli, C.; Matarazzo, S.; Ribatti, D.; Vermi, W.; et al. Long pentraxin-3 inhibits epithelial-mesenchymal transition in melanoma cells. *Mol. Cancer Ther.* **2013**, *12*, 2760–2771. [CrossRef]

24. Ronca, R.; Giacomini, A.; Di Salle, E.; Coltrini, D.; Pagano, K.; Ragona, L.; Matarazzo, S.; Rezzola, S.; Maiolo, D.; Torrella, R.; et al. Long-pentraxin 3 derivative as a small-molecule FGF trap for cancer therapy. *Cancer Cell* **2015**, *28*, 225–239. [CrossRef]

25. Leali, D.; Alessi, P.; Coltrini, D.; Ronca, R.; Corsini, M.; Nardo, G.; Indraccolo, S.; Presta, M. Long pentraxin-3 inhibits FGF8b-dependent angiogenesis and growth of steroid hormone-regulated tumors. *Mol. Cancer Ther.* **2011**, *10*, 1600–1610. [CrossRef] [PubMed]

26. Rodrigues, P.F.; Matarazzo, S.; Maccarinelli, F.; Foglio, E.; Giacomini, A.; Silva Nunes, J.P.; Presta, M.; Dias, A.A.M.; Ronca, R. Long pentraxin 3-mediated fibroblast growth factor trapping impairs fibrosarcoma growth. *Front. Oncol.* **2018**, *8*, 472. [CrossRef] [PubMed]

27. Matarazzo, S.; Melocchi, L.; Rezzola, S.; Grillo, E.; Maccarinelli, F.; Giacomini, A.; Turati, M.; Taranto, S.; Zammataro, L.; Cerasuolo, M.; et al. Long pentraxin-3 follows and modulates bladder cancer progression. *Cancers* **2019**, *11*, 1277. [CrossRef] [PubMed]

28. Ronca, R.; Ghedini, G.C.; Maccarinelli, F.; Sacco, A.; Locatelli, S.L.; Foglio, E.; Taranto, S.; Grillo, E.; Matarazzo, S.; Castelli, R.; et al. FGF trapping inhibits multiple myeloma growth through c-Myc degradation-induced mitochondrial oxidative stress. *Cancer Res.* **2020**. [CrossRef]

29. Giacomini, A.; Ghedini, G.C.; Presta, M.; Ronca, R. Long pentraxin 3: A novel multifaceted player in cancer. *Biochim. Biophys. Acta* **2018**, *1869*, 53–63. [CrossRef]

30. Kunova Bosakova, M.; Varecha, M.; Hampl, M.; Duran, I.; Nita, A.; Buchtova, M.; Dosedelova, H.; Machat, R.; Xie, Y.; Ni, Z.; et al. Regulation of ciliary function by fibroblast growth factor signaling identifies FGFR3-related disorders achondroplasia and thanatophoric dysplasia as ciliopathies. *Hum. Mol. Genet.* **2018**, *27*, 1093–1105. [CrossRef]

31. Neugebauer, J.M.; Amack, J.D.; Peterson, A.G.; Bisgrove, B.W.; Yost, H.J. FGF signalling during embryo development regulates cilia length in diverse epithelia. *Nature* **2009**, *458*, 651–654. [CrossRef] [PubMed]

32. Greenberg, N.M.; DeMayo, F.; Finegold, M.J.; Medina, D.; Tilley, W.D.; Aspinall, J.O.; Cunha, G.R.; Donjacour, A.A.; Matusik, R.J.; Rosen, J.M.; et al. Prostate cancer in a transgenic mouse. *Proc. Natl. Acad. Sci. USA* **1995**, *92*, 3439–3443. [CrossRef] [PubMed]

33. Capdevila, I.; Izpisua Belmonte, J.C. Knowing left from right: The molecular basis of laterality defects. *Mol. Med. Today* **2000**, *6*, 112–118. [CrossRef]

34. Long, S.; Ahmad, N.; Rebagliati, M. The zebrafish nodal-related gene southpaw is required for visceral and diencephalic left-right asymmetry. *Development* **2003**, *130*, 2303–2316. [CrossRef]

35. Komatsu, Y.; Mishina, Y. Establishment of left-right asymmetry in vertebrate development: The node in mouse embryos. *Cell. Mol. Life Sci.* **2013**, *70*, 4659–4666. [CrossRef]

36. Kramer-Zucker, A.G.; Olale, F.; Haycraft, C.J.; Yoder, B.K.; Schier, A.F.; Drummond, I.A. Cilia-driven fluid flow in the zebrafish pronephros, brain and Kupffer's vesicle is required for normal organogenesis. *Development* **2005**, *132*, 1907–1921. [CrossRef]

37. Ronca, R.; Alessi, P.; Coltrini, D.; Di Salle, E.; Giacomini, A.; Leali, D.; Corsini, M.; Belleri, M.; Tobia, C.; Garlanda, C.; et al. Long pentraxin-3 as an epithelial-stromal fibroblast growth factor-targeting inhibitor in prostate cancer. *J. Pathol.* **2013**, *230*, 228–238. [CrossRef]

38. Ronca, R.; Benzoni, P.; Leali, D.; Urbinati, C.; Belleri, M.; Corsini, M.; Alessi, P.; Coltrini, D.; Calza, S.; Presta, M.; et al. Antiangiogenic activity of a neutralizing human single-chain antibody fragment against fibroblast growth factor receptor 1. *Mol. Cancer Ther.* **2010**, *9*, 3244–3253. [CrossRef]

39. Kasahara, K.; Aoki, H.; Kiyono, T.; Wang, S.; Kagiwada, H.; Yuge, M.; Tanaka, T.; Nishimura, Y.; Mizoguchi, A.; Goshima, N.; et al. EGF receptor kinase suppresses ciliogenesis through activation of USP8 deubiquitinase. *Nat. Commun.* **2018**, *9*, 758. [CrossRef]

40. Hui, C.C.; Angers, S. Gli proteins in development and disease. *Annu. Rev. Cell. Dev. Biol.* **2011**, *27*, 513–537. [CrossRef]

41. Lu, J.; Liu, L.; Zheng, M.; Li, X.; Wu, A.; Wu, Q.; Liao, C.; Zou, J.; Song, H. MEKK2 and MEKK3 suppress Hedgehog pathway-dependent medulloblastoma by inhibiting GLI1 function. *Oncogene* **2018**, *37*, 3864–3878. [CrossRef] [PubMed]

42. Castelli, R.; Giacomini, A.; Anselmi, M.; Bozza, N.; Vacondio, F.; Rivara, S.; Matarazzo, S.; Presta, M.; Mor, M.; Ronca, R.; et al. Synthesis, structural elucidation, and biological evaluation of NSC12, an orally available Fibroblast Growth Factor (FGF) ligand trap for the treatment of FGF-dependent lung tumors. *J. Med. Chem.* **2016**, *59*, 4651–4663. [CrossRef]

43. Singla, V.; Reiter, J.F. The primary cilium as the cell's antenna: Signaling at a sensory organelle. *Science* **2006**, *313*, 629–633. [CrossRef]

44. Martin, L.; Kaci, N.; Estibals, V.; Goudin, N.; Garfa-Traore, M.; Benoist-Lasselin, C.; Dambroise, E.; Legeai-Mallet, L. Constitutively-active FGFR3 disrupts primary cilium length and IFT20 trafficking in various chondrocyte models of achondroplasia. *Hum. Mol. Genet.* **2018**, *27*, 1–13. [CrossRef]

45. Fabbri, L.; Bost, F.; Mazure, N.M. Primary cilium in cancer hallmarks. *Int. J. Mol. Sci.* **2019**, *20*, 1336. [CrossRef]

46. Yuan, K.; Frolova, N.; Xie, Y.; Wang, D.; Cook, L.; Kwon, Y.J.; Steg, A.D.; Serra, R.; Frost, A.R. Primary cilia are decreased in breast cancer: Analysis of a collection of human breast cancer cell lines and tissues. *J. Histochem. Cytochem.* **2010**, *58*, 857–870. [CrossRef]

47. Seeley, E.S.; Carriere, C.; Goetze, T.; Longnecker, D.S.; Korc, M. Pancreatic cancer and precursor pancreatic intraepithelial neoplasia lesions are devoid of primary cilia. *Cancer Res.* **2009**, *69*, 422–430. [CrossRef]

48. Kim, J.; Dabiri, S.; Seeley, E.S. Primary cilium depletion typifies cutaneous melanoma in situ and malignant melanoma. *PLoS ONE* **2011**, *6*, e27410. [CrossRef]

49. Hassounah, N.B.; Bunch, T.A.; McDermott, K.M. Molecular pathways: The role of primary cilia in cancer progression and therapeutics with a focus on Hedgehog signaling. *Clin. Cancer Res.* **2012**, *18*, 2429–2435. [CrossRef]

50. Higgins, M.; Obaidi, I.; McMorrow, T. Primary cilia and their role in cancer. *Oncol. Lett.* **2019**, *17*, 3041–3047. [CrossRef]

51. Helsten, T.; Elkin, S.; Arthur, E.; Tomson, B.N.; Carter, J.; Kurzrock, R. The FGFR landscape in cancer: Analysis of 4853 tumors by next-generation sequencing. *Clin. Cancer Res.* **2016**, *22*, 259–267. [CrossRef] [PubMed]

52. Patani, H.; Bunney, T.D.; Thiyagarajan, N.; Norman, R.A.; Ogg, D.; Breed, J.; Ashford, P.; Potterton, A.; Edwards, M.; Williams, S.V.; et al. Landscape of activating cancer mutations in FGFR kinases and their differential responses to inhibitors in clinical use. *Oncotarget* **2016**, *7*, 24252–24268. [CrossRef] [PubMed]

53. Ghedini, G.C.; Ronca, R.; Presta, M.; Giacomini, A. Future applications of FGF/FGFR inhibitors in cancer. *Expert Rev. Anticancer Ther.* **2018**, *18*, 861–872. [CrossRef] [PubMed]

54. Presta, M.; Chiodelli, P.; Giacomini, A.; Rusnati, M.; Ronca, R. Fibroblast growth factors (FGFs) in cancer: FGF traps as a new therapeutic approach. *Pharmacol. Ther.* **2017**, *179*, 171–187. [CrossRef]

55. Dai, S.; Zhou, Z.; Chen, Z.; Xu, G.; Chen, Y. Fibroblast Growth Factor Receptors (FGFRs): Structures and Small Molecule Inhibitors. *Cells* **2019**, *8*, 614. [CrossRef] [PubMed]

56. Giacomini, A.; Chiodelli, P.; Matarazzo, S.; Rusnati, M.; Presta, M.; Ronca, R. Blocking the FGF/FGFR system as a "two-compartment" antiangiogenic/antitumor approach in cancer therapy. *Pharmacol. Res.* **2016**, *107*, 172–185. [CrossRef]

3

The Aberrant Expression of the Mesenchymal Variant of FGFR2 in the Epithelial Context Inhibits Autophagy

Monica Nanni [1,†], Danilo Ranieri [1,†], Flavia Persechino [1], Maria Rosaria Torrisi [1,2,†] and Francesca Belleudi [1,*,†]

1   Laboratory affiliated to Istituto Pasteur Italia—Fondazione Cenci Bolognetti, Department of Clinical and Molecular Medicine, Sapienza University of Rome, 00185 Rome, Italy

2   S. Andrea University Hospital, 00189 Rome, Italy

*   Correspondence: francesca.belleudi@uniroma1.it

†   These authors contributed equally.

Abstract: Signaling of the epithelial splice variant of fibroblast growth factor receptor 2 (FGFR2b) triggers both differentiation and autophagy, while the aberrant expression of the mesenchymal FGFR2c isoform in epithelial cells induces impaired differentiation, epithelial mesenchymal transition (EMT) and tumorigenic features. Here we analyzed in the human keratinocyte cell line, as well as in primary cultured cells, the possible impact of FGFR2c forced expression on the autophagic process. Biochemical and quantitative immunofluorescence analysis, coupled to the use of autophagic flux sensors, specific substrate inhibitors or silencing approaches, showed that ectopic expression and the activation of FGFR2c inhibit the autophagosome formation and that AKT/MTOR is the downstream signaling mainly involved. Interestingly, the selective inhibition of AKT or MTOR substrates caused a reversion of the effects of FGFR2c on autophagy, which could also arise from the imbalance of the interplay between AKT/MTOR pathway and JNK1 signaling in favor of JNK1 activation, BCL-2 phosphorylation and possibly phagophore nucleation. Finally, silencing experiments of depletion of ESRP1, responsible for FGFR2 splicing and consequent FGFR2b expression, indicated that the switching from FGFR2b to FGFR2c isoform could represent the key event underlying the inhibition of the autophagic process in the epithelial context. Our results provide the first evidence of a negative impact of the out-of-context expression of FGFR2c on autophagy, suggesting a possible role of this receptor in the modulation of the recently proposed negative loop between autophagy and EMT during carcinogenesis.

Keywords: FGFR2c; autophagy; keratinocyte; MTOR; JNK1

1. Introduction

The fibroblast growth factor receptors (FGFR1-4) are four receptor tyrosine kinases regulating key processes, such as cell proliferation, differentiation, migration and survival [1,2]. The alternative splicing of the IgIII loop in FGFR1-3, generates the FGFRIIIb or the FGFRIIIc isoforms, which are mainly expressed in epithelial and mesenchymal tissues, respectively [3]. Deregulation of FGF/FGFR signaling can play either oncogenic or tumor suppressive roles [2,4]. In this regard, the epithelial isoform of FGFR2 (FGFR2b) is a well-recognized regulator of epidermal differentiation and skin homeostasis [5–7] exerting a tumor suppressive role in vitro and in vivo [8,9]. According to these studies, our group has demonstrated that FGFR2b controls the entire program of human keratinocyte differentiation [10–12] and that PKCδ and PKCα signaling downstream FGFR2b are involved in different steps of this process [12]. However, we also found that, in the same epidermal tissue context, the altered FGFR2 splicing and the aberrant expression of the mesenchymal FGFR2c isoform induces changes in the specificity for FGFs, leading to impairment of differentiation [13], epithelial mesenchymal transition (EMT) and early tumorigenic features [14]. The observation that FGFR2b/FGFR2c switching is also induced in keratinocytes by the E5 oncoprotein of human papillomavirus 16 (HPV16E5) [15], which is expressed in the early stages of virus infection, further supports the hypothesis that FGFR2c aberrant expression might be a precocious event in epithelial tumorigenesis.

At the light of growing evidences showing that FGFs would control cell differentiation by regulating autophagy in several tissues [16–18], we have also demonstrated that FGFR2b signaling triggers the autophagic process in human keratinocytes, showing that FGFR2b-induced autophagy and receptor-mediated early differentiation are interplaying events [19–21] and that JNK1 is the downstream signaling pathway at the crossroad between them [21,22].

However, autophagy not only regulates several biological functions, such as cell differentiation, but can also play either onco-suppressive or oncogenic roles in cancer, depending on its stage. In particular, during the initial steps of tumorigenesis autophagy, it appears to be linked to a negative loop to EMT, in established tumors this process has a pro-survival effect and its possible interplay with EMT remains still debated [23–25]. Therefore, keeping in mind the role of FGFR2c in driving EMT, here we pointed on to establish if and how the aberrant expression of FGFR2c could impact on autophagy. The results obtained showed that FGFR2c expression and signaling in epithelial context negatively interfere with the autophagic process, suggesting that this interference could significantly contribute to cancerogenesis.

2. Results

2.1. FGFR2c Expression and Signaling Inhibit Autophagy in Human Keratinocytes

To analyze the effect of FGFR2c expression and signaling on autophagy and to compare it to that previously described by us for FGFR2b [19–22], we took advantage of the human keratinocyte HaCaT clones stably transduced with pBp-FGFR2b or pBp-FGFR2c retroviral constructs or with empty pBp vector, as negative control [14]. Cells were left untreated or stimulated with FGF7, the specific ligand of FGFR2b, or with FGF2, which does not bind to FGFR2b, but is able to activate other FGFRs including FGFR2c. Western blot analysis showed that, while FGF7 stimulation increased the levels of the band corresponding to the lipidated form of the microtubule associated protein 1 light chain 3 (MAP1LC3/LC3-II) in all clones (Figure 1A), as expected [19–22], FGF2 treatment significantly decreased them only in cells ectopically expressing FGFR2c (Figure 1A). Both the FGFR2b-dependent induction and FGFR2c-mediated inhibition of autophagy were abolished by the specific FGFR2 tyrosine kinase inhibitor SU5402 (Figure 1A), demonstrating that, in both cases, the signaling of FGFR2 isoforms was required. Since it is widely accepted that autophagy is not only a post-translationally regulated, but also a transcriptionally controlled process [26] and we have previously shown that FGFR2b signaling plays a role in this transcriptional control [20,21], we wondered whether FGFR2c might impact on autophagy also by affecting the expression of LC3 gene. To this aim, the mRNA transcript levels of LC3, which we have previously demonstrated to be increased in response to FGF7 in keratinocytes [20,21], were estimated by real-time relative RT-PCR. The results showed that, while FGF7 stimulation increased the expression of this gene in all clones (Figure 1B), FGF2 treatment does not affect them, even in HaCaT-pBp FGFR2c (Figure 1B). The impact of FGFR2c aberrant expression and signaling on autophagy was also investigated by immunofluorescence approach. Quantitative immunofluorescence analysis showed that, while LC3 signal intensity, as well as the number of LC3 positive dots per cell, were increased by FGF7 stimulation in all clones (Figure 1C), they appeared reduced in response to FGF2 only in HaCaT pBp-FGFR2c cells (Figure 1C). Again, all the observed effects were abolished by SU5402 (Figure 1C), confirming the requirement of FGFR2 isoform activation. Thus, the ectopic expression of FGFR2c and its signaling, which is known to exert an oncogenic outcome in human keratinocytes, appear also to negatively impact on autophagy.

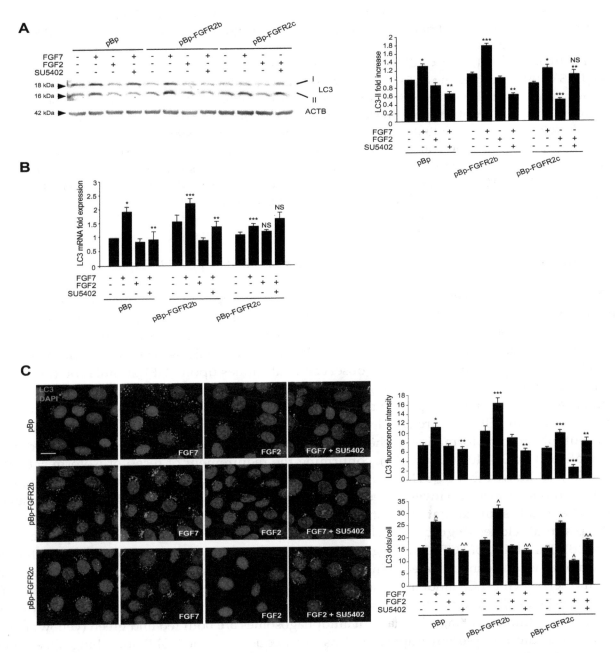

Figure 1. The ectopic expression of FGFR2c and its signaling inhibit autophagy in human keratinocytes. HaCaT cells, stably transduced with pBp-FGFR2b or pBp-FGFR2c constructs or with the empty pBp retroviral vector as control, were left untreated or stimulated with FGF7 or FGF2 in presence or absence of the FGFR2 tyrosine kinase inhibitor SU5402 as described in Material and Methods. (**A**) Western blot analysis showed that, while FGF7 stimulation increases the levels of LC3-II in all clones, FGF2 stimulation significantly decreased them only in HaCaT pBp-FGFR2c cells. All the observed effects were abolished by the presence of SU5402. Equal loading was assessed with the anti-ACTB antibody. For the densitometric analysis the values from three independent experiments were normalized and expressed as fold increases and are reported as mean values ± standard deviations (SD). Student's t test was performed, and significance levels are defined as $P < 0.05$. * $p < 0.05$ and *** $p < 0.001$ vs the corresponding FGF-unstimulated cells; ** $p < 0.05$ vs the corresponding SU5402-untreated cells; not significant (NS) vs the corresponding FGF-unstimulated, SU5402-untreated cells. (**B**) Real-time Reverse Transcriptase-Polymerase Chain Reaction (RT-PCR) analysis shows that while FGF7 stimulation induces

the increases of LC3 mRNA transcripts in all clones, FGF2 treatment does not affect them. The results observed in HaCaT pBp and pBp-FGFR2b upon FGF7 stimulation were abolished by SU5402. Results are expressed as mean values ± SE. Student's t test was performed, and significance levels were defined as $P < 0.05$. * $p < 0.01$, *** $p < 0.05$ and NS vs the corresponding FGF-unstimulated cells; ** $p < 0.05$ and NS vs the corresponding SU5402-untreated-cells. (**C**) Quantitative immunofluorescence analysis shows that LC3 signal intensity was increased by FGF7 stimulation in all clones, but it appears strongly reduced upon FGF2 treatment only in HaCaT pBp-FGFR2c cells. The observed effects were abolished by SU5402 treatment. Quantitative analysis of the fluorescence intensity and LC3 positive dots per cell were performed as described in Materials and Methods, and the results are expressed as mean values ± standard errors (SE). The student's t test was performed, and significance levels were defined as $P < 0.05$. * $p < 0.01$, *** $p < 0.001$ and ^ $p < 0.0001$, vs the corresponding FGF-unstimulated cells; ** $p < 0.001$ and ^^ $p < 0.0001$ vs the corresponding SU5402-untreated cells.

2.2. The Autophagosome Formation is the Autophagic Step Impaired by FGFR2c Expression and Signaling

The amount of intracellular autophagosomes usually depends on the balance between their formation and their lysosomal-mediated degradation. Therefore, in order to assess how the ectopic FGFR2c could impact on the autophagic flux, the levels of the well-known autophagy substrate SQSTM1/p62 (sequestosome 1) was estimated by Western blot analysis. The evident decrease of the 62 kDa band corresponding to SQSTM1, observed in all clones upon FGF7 stimulation (Figure 2A), confirmed the ability of FGFR2b signaling to trigger mainly the autophagosome assembly. In contrast, the significant increase of the SQSTM1 band, observed exclusively in HaCaT pBp-FGFR2c clones and only in response to FGF2 (Figure 2A), indicated that FGFR2c signaling might act via the inhibition of new autophagosome formation, rather than by accelerating their turnover. The observed effects were abolished by SU5402 (Figure 2A), confirming the requirement of receptor isoform activation. Since it is well known that SQSTM1 can be also transcriptionally regulated under conditions that modulate autophagy, we also investigated its mRNA expression levels in HaCaT clones stimulated as above. The results showed that FGF7 stimulation induced an evident decrease of SQSTM1 mRNA transcripts in all clones (Figure 2B), while FGF2 treatment did not significantly impact on them (Figure 2B). The ability of FGFR2c to negatively interfere with the phagosome formation, rather than their turnover, was also investigated using fluorescence approaches, transfecting HaCaT clones with a pDest-mCherry-EGFP-LC3 tandem construct [27]. In fact, mCherry-EGFP-LC3 is an autophagic flux sensor, since EGFP fluorescence (green) is quenched in acidic environments, whereas mCherry (red) is an acidic-stable fluorescent tag: The nascent autophagosomes are both red and green (yellow) labeled, whereas the acidic autolysosomes appear red, as a consequence of the EGFP quenching. Quantitative fluorescence analysis, performed on transfected cells left untreated or stimulated with FGFR2 ligands as above, showed that, while FGF7 stimulation increased both yellow and red dots (corresponding to autophagosomes and autophagolysosomes, respectively) (Figure 2C), FGF2 treatment significantly decreased them in HaCaT pBp-FGFR2c cells (Figure 2C). These results further confirmed that FGFR2c activation appear to inhibit new autophagosome assembly.

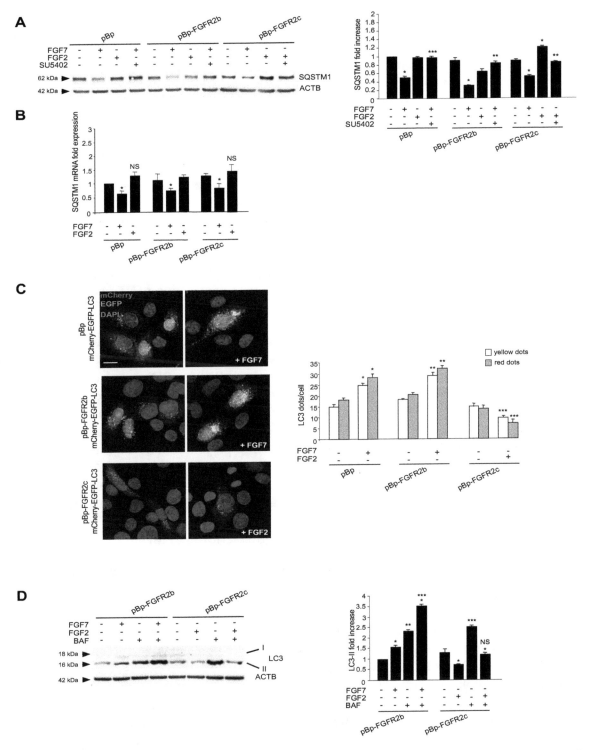

Figure 2. FGFR2c expression and signaling inhibits the autophagosome formation. (**A**) HaCaT pBp, pBp-FGFR2b or pBp-FGFR2c cells were left untreated or stimulated with FGF7 or FGF2 in presence or absence of the FGFR2 tyrosine kinase inhibitor SU5402 as above. Western blot analysis shows that, the 62 kDa band corresponding to SQSTM1 was significantly increased only in HaCaT pBp-FGFR2c clones after FGF2 stimulation, while it decreases in all clones upon FGF7 stimulation. All the observed effects were abolished by SU5402. Equal loading was assessed with anti-ACTB antibody. Densitometric analysis and the student's t test were performed as reported above. * $p < 0.05$ vs the corresponding FGF-unstimulated cells; ** $p < 0.05$ and *** $p < 0.001$ vs the corresponding SU5402-untreated cells. (**B**) HaCaT clones were stimulated with FGFR2 ligands as above. Real-time RT-PCR analysis shows that

FGF7 stimulation decreased SQSTM1 mRNA transcripts in all clones, while FGF2 treatment did not affect them. Results are expressed as mean values ± SE. The student's t test was performed, and significance levels were defined as $P < 0.05$. * $p < 0.01$ vs the corresponding FGF7-unstimulated cells; NS vs the corresponding FGF7-unstimulated cells. (C) HaCaT clones were transiently transfected with mCherry-EGFP-LC3 construct. Cells were then left untreated or stimulated with FGF7 or FGF2 as above. Quantitative fluorescence analysis showed that, while FGF7 stimulation increased both yellow and red dots in HaCaT pBp and HaCaT pBp-FGFR2b cells, FGF2 stimulation decreased them in HaCaT pBp-FGFR2c cells. Quantitative analysis was performed as described in Materials and Methods, and results were expressed as mean values ± SE. The student's t test was performed as reported in the legend to Figure 1B. * $p < 0.05$ and ** $p < 0.001$ vs the corresponding FGF7-unstimulated cells; *** $p < 0.01$ vs the corresponding FGF2-unstimulated cells. (D) HaCaT pBp-FGFR2b and pBp-FGFR2c clones were left untreated or stimulated with FGF7 or FGF2 in the presence or absence of bafilomycin A1 for the last 3 h. Western blot analysis showed that the increase in the levels of LC3-II upon FGF7 stimulation observed in HaCaT pBp-FGFR2b was further enhanced by bafilomycin while the decrease of LC3-II observed in HaCaT pBp-FGFR2c cells upon FGF2 stimulation was not recovered in the presence of the drug. Equal loading was assessed with anti-ACTB antibody. Densitometric analysis and the student's t test were performed as reported above. * $p < 0.05$ vs the corresponding FGF-unstimulated cells; ** $p < 0.05$, *** $p < 0.01$ and NS vs the corresponding bafilomycin-untreated cells.

Finally, we monitored the LC3-II levels in the presence or absence of the well-known inhibitor of the autophagosome-lysosome fusion bafilomycin A1. Western blot analysis showed that the effect of increase of LC3-II band generally induced by this drug (Figure 2D) was not found in HaCaT pBp-FGFR2c upon FGF2 stimulation (Figure 2D). These findings demonstrated that the observed decrease of LC3-II induced by FGF2 in HaCaT pBp-FGFR2c cannot be ascribed to an acceleration of the autophagic flux.

Since we have previously shown that FGFR2b signaling plays a role in the transcriptional control of several ATG genes, other than LC3 [20,21], we also assessed whether FGFR2c might impact on autophagy by affecting some of these genes To this aim, the mRNA transcript levels of the BECN1 and ATG5 were estimated by real-time RT-PCR. The results showed that FGF2 treatment did not affect them, even in HaCaT-pBp FGFR2c (Figure S1). Therefore, we feel confident to conclude that the interference exerted by FGFR2c on autophagy does not occur at the transcriptional level.

2.3. AKT/MTOR is the Downstream Pathway Responsible for FGFR2c-Mediated Inhibition of Autophagy

To search for the possible downstream signaling pathways responsible for FGFR2c-mediated autophagic repression, MAPK/ERK1/2 and AKT/MTOR signaling were first considered, since they represent the main pathways involved in FGFR-mediated inhibition of autophagy [16,28–30]. Western blot analysis demonstrated that, only in cells ectopically expressing FGFR2c, FGF2 stimulation triggered ERK1/2, as well as AKT and MTOR phosphorylation (Figure 3); these effects were abolished by the FGFR2 kinase inhibitor SU5402 (Figure 3), confirming their dependence from FGFR2c activation. Then, in order to assess the possible involvement of these two pathways in FGFR2c-mediated repression of autophagy, we took advantage of specific substrate inhibitors: The ERK1/2 upstream substrates MAP2K/MEK1/2 inhibitor PD0325901 [31], the AKT inhibitor AKT-I-1/2 [32] and the widely used MTOR inhibitor rapamycin. The efficiency of each inhibitor was first assayed by Western blot analysis (Figure S2A). Then, we analyzed their effects on LC3-II expression in HaCaT pBp controls and HaCaT pBp-FGFR2c left untreated or stimulated with FGF2, as above. Surprisingly, Western blot analysis showed that, in the presence of either AKT inhibitor or rapamycin, but not of MEK1/2 inhibitor, the decrease of LC3-II levels induced by FGF2 stimulation in pBp-FGFR2c cells was not only recovered, but significantly increased, compared to the corresponding unstimulated cells (Figure 4A). These results indicated that the selective block of AKT/MTOR pathway was able to revert the effects on autophagy. No effects of all the inhibitors were detectable in cells not stimulated with FGF2 or in pBp control cells (Figure 4A), suggesting that, at least in our keratinocyte model, the ERK1/2 or AKT/MTOR shut-off did not significantly interfere with basal autophagy. The crucial role of AKT/MTOR pathway in FGFR2c-mediated inhibition of autophagy was also confirmed by immunofluorescence approaches, demonstrating that FGF2 stimulation dampened LC3 signal intensity, as well as reduced the LC3 positive dots per cell, only in HaCaT pBp-FGFR2c clones (Figure 4B); however, in the presence of the AKT inhibitor or rapamycin, this effect was completely reversed, resulting in a visible increase of the staining and number of LC3 dots (Figure 4B). Thus, AKT/MTOR is the FGFR2c downstream pathway responsible for autophagy inhibition; nevertheless, if this pathway is selectively switched-off, the negative effect of FGFR2c signaling cascade on autophagic process is not only recovered, but even reversed in autophagy induction. In order to further assess the interesting outcome of AKT/MTOR shut-off on FGFR2c-mediated autophagic effects, we carried out specific protein depletion by siRNA. HaCaT pBp and HaCaT pBp-FGFR2c clones were transfected with MTOR siRNA or with an unrelated siRNA, as negative control, and the efficiency of MTOR depletion was checked by Western blot analysis (Figure S2B). After siRNA transfection, cells were left untreated or stimulated with FGF2 as above. Western blot analysis showed that the decrease of LC3-II, evident in HaCaT pBp-FGFR2c control siRNA cells upon FGF2 stimulation (Figure 4C), turned into a clear increase upon MTOR depletion (Figure 4C).

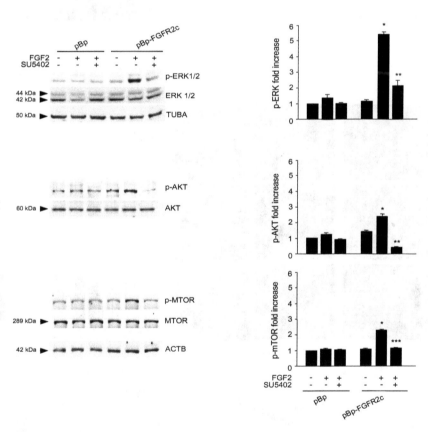

Figure 3. FGF2 stimulation activates ERK1/2 and AKT/MTOR signaling pathways in HaCaT pBp-FGFR2c. HaCaT pBp and HaCaT pBp-FGFR2c clones were left untreated or stimulated with FGF2 in the presence or absence of the FGFR2 tyrosine kinase inhibitor SU5402, as above. Western blot analysis performed using an antibody directed against the phosphorylated form of ERK1/2, AKT and MTOR demonstrates the activation of each substrates only in HaCaT pBp-FGFR2c upon FGF2 stimulation. These effects were abolished by SU5402 treatment. Equal loading was assessed with anti-ERK1/2, anti-AKT, and anti-MTOR antibodies. Densitometric analysis and Student's t test were performed as reported above. * $p < 0.05$ vs the corresponding FGF2-unstimulated cells; ** $p < 0.05$ and *** $p < 0.01$ vs the corresponding SU5402-untreated cells.

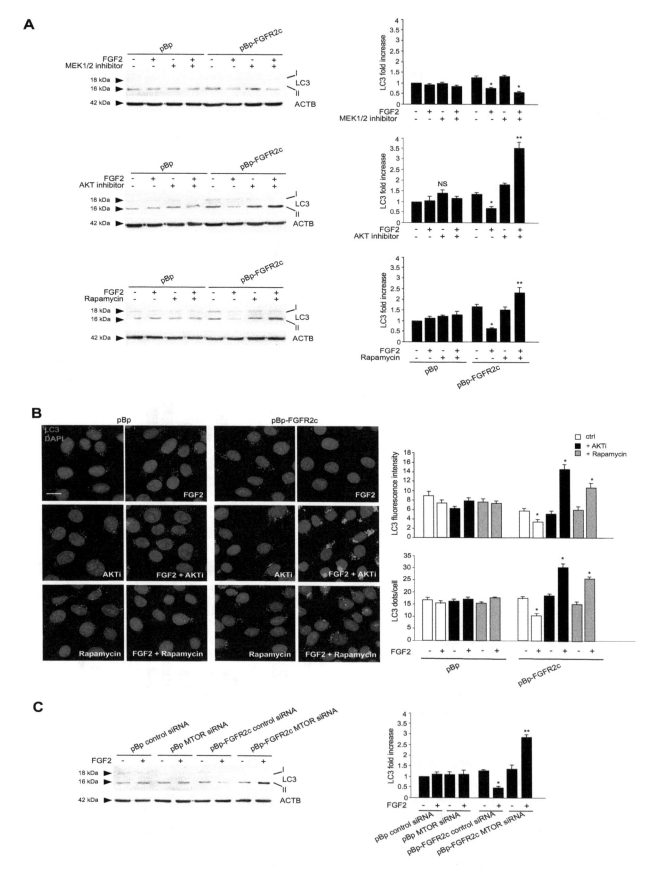

Figure 4. AKT/MTOR is the signaling pathway downstream FGFR2c responsible for FGFR2c-mediated inhibition of autophagy. (**A**) HaCaT pBp and HaCaT pBp-FGFR2c clones were left untreated or stimulated

with FGF2 in the presence or absence of the indicated substrate inhibitors as reported in Materials and Methods. Western blot analysis shows that both AKT inhibitor and rapamycin increased the level of LC3-II upon FGF2 stimulation in pBp-FGFR2c cells compared to the corresponding unstimulated cells, while MEK1/2 inhibitor shows no effects. No effects of all the inhibitors are detectable in pBp-FGFR2c cells not stimulated with FGF2 or in pBp control cells. Equal loading was assessed with anti-ACTB antibody. Densitometric analysis and the student's t test were performed as reported above. * $p < 0.05$ vs the corresponding FGF2-unstimulated cells; ** $p < 0.05$ and NS vs the corresponding substrate inhibitor-untreated cells. (B) HaCaT pBp and HaCaT pBp-FGFR2c clones were left untreated or stimulated with FGF2 in the presence or absence of AKT inhibitor or rapamycin as reported in Materials and Methods. Quantitative immunofluorescence analysis showed that both AKT inhibitor and rapamicyn reversed the inhibitory effect of FGF2 on LC3 signal intensity and on LC3 positive dots formation only in HaCaT pBp-FGFR2c clones. Quantitative analysis was performed as described in Materials and Methods, and results are expressed as mean values ± SE. The student's t test was performed as reported above. * $p < 0.001$ vs the corresponding FGF2-unstimulated cells. (C) HaCaT pBp and HaCaT pBp-FGFR2c clones were transiently transfected with MTOR siRNA or an unrelated siRNA as a control. Cells were then left untreated or stimulated with FGF2 as described in Materials and Methods. Western blot analysis shows that the decrease of LC3-II, observed in HaCaT pBp-FGFR2c control siRNA cells upon FGF2 stimulation, was reversed upon MTOR depletion. Equal loading was assessed with anti-ACTB antibody. Densitometric analysis and the student's t test were performed as reported in the legend to Figure 1A. * $p < 0.05$ and ** $p < 0.01$ vs the corresponding FGF2-unstimulated cells.

2.4. The Reversion of the Effects of FGFR2c on Autophagy Upon the Selective Block of AKT/MTOR Pathway is Accompanied by JNK1 Activation

Searching for the signaling events possibly involved in the reversion of autophagic response to FGFR2c, we focused our attention on MAPK8/JNK1 pathway. In fact, it has been recently proposed that AKT/MTOR signaling exerts an inhibitory function on JNK [33], which is the pathway involved in the induction of autophagy mediated by various FGFRs [17], including FGFR2b [21,22]. In fact, PI3K/AKT/MTOR and JNK signaling are not independent pathways, but a complex network playing important biological roles in cancer [33] and cooperating in the control of different events, including autophagy [34]. Therefore, it is reasonable to suppose that, in our cellular model of keratinocytes ectopically expressing FGFR2c, the induction of autophagy upon AKT/MTOR shut down could be the consequence of the imbalance of the interplay between AKT/MTOR and JNK pathways in favor of JNK1 activation. In order to ascertain it, we checked the phosphorylation levels of JNK1 in the presence of AKT inhibitor: Western blot analysis showed that, in cells expressing FGFR2c, the basal phosphorylation of JNK1 was significantly decreased by FGF2 stimulation (Figure 5), but this effect was reversed by the presence of the AKT inhibitor (Figure 5). Overall, these results suggested that FGFR2c-induced repression of autophagy involves the AKT/MTOR pathway, which also inhibits JNK1 signaling. In fact, upon the selective block of AKT/MTOR pathway, JNK1 phosphorylation/activation increases and contributes to autophagy stimulation.

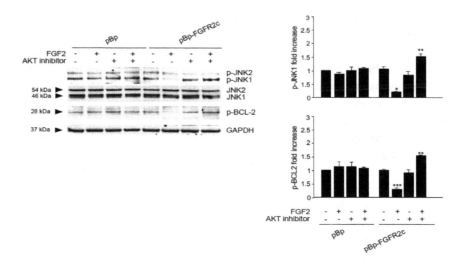

Figure 5. AKT signaling is required for FGFR2c-mediated inhibition of JNK e BCL-2 phosphorylation. HaCaT pBp and HaCaT pBp-FGFR2c clones were left untreated or stimulated with FGF2 in the presence or absence of AKT inhibitor as reported in Materials and Methods. Western blot analysis performed using antibody directed against the phosphorylated form of JNK and BCL-2, demonstrated that FGF2 stimulation decreases the basal phosphorylation of both substrates, while AKT inhibitor reverses this effect. Densitometric analysis and the student's t test were performed as reported in the legend to Figure 1A. * $p < 0.05$ and *** $p < 0.01$ vs the corresponding FGF2-unstimulated cells; ** $p < 0.05$ vs the corresponding AKT inhibitor-untreated cells.

It is well known that JNK1 signaling triggers the autophagic step of phagophore nucleation via BCL-2 phosphorylation and consequent BECN1 release from the BCL-2/BECN1 inhibitory complex [35–37]. This mechanism has been described for the autophagy triggered by FGFR4 [17] and FGFR2b [21]. Therefore, we wondered if FGFR2c activation, which inhibits JNK1 downstream signaling, could also negatively impact on BCL-2 phosphorylation. Western blot analysis showed that the basal phosphorylation of BCL-2 in HaCaT pBp-FGFR2c clones appeared decrease by FGF2 stimulation (Figure 5), but the presence of AKT inhibitor reversed this effect (Figure 5). Thus, while FGFR2c signaling appears to repress autophagy via AKT/MTOR phosphorylation/activation, which also negatively impact on JNK1-mediated phosphorylation of BCL-2, the selective block of AKT/MTOR pathway appears to redirect it toward JNK1 activation, BCL-2 phosphorylation and possibly phagophore nucleation.

2.5. Switching From the Epithelial FGFR2b to the Mesenchymal FGFR2c Isoform Underlies the FGF2-Mediated Inhibition of Autophagy in Epithelial Context

The epithelial splicing regulatory proteins (ESRPs), and in particular, the ESRP1 isoform are responsible for the FGFR2 splicing and consequent expression of the epithelial FGFR2b isoform [15,38]. Therefore, in order to assess if the FGFR2b versus FGFR2c isoform switching could represent a key event responsible for FGF2-induced inhibition of the autophagic process in epithelial context, we forced this event in keratinocytes, performing ESRP1 depletion by siRNA approach. HaCaT cells where transfected with ESRP1 siRNA (HaCaT ESRP1 siRNA) or with an unrelated siRNA (HaCaT control siRNA), as control, and then stimulated with FGF7 or FGF2 in the presence or not of the FGFR2 kinase inhibitor SU5402, as reported above. The efficiency of ESRP1 depletion was verified through either molecular (Figure 6A, left panel) and biochemical approaches (Figure 6A, right panel), while its effect on FGFR2 isoform expression was quantitated by real-time relative RT–PCR, using human fibroblasts (HFs) as positive control for FGFR2c expression. The results showed that ESRP1 depletion led to a significant decrease of FGFR2b expression (Figure 6B, left panel) and to the appearance of FGFR2c (Figure 6B, right panel), indicating that the correct splicing of the FGFR2 gene, occurring in epithelial context, had been impaired. Then, we focused our attention on the autophagic events. Western blot analysis showed that, the increase of LC3-II in response to FGF7, visible in HaCaT control siRNA cells (Figure 6C) and attributable to FGFR2b activation and signaling, was abolished by ESRP1 depletion (Figure 6C). In contrast, in response to FGF2 stimulation, these HaCaT ESRP1 siRNA cells showed a rate of LC3-II decrease (Figure 6C), as well as of MTOR phosphorylation/activation (Figure 6D) comparable to that observed in HaCaT pBp-FGFR2c cells (see Figures 1A and 3). These results were also confirmed in parallel experiments, using skin-derived primary human keratinocytes (HKs) transfected with ESRP1 siRNA or control siRNA and stimulated with FGF7 or FGF2 as above. Molecular and biochemical approaches were used to verify the efficiency of ESRP1 depletion (Figure 6E), as well as its ability to lead FGFR2b down-regulation and ex-novo expression of FGFR2c (Figure 6F). Finally, Western blot analysis confirmed that ESRP1 depletion was able to abolish the autophagic response of HKs to FGF7 and to sensitize them to FGF2, inducing an LC3-II decrease (Figure 6G) and a MTOR phosphorylation/activation (Figure 6H) comparable to that observed in HaCaT ESRP1 siRNA cells. These results suggested that an altered FGFR2 splicing and the consequent switch from FGFR2b to FGFR2c in epithelial context might drive to autophagy inhibition.

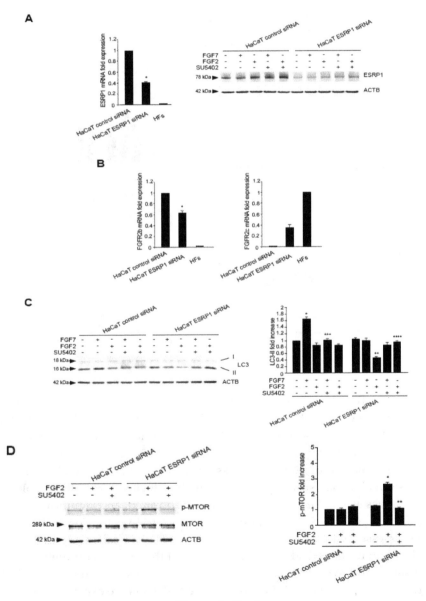

Figure 6. *Cont.*

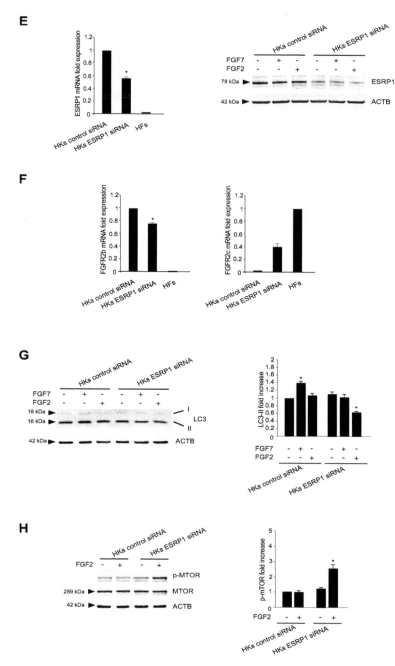

Figure 6. ESRP1 depletion by siRNA in haCaT cells results in aberrant FGFR2 splicing and FGF2-mediated inhibition of autophagy. (**A**) HaCaT cells were transiently transfected with ESRP1 siRNA or an unrelated siRNA as a control. Cells were then left in complete medium (left panel). Alternatively, cells were left untreated or stimulated with FGF7 or FGF2 in the presence or absence of the FGFR2 tyrosine kinase inhibitor SU5402 as above (right panel). Real-time RT-PCR analysis (left panel) and Western blot analysis (right panel) show the efficiency of ESRP1 depletion. Results are expressed as mean values ± SE. The student's t test was performed, and significance levels are defined as P values of ± 0.05. * $p < 0.01$ vs the corresponding control siRNA cells. For the Western blot analysis, equal loading was assessed with anti-ACTB antibody. (**B**) HaCaT cells were transiently transfected with ESRP1 siRNA or an unrelated siRNA as a control and then left in complete medium. Real-time RT-PCR analysis shows that ESRP1 depletion leads to a significant decrease of FGFR2b expression (left panel) and to the appearance of FGFR2c (right panel). Results are expressed as mean values ± SE. The student's t test was performed, and significance levels are defined as P values of 0.05. * $p < 0.05$ vs

the corresponding control siRNA cells. (C) HaCaT cells were transiently transfected with ESRP1 siRNA or an unrelated siRNA as a control. Cells were left untreated or stimulated with FGF7 or FGF2 in the presence or absence of the FGFR2 tyrosine kinase inhibitor SU5402 as above. Western blot analysis shows that ESRP1 depletion abolishes the increase of LC3 induced by FGF7 while it induces a decrease of LC3 upon FGF2 stimulation. Equal loading was assessed with anti-ACTB antibody. Densitometric analysis and the student's t test were performed as reported in the legend to Figure 1A. * $p < 0.05$ and ** $p < 0.01$ vs the corresponding FGF-unstimulated cells; *** $p < 0.01$ and **** $p < 0.05$ vs the corresponding SU5402-untreated cells. (D) HaCaT cells were transiently transfected with ESRP1 siRNA or an unrelated siRNA as a control and then left untreated or stimulated with FGF2 in presence or absence of SU5402 as above. Western blot analysis performed using antibody directed against the phosphorylated form of MTOR shows the phosphorylation of this substrates upon FGF2 stimulation in HaCaT ESRP1 siRNA cells. Equal loading was assessed with anti-ACTB antibody. Densitometric analysis and the student's t test were performed as reported in the legend to Figure 1A. * $p < 0.05$ vs the corresponding FGF-unstimulated cells; ** $p < 0.001$ vs the corresponding SU5402-untreated cells. (E) HKs cells were transiently transfected with ESRP1 siRNA or an unrelated siRNA as a control and left in complete medium (left panel) or stimulated with FGF7 or FGF2 as above (right panel). Real-time RT-PCR analysis (left panel) and Western blot analysis (right panel) showed the efficiency of ESRP1 depletion. Results are expressed as mean values ± SE. The student's t test was performed, and significance levels are defined as P values of ± 0.05. * $p < 0.05$ vs the corresponding control siRNA cells. For the Western blot analysis, equal loading was assessed with anti-ACTB antibody. (F) HKs cells were transiently transfected with ESRP1 siRNA or an unrelated siRNA as a control and then left in complete medium. Real-time RT-PCR analysis showed that ESRP1 depletion lead to a significant decrease of FGFR2b expression (left panel) and to the appearance of FGFR2c (right panel). Results are expressed as mean values ± SE. The student's t test was performed, and significance levels are defined as P values of ± 0.05. * $p < 0.01$ vs the corresponding control siRNA cells. (G) HKs cells were transiently transfected with ESRP1 siRNA or an unrelated siRNA as a control. Cells were left untreated or stimulated with FGF7 or FGF2 as above. Western blot analysis shows that ESRP1 depletion abolished the increase of LC3 induced by FGF7 while it induced a decrease of LC3 upon FGF2 stimulation. Equal loading was assessed with anti-ACTB antibody. Densitometric analysis and the student's t test were performed as reported in the legend to Figure 1A. * $p < 0.05$ vs the corresponding FGF-unstimulated cells. (H) HKs cells were transiently transfected with ESRP1 siRNA or an unrelated siRNA as a control and then left untreated or stimulated with FGF2. Western blot analysis performed using antibody directed against the phosphorylated form of MTOR shows that ESRP1 depletion induces the phosphorylation of this substrates upon FGF2 stimulation. Equal loading was assessed with anti-ACTB antibody. Densitometric analysis and the student's t test were performed as reported in the legend to Figure 1A. * $p < 0.05$ vs the corresponding control siRNA cells.

3. Discussion

The human genome consists of 20,000–25,000 genes, 95% of which undergo alternative splicing, ensuring the higher proteome diversity [39]. Isoforms arising from the splicing events show distinct and often opposing functions and several studies have suggested that the aberrant splicing represents a crucial event in cancer [39]. The high frequency of splicing aberrations in neoplastic diseases has made necessary new strategies for therapeutic approaches pointed on targeting the aberrant variants and their signaling.

In agreement with the proposed central role of the aberrant splicing in tumorigenesis, the altered splicing of FGFR2 and the consequent appearance of the mesenchymal FGFR2c isoform was observed in several carcinomas [40–44]. Consistent with this clinical evidence, recent studies from our group have demonstrated that the FGFR2 isoform switching and aberrant expression of the mesenchymal FGFR2c isoform in human keratinocytes induces impaired differentiation [13] and EMT [14,15]. The initiation of a pathological type III EMT is also accompanied by the appearance of tumorigenic features [14] indicating that FGFR2 aberrant splicing might represent the precocious event driving the early step of carcinogenesis.

Very recently it is emerging that EMT and autophagy are tumor cell responses to microenvironment stresses occurring in different stages of cancer progression [24,25]. These responses appear to be mutually exclusive and they regulate each other in a complex negative loop [24,25]. However, still many open questions remain to be answered to clarify how tumor cells decide whether to enter one or the other cell stress response. To address tumor cell heterogeneity, the identification of hub molecules able to regulate this intricate interplay appears a promising goal for cancer therapy.Starting from our recent results dealing with the ability of the epithelial FGFR2b isoform in promoting autophagy [19–22], we speculated that the ectopic FGFR2c might play a central role in the regulation of the negative crosstalk between EMT and the autophagic process in epithelial context. Consistent with this hypothesis, using biochemical, molecular and immunofluorescence approaches, we demonstrated here that the ectopic expression of FGFR2c in normal human keratinocytes efficiently counteracts autophagy. Moreover, forcing the aberrant splicing of FGFR2 in keratinocytes via ESRP1 depletion by siRNA approach, we found that the switching from the epithelial FGFR2b to the mesenchymal FGFR2c isoforms could be the specific event underlying the negative impact on autophagy in epithelial context. This is consistent with the fact that dysregulations in RNA alternative splicing is linked to EMT induction and tumor development [39]. In this regard, RNA splicing regulators, such as ESRPs, are emerging as key proteins playing both oncogenic and tumor suppressive roles via the modulation of RNA isoforms involved in oncogenic signaling pathways [45]. Therefore, we might speculate that, at least in the epidermal context, the down-regulation of ESPRs proteins, responsible for FGFR2 isoform switching, would be the molecular event crucial not only for EMT induction [15] but also for autophagy repression.

Since our results also provide elements to assume that FGFR2c-induced inhibition of the autophagic process is not transcriptionally regulated, we further progressed on the identification of the molecular mechanisms underlying FGFR2c-mediated inhibition of autophagy, identifying AKT/MTOR as the crucial pathway involved (Figure 7). We also observed that the selective inhibition of AKT or MTOR, but not that of MEK1/2, was able not just to dampen, but even to reverse the effects of FGFR2c signaling on autophagy (Figure 7). Interestingly, this unexpected event was accompanied by JNK1 activation. JNK1 signaling is known to be involved in the induction of autophagy mediated by various FGFRs [17], including FGFR2b [21,22] and proceeds via BCL-2 phosphorylation, which in turn allows phagophore nucleation [35–37]. Indee, our results showed that both JNK1 and BCL-2 phosphorylation appeared repressed by FGFR2c activation, but strongly activated upon AKT/MTOR shut-off. These findings are in agreement with previous observations of a cooperative crosstalk between PI3K/AKT/MTOR and JNK pathways in the regulation of autophagy also in different cellular contexts, such as PC12 cells and various cancer cell lines [34,46]. In addition, increasing evidences revealed that AKT and JNK pathways interact with each other and that AKT signaling inhibits JNK and different mechanisms have been proposed [33]: In fact, AKT would counteract JNK activation antagonizing the formation of the MAPK8IP1/JIP1-JNK complex [47,48], as well as interfering with the activation of JNK upstream kinases, such as MAP3K5/ASK1, MAP2K4/7/MKK4/7 and MLK [33]. However, since the PI3K/AKT pathway plays essential roles in cancer [33], understanding the molecular mechanisms regulating its crosstalk with JNK may essentially contribute to clarify the specific involvement of each of these pathways in tumor development. Moreover, since autophagy and EMT negatively regulate each other in tumor cells, it is reasonable to suppose that AKT/MTOR inhibition would be able not only to restore autophagy, but also to reverse FGFR2c-induced EMT and tumorigenic features: Further future investigations will be performed in order to address this topic.

Overall, since FGFR2c is simultaneously responsible for the unbalance of the negative crosstalk between autophagy and EMT in favor of this latter, as well as for impairment of differentiation and induction of tumorigenic features, we can conclude that this receptor, expressed as a consequence of an altered splicing event, could be one of the crucial molecular drivers of epithelial deregulation during tumorigenesis. On the other hand, the selective inhibition of specific pathways downstream FGFR2c, such as AKT/MTOR, could represent an effective tool to interfere with the oncogenic autophagy/EMT crosstalk and consequently to counteract the early steps of carcinogenesis.

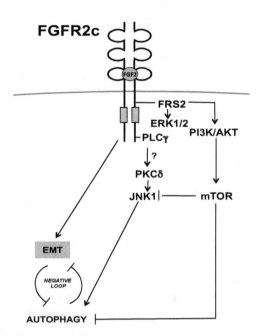

Figure 7. Schematic drawing of the proposed role of FGFR2c and its signaling in the regulation of autophagy and EMT interplay. FGFR2c inhibits autophagy via AKT/MTOR pathway, which also negatively interferes with JNK1 signaling.

4. Materials and Methods

4.1. Cells and Treatments

The human keratinocyte cell line HaCaT, stably expressing FGFR2c (pBp-FGFR2c), overexpressing FGFR2b (pBp-FGFR2b) or the empty vector (pBp) were cultured in Dulbecco's modified eagle's medium (DMEM), supplemented with 10% fetal bovine serum (FBS) plus antibiotics. Primary cultures of human keratinocytes and human fibroblasts derived from healthy skin (HKs and HFs, respectively) were obtained from patients attending the Dermatology Unit of the Sant'Andrea Hospital of Rome; all patients were extensively informed and their consent for the investigation was given and collected in written form in accordance with guidelines approved by the management of the Sant'Andrea Hospital. The research was done in agreement with the guidelines of the Helsinki declaration, according to a protocol study approved by the Ethical Committee of Sant'Andrea University Hospital (Prot. CE n. 1591/2013). Primary cells were isolated and cultured as previously described [49,50].

HaCaT clones were transiently transfected with the pDest-mCherry-EGFP tandem expression vector containing LC3 (HaCaT mCherry-EGFP-LC3) [27].

For RNA interference and MTOR or ESRP1 silencing, cells were transfected with MTOR small interfering RNA (MTOR siRNA) (Santa Cruz Biotechnology, Inc., Santa Cruz, CA, USA; SC35409), ESRP1 siRNA (Santa Cruz Biotechnology, SC77526), or an unrelated siRNA as a control, using Lipofectamine 2000 transfection reagent (Life Technologies, Carlsbad, CA, USA; 11668-019) or Fugene HD (Promega, Madison, WI, E2311) according to the manufacturer's protocol.

For growth factors stimulation, cells were left untreated or incubated with FGF7 (Upstate Biotechnology, Lake Placid, NY, 01-118) or with FGF2 (PeproTech, London, BFGF 100-188) 100 ng/mL for 24 h at 37 °C. To induce activation and signaling of FGFR2 isoforms, cells were serum starved and incubated with FGF7 or FGF2 100 ng/mL for 10 min at 37 °C. For inhibition of FGFR2b and FGFR2c tyrosine kinase activity, cells were pre-incubated with a specific FGFR2 tyrosine kinase inhibitor, SU5402 25 µM (Calbiochem, Nottingham, UK; 572630) for 1 h before treatments with growth factors (GFs).

To inhibit ERK, AKT, or MTOR cells were incubated with the MEK1/2-specific inhibitor PD0325901 (1 µM; Sigma-Aldrich, Saint Louis, MO, USA; PZ0162) AKT-specific inhibitor Akt-I-1/2, (1 µM; Calbiochem, 124005), or with the specific MTOR inhibitor rapamycin (100 nM; Cell Signaling Technology, Beverly, MA, USA; 9904) respectively, for 1 h at 37 °C before being treated with FGF2 in the presence of each inhibitor.

To irreversibly block the fusion between autophagosomes and lysosomes, cells were incubated with bafilomycin A1 (20 nM; Sigma-Aldrich, B1793) for 3 h at 37 °C after treatment with GFs in the presence of the inhibitor.

4.2. Immunoflurescence

HaCaT clones, grown on coverslips, were fixed with 4% paraformaldehyde in phosphate-buffered saline (PBS) for 30 min at 25 °C, followed by treatment with 0.1 M glycine for 20 min at 25 °C and with 0.1% Triton X-100 for an additional 5 min at 25 °C to allow permeabilization. Cells were then incubated for 1 h at 25 °C with the following primary antibodies: Mouse monoclonal anti-LC3 (1:100 in PBS, 5F10 Nanotools, Teningen, Germany, 0231). The primary antibodies were visualized using goat anti-mouse IgG-Alexa Fluor 488 (1:200 in PBS, Life Technologies, Carlsbad, CA, A11001) for 30 min at 25 °C. Nuclei were stained with DAPI (1:1000 in PBS; Sigma-Aldrich, D9542). Coverslips were finally mounted with Mowiol (Sigma) for observation. Fluorescence signals were analyzed by scanning cells in a series of sequential sections with an ApoTome System (Zeiss) connected with an Axiovert 200 inverted microscope (Zeiss); image analysis was performed by the Axiovision software (Zeiss) and images were obtained by 3D reconstruction of the total number of the serial optical sections. Quantitative analysis of the fluorescence intensity was performed by the Axiovision software (Zeiss), analyzing 10 different fields randomly taken from 3 independent experiments. Quantitative analysis of LC3-positive dots per cell was performed analyzing 100 cells for each sample in 5 different microscopy fields from 3 different experiments. Results are shown as means ± standard error (SE). The student's t test was performed and significance levels have been defined as $p < 0.05$.

4.3. Western Blot Analysis

Cells were lysed in a buffer containing 50 mM HEPES, pH 7.5, 150 mM NaCl, 1% glycerol, 1% Triton X-100, 1.5 mM MgCl2, 5 mM EGTA, supplemented with protease inhibitors (10 g/mL aprotinin, 1 mM phenylmethylsulfonyl fluoride [PMSF], 10 µg/mL leupeptin) and phosphatase inhibitors (1 mM sodium orthovanadate, 20 mM sodium pyrophosphate, 0.5 M NaF). A range of 20 to 50 µg of total protein was resolved under reducing conditions by 8 or 12% SDS-PAGE and transferred to reinforced nitrocellulose (BA-S 83; Schleicher & Schuell, Keene, NH, USA; BA-S83). The membranes were blocked with 5% nonfat dry milk (Bio-Rad Laboratories, Hercules, CA, USA, 170-6404) in PBS 0.1% Tween 20 (Bio-Rad, 170-6531) and incubated with anti-SQSTM1 (BD Bioscience, San Josè, CA, USA, 610833), anti-phospho-JNK (anti-p-JNK) (Thr183/Tyr185, Cell Signaling, 9255S), anti-p-MTOR (Ser 2448, Cell Signaling, 5536S), monoclonal antibodies or with anti-LC3 (MBL, Woburn, MA, PD014), anti ESRP1 (Sigma-Aldrich, HPA023719), anti-p-p44/42 mitogen-activated protein kinase (MAPK) (p-ERK1/2) (Thr202/Tyr204; Cell Signaling, 9101S), anti-p-AKT (Ser 473; Cell Signaling, 9271), anti-p-BCL2 (Ser 70; Cell Signaling, 2827) polyclonal antibodies, followed by enhanced chemiluminescence (ECL) detection (Thermo Scientific, Rockford, IL, USA; 34580).

The membranes were rehydrated by washing in PBS/Tween-20, stripped with 100 mM mercaptoethanol and 2% SDS for 30 min at 55°C and probed again with, anti-AKT (H-136; Santa Cruz Biotechnology, sc-8312), anti-p44/42 MAPK (ERK1/2) (137F5, Cell Signaling, 4695S), anti-JNK (Cell Signaling, 9252S), anti-α-TUBA (Cell Signaling, 2148S) polyclonal antibodies or with anti-MTOR (7C10, Cell Signaling, 2983S), anti-ACTB (Sigma-Aldrich, A5441) monoclonal antibody to estimate the protein equal loading.

Densitometric analysis was performed using Quantity One Program version 4.6.8 (Bio-Rad). The resulting values from three different experiments were normalized, expressed as fold increase respect to the control value and reported in graph as mean values ± standard deviation (SD). The student's t test was performed and significance levels have been defined as $p < 0.05$.

4.4. Primers

Oligonucleotide primers necessary for target genes and the housekeeping gene were chosen by using the online tool Primer-BLAST [51] and purchased from Invitrogen (Invitrogen, Carlsbad, CA, USA). The following primers were used: For the MAP1LC3B target gene: 5'-CGCACCTTCGAACA AAGAG-3' (sense) and 5'-CTCACCCTTGTATCGTTCTATTATCA-3' (antisense); for the BECN1 target gene: 5'-GGATGGTGTCTCTCGCAGAT-3' (sense) and 5'-TTGGCACTTTCTGTGGACAT-3' (antisense); for the ATG5 target gene: 5'-CAACTTGTTTCACGCTATATCAGG-3' (sense) and 5'-CAC TTTGTCAGTTACCAACGTCA-3' (antisense); for ESRP1 target gene: 5'-GGCTCGGATGAGAAGGA GTT-3' (sense), 5'-GCACTTCGTGCAACTGTCC-3' (antisense), for FGFR2b target gene: 5'-CGTGGA AAAGAACGGCAGTAAATA-3' (sense), 5'-GAACTATTTATCCCCGAGTGCTTG-3' (antisense); for FGFR2c target gene: 5'- TGAGGACGCTGGGGAATATACG-3 (sense), 5'-TAGTCTGGGGAAGCTGT AATCTCCT 3' (antisense); for the SQSTM1 target gene: 5'-AGCTGCCTTGTACCCACATC-3 (sense), 5'- CAGAGAAGCCCATGGACAG-3' (antisense); for the 18S rRNA housekeeping gene: 5'-CGAGCCGCCTGGATACC-3' (sense) and 5'-CATGGCCTCAGTTCCGAAAA-3' (antisense). For each primer pair, we performed no-template control and no-reverse-transcriptase control (reverse transcription [RT]-negative) assays, which produced negligible signals

4.5. RNA Extraction and cDNA Synthesis

RNA was extracted using the TRIzol method (Invitrogen, Carlsbad, CA, USA; 15596018) according to the manufacturer's instructions and eluted with 0.1% diethylpyrocarbonate (DEPC)-treated water. Each sample was treated with DNase I (Invitrogen, 18068-015). The total RNA concentration was quantitated by spectrophotometry; 1 µg of total RNA was used for reverse transcription using the iScriptTM cDNA synthesis kit (Bio-Rad, 170-8891) according to the manufacturer's instructions.

4.6. PCR Amplification and Real-Time Quantitation

Real-time RT-PCR was performed using the iCycler real-time detection system (iQ5 Bio-Rad) with optimized PCR conditions. The reactions were carried out in a 96-well plate using iQ SYBR green supermix (Bio-Rad, 1708882), adding forward and reverse primers for each gene and 1 µl of diluted template cDNA to a final reaction mixture volume of 15 µl. All assays included a negative control and were replicated three times. The thermal cycling program was performed as described previously [52]. Real-time quantitation was performed with the help of the iCycler IQ optical system software, version 3.0a (Bio-Rad), according to the manufacturer's manual. Results are reported as mean values ± SE from three different experiments in triplicate. The student's t test was performed, with significance levels defined as P values < 0.05.

Supplementary Materials:
Figure S1: FGFR2c-induced inhibition of autophagy is not transcriptionally regulated, Figure S2: Biochemical evaluation of the efficiency of specific signaling pathway substrate inhibitors and siRNAs.

Author Contributions: M.R.T and F.B. have made substantial contribution to conception and design of the project, M.N., D.R., M.R.T and F.B. have been involved in drafting the manuscript and has given final approval of the version to be published. M.N., D.R. and F.P. have made substantial contributions to acquisition and analysis of the data and have given final approval of the version to be published.

Abbreviations

ATG5	autophagy-related 5
BCL-2	BCL2 apoptosis regulator
BECN1	Beclin 1, autophagy-related
DAPI	4′,6-diamidino-2-phenylindole
EGFP	enhanced green fluorescent protein
EMT	epithelial-mesenchymal transition
ESRP	Epithelial Splicing Regulatory Protein
FGF2	fibroblast growth factor 2 (basic)
FGF7/KGF	fibroblast growth factor 7
FGFR	fibroblast growth factor receptor
GAPDH	glyceraldehyde-3-phosphate dehydrogenase
HaCaT	human adult skin keratinocytes propagated under low calcium
HFs	Human fibroblasts
HKs	human keratinocytes
HPV16	human papillomavirus type 16
MAP1LC3/LC3	microtubule-associated protein 1 light chain 3
MAP2K/MEK	mitogen-activated protein kinase kinase
MAP2K4/MKK4	mitogen-activated protein kinase kinase 4
MAP2K7/MKK7	mitogen-activated protein kinase kinase 7
MAPK/ERK	mitogen-activated protein kinase
MAPK8/JNK	mitogen-activated protein kinase 8
ACTB	actin beta
MAP3K5/ASK1	mitogen-activated protein kinase kinase kinase 5
MAPK8IP1/JIP1	mitogen-activated protein kinase 8 interacting protein 1
MLK	mixed-lineage kinase
MTOR	mechanistic target of rapamycin
NS	not significant
PI3K	class I phosphoinositide 3 kinase
PKC	protein kinase C
SD	standard deviation
SE	standard error
SDS-PAGE	sodium dodecyl sulphate-polyacrylamide gel electrophoresis
RT-PCR	Reverse Transcriptase-Polymerase Chain Reaction
siRNA	small interfering RNA
SQSTM1/P62	sequestosome 1
TUBA	α-tubulin

References

1.   Brewer, J.R.; Mazot, P.; Soriano, P. Genetic insights into the mechanisms of FGF signaling. *Genes Dev.* **2016**, *30*, 751–771. [CrossRef] [PubMed]

2.   Tanner, Y.; Grose, R.P. Dysregulated FGF signalling in neoplastic disorders. *Semin. Cell Dev. Biol.* **2016**, *53*, 126–135. [CrossRef] [PubMed]

3.   Turner, N.; Grose, R. Fibroblast growth factor signalling: From development to cancer. *Nat. Rev. Cancer* **2010**, *10*, 116–129. [CrossRef]

4.   Brooks, A.N.; Kilgour, E.; Smith, P.D. Molecular pathways: Fibroblast growth factor signaling: A new therapeutic opportunity in cancer. *Clin. Cancer Res.* **2012**, *18*, 1855–1862. [CrossRef] [PubMed]

5.   Petiot, A.; Conti, F.J.; Grose, R.; Revest, J.M.; Hodivala-Dilke, K.M.; Dickson, C. A crucial role for FGFR-IIIb signalling in epidermal development and hair follicle patterning. *Development* **2003**, *130*, 5493–5501. [CrossRef] [PubMed]

6.   Grose, R.; Fantl, V.; Werner, S.; Chioni, A.M.; Jarosz, M.; Rudling, R.; Cross, B.; Hart, I.R.; Dickson, C. The role of fibroblast growth factor receptor 2b in skin homeostasis and cancer development. *EMBO J.* **2007**, *26*, 1268–1278. [CrossRef]

7. Yang, J.; Meyer, M.; Müller, A.K.; Böhm, F.; Grose, R.; Dauwalder, T.; Verrey, F.; Kopf, M.; Partanen, J.; Bloch, W. Fibroblast growth factor receptors 1 and 2 in keratinocytes control the epidermal barrier and cutaneous homeostasis. *J. Cell Biol.* **2010**, *188*, 935–952. [CrossRef]

8. Feng, S.; Wang, F.; Matsubara, A.; Kan, M.; McKeehan, W.L. Fibroblast growth factor receptor 2 limits and receptor 1 accelerates tumorigenicity of prostate epithelial cells. *Cancer Res.* **1997**, *58*, 1509–1514.

9. Zhang, Y.; Wang, H.; Toratani, S.; Sato, J.D.; Kan, M.; McKeehan, W.L.; Okamoto, T. Growth inhibition by keratinocyte growth factor receptor of human salivary adenocarcinoma cells through induction of differentiation and apoptosis. *Proc. Natl. Acad. Sci. USA.* **2001**, *98*, 11336–11340. [CrossRef]

10. Belleudi, F.; Purpura, V.; Torrisi, M.R. The receptor tyrosine kinase FGFR2b/KGFR controls early differentiation of human keratinocytes. *PLoS ONE* **2011**, *6*, e24194. [CrossRef]

11. Purpura, V.; Belleudi, F.; Caputo, S.; Torrisi, M.R. HPV16 E5 and KGFR/ FGFR2b interplay in differentiating epithelial cells. *Oncotarget* **2013**, *4*, 192–205. [CrossRef] [PubMed]

12. Rosato, B.; Ranieri, D.; Nanni, M.; Torrisi, M.R.; Belleudi, F. Role of FGFR2b expression and signaling in keratinocyte differentiation: Sequential involvement of PKCδ and PKCα. *Cell Death Dis.* **2018**, *9*, 565. [CrossRef] [PubMed]

13. Ranieri, D.; Rosato, B.; Nanni, M.; Belleudi, F.; Torrisi, M.R. Expression of the FGFR2c mesenchymal splicing variant in human keratinocytes inhibits differentiation and promotes invasion. *Mol. Carcinog.* **2018**, *57*, 272–283. [CrossRef] [PubMed]

14. Ranieri, D.; Rosato, B.; Nanni, M.; Magenta, A.; Belleudi, F.; Torrisi, M.R. Expression of the FGFR2 mesenchymal splicing variant in epithelial cells drives epithelial-mesenchymal transition. *Oncotarget* **2016**, *7*, 5440–5460. [CrossRef] [PubMed]

15. Ranieri, D.; Belleudi, F.; Magenta, A.; Torrisi, M.R. HPV16 E5 expression induces switching from FGFR2b to FGFR2c and epithelial-mesenchymal transition. *Int. J. Cancer* **2015**, *137*, 61–72. [CrossRef] [PubMed]

16. Zhang, J.; Liu, J.; Huang, Y.; Chang, J.Y.F.; Liu, L.; McKeehan, W.L.; Martin, J.F.; Wang, F. FRS2α-mediated FGF signals suppress premature differentiation of cardiac stem cells through regulating autophagy activity. *Circ. Res.* **2012**, *110*, e29–e39. [CrossRef] [PubMed]

17. Cinque, L.; Forrester, A.; Bartolomeo, R.; Svelto, M.; Venditti, R.; Montefusco, S.; Polishchuk, E.; Nusco, E.; Rossi, A.; Medina, D. FGF signalling regulates bone growth through autophagy. *Nature* **2015**, *528*, 272–275. [CrossRef]

18. Wang, X.; Qi, H.; Wang, Q.; Zhu, Y.; Wang, X.; Jin, M.; Tan, Q.; Huang, Q.; Xu, W.; Li, X. FGFR3/fibroblast growth factor receptor 3 inhibits autophagy through decreasing the ATG12-ATG5 conjugate, leading to the delay of cartilage development in achondroplasia. *Autophagy* **2015**, *11*, 1998–2013. [CrossRef]

19. Belleudi, F.; Purpura, V.; Caputo, S.; Torrisi, M.R. FGF7/KGF regulates autophagy in keratinocytes: A novel dual role in the induction of both assembly and turnover of autophagosomes. *Autophagy* **2014**, *10*, 803–821. [CrossRef]

20. Belleudi, F.; Nanni, M.; Raffa, S.; Torrisi, M.R. HPV16 E5 deregulates the autophagic process in human keratinocytes. *Oncotarget* **2015**, *6*, 9370–9386. [CrossRef]

21. Nanni, M.; Ranieri, D.; Rosato, B.; Torrisi, M.R.; Belleudi, F. Role of fibroblast growth factor receptor 2b in the cross talk between autophagy and differentiation: Involvement of Jun N-terminal protein kinase signaling. *Mol. Cell. Biol.* **2018**, *38*, e00119-18.

22. Nanni, M.; Ranieri, D.; Raffa, S.; Torrisi, M.R.; Belleudi, F. Interplay between FGFR2b-induced autophagy and phagocytosis: Role of PLCγ-mediated signaling. *J. Cell Mol. Med.* **2018**, *22*, 668–683. [CrossRef] [PubMed]

23. Mowers, E.E.; Sharifi, M.N.; Macleod, K.F. Autophagy in cancer metastasis. *Oncogene* **2017**, *36*, 1619–1630. [CrossRef] [PubMed]

24. Marcucci, F.; Ghezzi, P.; Rumio, C. The role of autophagy in the cross-talk between epithelial-mesenchymal transitioned tumor cells and cancer stem-like cells. *Mol. Cancer* **2017**, *16*, 3. [CrossRef] [PubMed]

25. Marcucci, F.; Rumio, C. How tumor cells choose between epithelial-mesenchymal transition and autophagy to resist stress-therapeutic implications. *Front. Pharmacol.* **2018**, *9*, 714. [CrossRef] [PubMed]

26. Füllgrabe, J.; Ghislat, G.; Cho, D.H.; Rubinszteinet, D.C. Transcriptional regulation of mammalian autophagy at a glance. *J. Cell Sci.* **2016**, *129*, 3059–3066. [CrossRef] [PubMed]

27. Pankiv, S.; Clausen, T.H.; Lamark, T.; Brech, A.; Bruun, J.A.; Outzen, H.; Øvervatn, A.; Bjørkøy, G.; Johansen, T. p62/ SQSTM1 binds directly to Atg8/LC3 to facilitate degradation of ubiquitinated protein aggregates by autophagy. *J. Biol. Chem.* **2007**, *282*, 24131–24145. [CrossRef] [PubMed]

28. Lin, X.; Zhang, Y.; Liu, L.; McKeehan, W.L.; Shen, Y.; Song, S.; Wang, F. FRS2α is essential for the fibroblast growth factor to regulate the MTOR pathway and autophagy in mouse embryonic fibroblasts. *Int. J. Biol. Sci.* **2011**, *7*, 1114–1121. [CrossRef]

29. Chen, Y.; Xie, X.; Li, X.; Wang, P.; Jing, Q.; Yue, J.; Liu, Y.; Cheng, Z.; Li, J.; Song, H. FGFR antagonist induces protective autophagy in FGFR1-amplified breast cancer cell. *Biochem. Biophys. Res. Commun.* **2016**, *474*, 1–7. [CrossRef]

30. Yuan, H.; Li, Z.M.; Shao, J.; Ji, W.X.; Xia, W.; Lu, S. FGF2/FGFR1 regulates autophagy in FGFR1-amplified non-small cell lung cancer cells. *J. Exp. Clin. Cancer Res.* **2017**, *36*, 72. [CrossRef]

31. Nakanishi, Y.; Mizuno, H.; Sase, H.; Fujii, T.; Sakata, K.; Akiyama, N.; Aoki, Y.; Aoki, M.; Ishii, N. ERK signal suppression and sensitivity to CH5183284/Debio 1347, a selective FGFR inhibitor. *Mol. Cancer Ther.* **2015**, *14*, 2831–2839. [CrossRef] [PubMed]

32. Bain, J.; Plater, L.; Elliott, M.; Shpiro, N.; Hastie, C.J.; McLauchlan, H.; Klevernic, I.; Arthur, J.S.; Alessi, D.R.; Cohen, P. The selectivity of protein kinase inhibitors: A further update. *Biochem. J.* **2007**, *408*, 297–315. [CrossRef] [PubMed]

33. Zhao, H.F.; Wang, J.; Tony To, S.S. The phosphatidylinositol 3-kinase/Akt and c-Jun N-terminal kinase signaling in cancer: Alliance or contradiction? *Int. J. Oncol.* **2015**, *47*, 429–436. [CrossRef] [PubMed]

34. Rodríguez-Blanco, J.; Martín, V.; García-Santos, G.; Herrera, F.; Casado-Zapico, S.; Antolín, I.; Rodriguez, C. Cooperative action of JNK and AKT/MTOR in 1-methyl-4-phenylpyridinium-induced autophagy of neuronal PC12 cells. *J. Neurosci. Res.* **2012**, *90*, 1850–1860. [CrossRef] [PubMed]

35. Zhou, Y.Y.; Li, Y.; Jiang, W.Q.; Zhou, L.F. MAPK/JNK signalling: A potential autophagy regulation pathway. *Biosci. Rep.* **2015**, *35*, e00199.

36. Russell, R.C.; Yuan, H.X.; Guan, K.L. Autophagy regulation by nutrient signaling. *Cell Res.* **2014**, *24*, 42–57. [CrossRef]

37. Wei, Y.; Pattingre, S.; Sinha, S.; Bassik, M.; Levine, B. JNK1-mediated phosphorylation of Bcl-2 regulates starvation-induced autophagy. *Mol. Cell.* **2008**, *30*, 678–688. [CrossRef]

38. Warzecha, C.C.; Sato, T.K.; Nabet, B.; Hogenesch, J.B.; Carstens, R.P. ESRP1 and ESRP2 are epithelial cell-type-specific regulators of FGFR2 splicing. *Mol. Cell.* **2009**, *33*, 591–601. [CrossRef]

39. Wang, B.D.; Lee, N.H. Aberrant RNA Splicing in Cancer and Drug Resistance. *Cancers* **2018**, *10*, 458. [CrossRef]

40. Kawase, R.; Ishiwata, T.; Matsuda, Y.; Onda, M.; Kudo, M.; Takeshita, T.; Naito, Z. Expression of fibroblast growth factor receptor 2 IIIc in human uterine cervical intraepithelial neoplasia and cervical cancer. *Int. J. Oncol.* **2010**, *36*, 331–340.

41. Matsuda, Y.; Hagio, M.; Seya, T.; Ishiwata, T. Fibroblast growth factor receptor 2 IIIc as a therapeutic target for colorectal cancer cells. *Mol. Cancer Ther.* **2012**, *11*, 2010–2020. [CrossRef] [PubMed]

42. Ishiwata, T.; Matsuda, Y.; Yamamoto, T.; Uchida, E.; Korc, M.; Naito, Z. Enhanced expression of fibroblast growth factor receptor 2 IIIc promotes human pancreatic cancer cell proliferation. *Am. J. Pathol.* **2012**, *180*, 1928–1941. [CrossRef] [PubMed]

43. Zhao, Q.; Caballero, O.L.; Davis, I.D.; Jonasch, E.; Tamboli, P.; Yung, W.K.; Weinstein, J.N.; Kenna Shaw for TCGA research network; Strausberg, R.L.; Yao, J. Tumor-specific isoform switch of the fibroblast growth factor receptor 2 underlies the mesenchymal and malignant phenotypes of clear cell renal cell carcinomas. *Clin. Cancer Res.* **2013**, *19*, 2460–2472. [CrossRef] [PubMed]

44. Peng, W.X.; Kudo, M.; Fujii, T.; Teduka, K.; Naito, Z. Altered expression of fibroblast growth factor receptor 2 isoform IIIc: Relevance to endometrioid adenocarcinoma carcinogenesis and histological differentiation. *Int. J. Clin. Exp. Pathol.* **2014**, *7*, 1069–1076. [PubMed]

45. Urbanski, L.M.; Leclair, N.; Anczuków, O. Alternative-splicing defects in cancer: Splicing regulators and their downstream targets, guiding the way to novel cancer therapeutics. *Wiley Interdiscip. Rev. RNA* **2018**, *9*, e1476. [CrossRef]

46. Liu, J.; Zheng, L.; Zhong, J.; Wu, N.; Liu, G.; Lin, X. Oleanolic acid induces protective autophagy in cancer cells through the JNK and MTOR pathways. *Oncol. Rep.* **2014**, *32*, 567–572. [CrossRef] [PubMed]

47. Kim, A.H.; Yano, H.; Cho, H.; Meyer, D.; Monks, B.; Margolis, B.; Birnbaum, M.J.; Chao, M.V. Akt1 regulates a JNK scaffold during excitotoxic apoptosis. *Neuron* **2002**, *35*, 697–709. [CrossRef]

48. Pan, J.; Pei, D.S.; Yin, X.H.; Hui, L.; Zhang, G.Y. Involvement of oxidative stress in the rapid Akt1 regulating a JNK scaffold during ischemia in rat hippocampus. *Neurosci. Lett.* **2006**, *392*, 47–51. [CrossRef]

49. Cardinali, G.; Ceccarelli, S.; Kovacs, D.; Aspite, N.; Lotti, L.V.; Torrisi, M.R.; Picardo, M. Keratinocyte growth factor promotes melanosome transfer to keratinocytes. *J. Investig. Dermatol.* **2005**, *125*, 1190–1199. [CrossRef]

50. Raffa, S.; Leone, L.; Scrofani, C.; Monini, S.; Torrisi, M.R.; Barbara, M. Cholesteatoma-associated fibroblasts modulate epithelial growth and differentiation through KGF/FGF7 secretion. *Histochem. Cell Biol.* **2012**, *138*, 251–269. [CrossRef]

51. Ye, J.; Coulouris, G.; Zaretskaya, I.; Cutcutache, I.; Rozen, S.; Madden, T.L. Primer-BLAST: A tool to design target-specific primers for polymerase chain reaction. *BMC Bioinform.* **2012**, *13*, 134. [CrossRef] [PubMed]

52. Avitabile, D.; Genovese, L.; Ponti, D.; Ranieri, D.; Raffa, S.; Calogero, A.; Torrisi, M.R. Nucleolar localization and circadian regulation of Per2S, a novel splicing variant of the Period 2gene. *Cell Mol. Life Sci.* **2014**, *71*, 2547–2559. [CrossRef] [PubMed]

4

Cross-Talk between Fibroblast Growth Factor Receptors and Other Cell Surface Proteins

Marta Latko, Aleksandra Czyrek, Natalia Porębska, Marika Kucińska, Jacek Otlewski, Małgorzata Zakrzewska and Łukasz Opaliński *

Department of Protein Engineering, Faculty of Biotechnology, University of Wroclaw, Joliot-Curie 14a, 50-383 Wroclaw, Poland; marta.latko2@uwr.edu.pl (M.L.); aleksandra.czyrek@uwr.edu.pl (A.C.); natalia.porebska2@uwr.edu.pl (N.P.); kucinska.marika@gmail.com (M.K.); jacek.otlewski@uwr.edu.pl (J.O.); malgorzata.zakrzewska@uwr.edu.pl (M.Z.)
* Correspondence: lukasz.opalinski@uwr.edu.pl

Abstract: Fibroblast growth factors (FGFs) and their receptors (FGFRs) constitute signaling circuits that transmit signals across the plasma membrane, regulating pivotal cellular processes like differentiation, migration, proliferation, and apoptosis. The malfunction of FGFs/FGFRs signaling axis is observed in numerous developmental and metabolic disorders, and in various tumors. The large diversity of FGFs/FGFRs functions is attributed to a great complexity in the regulation of FGFs/FGFRs-dependent signaling cascades. The function of FGFRs is modulated at several levels, including gene expression, alternative splicing, posttranslational modifications, and protein trafficking. One of the emerging ways to adjust FGFRs activity is through formation of complexes with other integral proteins of the cell membrane. These proteins may act as coreceptors, modulating binding of FGFs to FGFRs and defining specificity of elicited cellular response. FGFRs may interact with other cell surface receptors, like G-protein-coupled receptors (GPCRs) or receptor tyrosine kinases (RTKs). The cross-talk between various receptors modulates the strength and specificity of intracellular signaling and cell fate. At the cell surface FGFRs can assemble into large complexes involving various cell adhesion molecules (CAMs). The interplay between FGFRs and CAMs affects cell–cell interaction and motility and is especially important for development of the central nervous system. This review summarizes current stage of knowledge about the regulation of FGFRs by the plasma membrane-embedded partner proteins and highlights the importance of FGFRs-containing membrane complexes in pathological conditions, including cancer.

Keywords: fibroblast growth factor receptors; signaling; receptor cross-talk; coreceptor; membrane proteins

---

1. Introduction

Fibroblast growth factor receptors 1–4 (FGFR1–4) form a group of receptor tyrosine kinases (RTKs) that are present on the surface of various cell types. FGFRs govern plethora of key cellular processes, including proliferation, migration, differentiation, and apoptosis, and their proper functioning is critical for development of the human body and homeostasis [1]. Alterations in FGFR1–4 are frequently detected in variety of developmental diseases and cancers, like prostate, breast, lung, and ovarian cancers [2,3]. The overall structure of FGFRs is typical for RTKs with an N-terminal region including three immunoglobulin-like domains D1–D3 exposed to the extracellular space, a single transmembrane span and a cytosolic tyrosine kinase domain (Figure 1a) [1,4]. The extracellular part of FGFRs constitutes binding sites for their natural ligands, FGFs, heparan cofactors, and a number of partner proteins [5,6]. Additionally, the ectodomain of FGFRs includes several motifs that prevent receptor autoactivation in the absence of growth factors [7–10]. The transmembrane helix of FGFRs anchors the receptors in the membrane and facilitates dimerization [11]. In the cytosol, the juxtamembrane (JM) region of FGFRs is involved in receptor dimerization and moderates transmission of signals [12–14]. The initiation of intracellular signaling circuits requires activation of FGFRs split kinase domain [1,5]. FGFR1–3 are subjected to alternative splicing in their extracellular region, yielding b and c isoforms of the receptors that differ in expression pattern and ligand specificity [15–17]. The FGFR family includes also fifth member—FGFRL1 (FGFR5)—which is homologous to FGFRs in the extracellular region, but lacks the cytosolic tyrosine kinase domain [18,19].

Classically, the transmission of signals through the plasma membrane via FGFRs requires binding of appropriate growth factors and subsequent receptor activation. The canonical FGFs (FGF1–FGF10, FGF16, FGF17, FGF18, FGF20, and FGF22) are effective ligands in FGFRs binding and activation. In an inactive state monomeric FGFRs bind canonical FGFs, which triggers conformational changes in the receptor, resulting in dimerization and transactivation of cytosolic tyrosine kinases [1,20]. Sequential phosphorylation of tyrosine residues within the cytosolic tail of FGFRs creates docking sites for downstream signaling proteins [1,21]. The signals are further propagated through several pathways: Ras/Raf-mitogen-activated protein kinase/extracellular signal regulated kinase kinase (MEK)–extracellular signal regulated kinase (ERK), phosphoinositide 3-kinase (PI3K)/protein kinase B (AKT)/mammalian target of rapamycin (mTOR), phospholipase Cγ (PLCγ), and signal transducer and activator of transcription (STAT) [1,20].

FGFR-dependent signaling can be adjusted in several ways, including the diversified tissue distribution, different expression level of signaling components and their alternative splicing, which influences tissue development and disease progression [1]. Transmission of signals can be further modulated by ligand type, as FGFR complexes with different FGFs may vary in the strength and duration of propagated signals, which in turn decides cell fate [20,22]. FGFRs signaling can be modified as well by spontaneous receptor dimerization in the absence of ligands [23]. The posttranslational modifications, like glycosylation, ubiquitination, and phosphorylation, influence ligand binding and constitute negative feedback mechanisms for inhibition of FGFRs signaling [24–28]. Additionally, the cellular trafficking of FGFRs may regulate signals specificity, intensity, and timing [29–31].

One of the emerging means to modulate FGFRs activity is via formation of complexes with other plasma membrane proteins. Assembly of such complexes can be critical for transmission of signals, which is the case for endocrine FGFs (FGF19, FGF21, and FGF23) [32]. Partner proteins may deliver cofactors that facilitate formation of productive signaling modules or regulate the cellular transport of FGFRs [1]. Distinct types of cell surface receptors interact with FGFRs, leading to integration of different signaling routes or modulation of signal transmission. Several high throughput studies led to the discovery of numerous potential interaction partners of FGFRs within the plasma membrane [33–35]. However, the biological significance for most of them still needs to be elucidated.

In the next chapters we focus on the interplay between FGFRs and their binding partners in the regulation of signaling and cell behavior.

2. Cross-Talk between FGFRs and G-Protein-Coupled Receptors in Regulation of the Central Nervous System

G-protein-coupled receptors (GPCRs) constitute one of the largest groups of receptors responsible for signal transmission [36–38]. GPCRs are composed of an N-terminal extracellular domain, seven transmembrane helices, and a C-terminal region directed to the cytosol. Stimulation of GPCRs by extracellular ligands induces conformational changes within GPCRs, triggering intracellular signaling pathways modulated by heterotrimeric G proteins [39,40]. Due to their wide diversity GPCRs modulate numerous processes, including, among others, nervous system transmission, visual, gustatory and smell sensing, inflammation, and recognition of cell density [41].

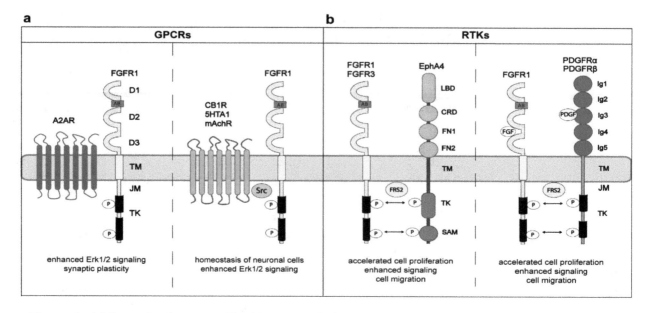

Figure 1. (a) Interplay between fibroblast growth factor receptors (FGFRs) and G-protein-coupled receptors (GPCRs) (a) and other receptor tyrosine kinases (RTKs) (b) in the regulation of downstream signaling. The extracellular region of FGFRs is composed of immunoglobulin like domains D1–D3 (gray) and the acidic box (AB; red). FGFRs are anchored in the plasma membrane by a single transmembrane helix (yellow). The cytosolic part of FGFRs consists of the juxtamembrane domain (JM) and the split tyrosine kinase domain (TK; black). GPCR–FGFR complexes may involve Src as a mediator between receptors or form functional heterocomplexes without involvement of Src. (b) FGFRs interact with other RTK members in the plasma membrane and can be directly activated by intracellular tyrosine kinase domains of partner proteins like Eph receptors or PDGFRs. EphA4 receptor contains the N-terminal ligand binding domain (LBD) followed by the cysteine rich domain (CDR) and two fibronectin type III domains (FN1–2). EphA4 is embedded in the membrane by a single transmembrane domain (TM). The cytosol-oriented region of EphA4 is composed of the tyrosine kinase domain (TK) and the sterile alpha motif (SAM). The TK domain of EphA4 interacts with JM region of FGFRs. PDGFRs contain five immunoglobulin-like domains (Ig1–Ig5) in their extracellular region, a single transmembrane span (TM), and intracellular juxtamembrane (JM) and tyrosine kinase (TK) domains. TK of PDGFRs directly phosphorylates FGFRs.

Various members of GPCRs and RTKs form heterocomplexes, which trigger intracellular signaling and cellular response different from that induced by RTKs or GPCRs alone [42]. The alterations in transmitted signals by GPCRs-RTKs heterocomplexes is achieved by the transactivation of RTKs by GPCRs which may occur via two distinct mechanisms: one relying on GPCRs activation and signaling that results in release of RTKs ligands and subsequent RTKs activation and second mechanism that involves a direct interaction and subsequent activation of RTKs by GPCRs [42]. The transactivation of RTKs by GPCRs was already demonstrated for a large number of RTKs, including epidermal growth factor receptors (EGFRs), platelet-derived growth factor receptors (PDGFRs), and insulin-like growth factor receptors (IGFRs) [42].

In the central nervous system (CNS), GPCR-dependent signaling controls proliferation, migration, survival, and differentiation of neurons [43]. FGFRs are expressed in different areas of brain. While FGFR1 is widely found in the hippocampus and in various parts of the cortex, FGFR2 and FGFR3 proteins are scattered throughout the CNS, and their expression profile changes with the brain development. FGFR4 is less abundant than other FGFRs and is mainly localized to the medial habenular nucleus [44–48]. The FGFRs are involved in the development, function and maintenance of the CNS [49]. Yeast two-hybrid (Y2H) screens revealed FGFR1 as a binding partner of G-protein-coupled receptor (GPCR)–adenosine receptor A2AR. The FGFR1-A2AR interaction was further confirmed by pull-down and coimmunoprecipitation [50]. The simultaneous stimulation of PC12 cells with A2AR agonist and FGF2 results in enhanced activation of downstream signaling pathways in comparison to single treatments, pointing on the synergistic effect of both receptors on cellular signaling. The enhanced activation of extracellular regulated kinases 1/2 (ERK1/2) requires assembly of the FGFR1-A2AR complex, pointing on the functional relevance of this interaction. The modulation of signaling by FGFR1-A2AR heterocomplexes was found to be important for regulation of the synaptic plasticity (Figure 1a) [50].

Cannabinoid receptor 1 (CB1R) is GPCR-ubiquitous in neurons, mediates the biological action of endogenous and synthetic cannabinoids, and regulates homeostasis of neuronal cells [51]. CB1R-FGFR1 interaction in neurons was demonstrated by means of coimmunoprecipitation. CB1R induces the transactivation of FGFR1 via protein kinase C (PKC) that in turn activates Fyn and Src. The latter proteins trigger activation of FGFR1 by phosphorylating key tyrosine residues of the receptor kinase domain [52]. The formation of CB1R-FGFR1 complexes occurs in lipid rafts of the plasma membrane, leads to activation of ERK1/2, and is important for neuronal differentiation (Figure 1a).

Using the proximity ligation assay (PLA) the interaction of FGFR1 with muscarinic acetylocholine receptor (mAChR) subtype M1R was visualized [53]. Upon stimulation of hippocampal neurons with M1R agonist oxotremorine-M the activation of FGFR1 was observed. The exact mechanism of FGFR1 transactivation is not clear, however it involves Src tyrosine kinase that phosphorylates FGFR1 [53]. The cross-talk between mAChR and FGFR1 enhances neurite growth (Figure 1a) [53].

Binding between FGFR1 and 5-hydroxytriptamine receptor 1A (5-HT1A) was also demonstrated with PLA, but it was further confirmed by coimmunoprecipitation and bioluminescence resonance energy transfer (BRET) in a wide variety of cell types [54–56]. The number of FGFR1-5-HT1A complexes increases upon stimulation of cells with the FGF2 and 5-HT1A agonist 7-(Dipropylamino)-5,6,7,8-tetrahydronaphthalen-1-ol (8-OH-DPAT), confirming the functional interplay between these receptors [55]. Activation of 5-HTA1 with 8-OH-DPAT causes subsequent FGFR1 phosphorylation mediated by Src [55]. The simultaneous activation of FGFR1 and 5-HTA1 results in synergistically enhanced signaling that induces growth and controls homeostasis of neuronal cells (Figure 1a) [55]. Interestingly, the FGFR1–5-HT1A heterocomplexes display anti-depressive effects and thus may constitute targets for treatment of mood disorders [55,57–59].

Mu-opoid receptor (MOR) binds with high affinity to enkephalins and endorphins that modulate neuronal excitability. In rat glioma C6 cells MOR induces rapid activation of ERK1/2 via the transactivation of FGFR1. Again, the exact mechanism of this transactivation is unknown. Also the direct interaction between MOR and FGFR1 has not been yet demonstrated [60].

Summarizing, various members of GPCRs affect activity of FGFRs through the transactivation, which usually requires formation of the direct interaction between these receptors and involves Src as a bridging factor. The cross-talk between GPCRs and FGFRs is especially relevant for the development and functioning of neurons. GPCRs constitute large group of receptors, however only few members of the GPCRs family were demonstrated to bind FGFR1. The function of one type of receptors can be modulated by binding to other group of receptors. Since GPCRs play diverse pivotal functions in cells, the involvement of FGFRs in the regulation of GPCRs needs to be elucidated.

3. Interplay between FGFRs and Other RTKs

Diversification of signals transmitted by FGFRs can be also achieved by the interplay with other members of RTK family. The cross-talk between RTKs can occur via formation of receptor heterocomplexes and subsequent tyrosine phosphorylation of one receptor by tyrosine kinase of the other one. Alternatively, the transphosphorylation of RTKs in the complex can be mediated by the cytosolic kinase, like Src [61].

Eph receptors are activated by ephrin ligands and constitute the largest family of RTKs [62,63]. Based on sequence similarity and preference for ephrins A or B, Eph receptors are divided into EphA (EphA1–EphA10) and EphB (EphB1–EphB6) receptors [64]. The Eph receptors contain structural features characteristic for RTKs: an extracellular ligand binding region, a transmembrane domain, and an intracellular tyrosine kinase module [65]. The N-terminal extracellular part of Eph receptors is composed of ephrin binding domain followed by the cysteine rich EGF-like motif and two fibronectin type III repeats (FN3) FN1 and FN2. The cytosolic region of Eph receptors includes the juxtamembrane domain, the tyrosine kinase and the sterile alpha motif (SAM) (Figure 1b) [66]. Remarkably, activation of Eph receptors by ephrins requires the assembly of cell to cell contacts, as ephrins are embedded in the plasma membrane by the glycosylphosphatidylinositol (GPI) anchor (ephrins A) or the transmembrane helix (ephrins B) [64]. Binding of Eph receptor to ephrin present on the surface of aligned cell is followed by the juxtaposition of cytoplasmic kinase domain that evokes the transphosphorylation of receptor tyrosine residues initiating downstream signaling cascades [67]. The Eph receptor–ephrin complexes can be further arranged into high order assemblies that modulate cellular signaling [68,69]. The Eph receptor–ephrin complexes adjust cell adhesion, organization of cytoskeleton, angiogenesis, neural development, and plasticity [70].

EphA4 receptor emerged as binding partner of FGFR3 in Y2H screens [71]. Further experiments, including coimmunoprecipitation revealed that the tyrosine kinase domain of Eph4 directly interacts with the JM domain of FGFR1–4 [71]. The formation of EphA4-FGFR complexes requires phosphorylation of tyrosine residues within JM domain of Eph4. Kinase domains of EphA4 and FGFRs can transphosphorylate each other. Furthermore, EphA4 ligand ephrin-A1 enhances FGFRs signaling, indicating significance of the FGFRs transactivation by EphA4 for the modulation of intracellular signal propagation [72]. Signals transmitted via FGF2/FGFR1/EphA4 complexes are enhanced in relation to FGF2/FGFR1, resulting in accelerated cell proliferation and migration [67]. In addition, the interaction between EphA4 and the fibroblast growth factor receptor substrate 2 alpha (FRS2α), a protein required for FGFRs signaling [73] was demonstrated with Y2H and pull down experiments. Noteworthy, the ternary complex, involving FGFR1, EphA4, and FRS2α was detected. Thus, FRS2α acts as a tethering molecule that integrates signals from both receptors and regulates self-renewal, differentiation, and proliferation of neural stem/progenitor cells [74,75]. The cross-talk between Eph and FGFRs and Eph receptors was further confirmed by the observation that FGFRs phosphorylate EphA receptor target molecule, ephexin-1 [76]. Furthermore, Dlg-1, a scaffolding protein directly interacting with EphA receptors, can modulate FGFRs signaling (Figure 1b) [77,78].

Platelet-derived growth factor receptors alpha and beta (PDGFRα and PDGFRα) are RTKs that are activated by five different platelet-derived growth factors (PDGF): PDGF-AA, PDGF-BB, PDGF-AB, PDGF-CC, and PDGF-DD [79,80]. Through regulation of cellular signaling PDGFRs influence cell motility, proliferation, and angiogenesis and aberrant PDGFRs are implicated in cancer [79]. PDGFRs are composed of the extracellular region divided into five Ig-like domains, from which Ig2 and Ig3 form the PDGF binding site, a single transmembrane span, and the intracellular tyrosine kinase domain (Figure 1b) [81,82]. In vitro and in vivo experiments using solid-phase assay (SPA), coimmunoprecipitation, and Förster Resonance Energy Transfer (FRET) revealed that PDGFRα interacts with high affinity with FGFR1 [83]. The formation of PDGFRα-FGFR1 complexes is facilitated by the presence of ligands for both receptors [83]. The interaction between PDGFRβ and FGFR1 was demonstrated by means of coimmunoprecipitation [84]. In this receptor heterocomplex PDGFRβ directly phosphorylates FGFR1 on tyrosine residues [84]. Interestingly, FRS2 functions as a bridging molecule between PDGFRβ and FGFR1 (Figure 1b) [84]. The interplay is not only observed between the receptors but also at the level of their ligands. PDGF-BB and FGF2 interact with each other and activity of individual ligands in PDGF-BB-FGF2 complex is altered [85–87]. Remarkably, PDGFRs and FGFRs are often dysregulated in cancer and are targets of numerous therapeutic approaches [88].

Summarizing, FGFRs assemble into large multiprotein complexes with other RTK members and accessory proteins. The tyrosine kinase domains of different RTKs are able to transphosphorylate each other, initiating signals and adjusting their strength and specificity. Importantly, the interplay between RTKs is often coordinated at the level of FRS2. The fact that different members of RTKs can transactivate each other suggests the presence of an additional level of complexity in RTKs signaling. The family of RTKs is composed of 58 members; however, to date only few RTKs have been implicated in the FGFRs transactivation. Further studies on the interplay of FGFRs with other RTKs may uncover novel cellular regulatory mechanisms. Numerous FGFR-targeted anticancer therapies aim on the inhibition of FGFs interaction with FGFRs. Since FGFRs can be activated by other receptors in the absence of ligands, the detailed knowledge about FGFRs interplay with other RTKs may help in the development of novel therapeutics downregulating FGFRs signaling.

4. Modulation of FGFRs Activity by Cell-Surface Proteins Involved in Adhesion

Establishing cell-cell contacts requires an extensive remodeling of cellular components. Communication between cells involves interactions that are mediated by various cell adhesion molecules (CAMs). At the cell-cell interface extensive signaling is triggered, which coordinates remodeling of cellular structures. Noteworthy, FGFRs emerged as CAMs binding partners that participate in the signaling initiated by CAMs at cell-cell contacts (Figure 2).

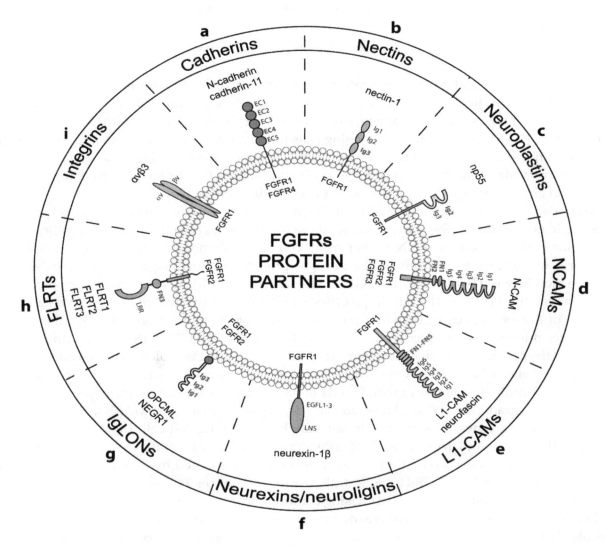

Figure 2. Cross-talk between FGFRs and various cell adhesion molecules. The interaction of particular FGFR with members of CAMs subgroup is indicated. The domain architecture of FGFR partner proteins is shown. Domains (where identified) responsible for the interaction between the partner protein and FGFR are indicated in red. (**a**) Cadherins reported to interact with FGFR1 and FGFR4 contain five EC domains in their extracellular region, a single transmembrane helix, and a cytosoilc tail interacting with several signaling proteins. (**b**) Nectins are composed of three immunoglobulin-like domains Ig1–Ig3, a single transmembrane domain, and a cytosolic region. Nectins bind FGFR1 using the Ig3 domain (**c**) Neuroplastin (Np55) contains two immunoglobulin-like Ig1–Ig2 domains in their extracellular region

and are embedded in the membrane by a single transmembrane helix, exposing short tail into the cytosol. Np55-FGFR1 interaction involves the Ig2 domain of Np55 (**d**) NCAMs expose on the surface of the cells five immunoglobulin-like domains Ig1–Ig5 and two fibronectin type III domains FN1 and FN2. The cytosolic tail of NCAMs varies in length. NCAMs bind FGFR1-FGFR3 using FN1–FN2 domains (**e**) L1-CAM is a single spanning plasma membrane protein with six Ig-like domains (Ig1–Ig6) and five fibronectin type III domains (FN1–FN5) in its extracellular region. FGFR1 binding requires the FN1–FN5 region of L1-CAM (**f**) Neurexins contain different numbers of the laminin-neurexin-sex hormone binding globulin domains (LNS) and three EGF-like domains (EGFL1–3), a single transmembrane span and the cytosolic tail interacting with cytoskeletal and signaling proteins. The extracellular region of neurexin 1-β interacts with FGFR1. (**g**) Ig-LON family members: OPCML and NEGR1 interact with FGFR1 and FGFR2. Ig-LON proteins contain three immunoglobulin-like domains Ig1–Ig3 that are implicated in FGFR binding. (**h**) FLRTs are single spanning transmembrane proteins containing the leucine-rich repeat domain (LRR) and the FN3 domain in their extracellular region. FLRTs employ the FN3 domain for FGFR1 and FGFR2 binding. (**i**) Integrins are composed of different α and β subunits. Integrin αvβ3 forms complexes with FGFR1.

4.1. Cadherins

Cadherins are integral membrane proteins that are involved in the formation of specific cell-cell contacts, the adherens junctions (AJs) [89]. AJs are regulated by the alternative splicing of cadherins and are important for tissue development, homeostasis of epithelium and are implicated in different types of cancer [90–92]. Cadherins on opposing cells interact with each other via extracellular regions composed of five domains (EC1–EC5) in a calcium-dependent manner (Figure 2a) [92]. The cytosolic tail of cadherins binds catenins and other intracellular factors that link cadherin complexes to the cytoskeleton and forms signaling platforms at the cell-cell interface [90].

Neuronal cadherin (N-cadherin, cadherin-2) is expressed in various cell types, but its highest level is detected in neuronal and mesenchymal cells, where it coordinates cell migration and proliferation [93]. The functional interaction between N-cadherin and FGFRs was demonstrated in numerous cells, where N-cadherin was shown to activate FGFRs and receptor-downstream signaling (Figure 2a) [94–96]. The interaction between N-cadherin and FGFR1 was demonstrated by means of coimmunoprecipitation in different cell lines [97,98]. The binding studies with truncated variants of FGFR1 revealed that the acidic box of the receptor extracellular region is required for the interaction with N-cadherin [97,98]. Fluorescence microscopy analyses revealed that in transfected NIH3T3 cells N-cadherin and FGFR1 colocalize at the plasma membrane, however the N-cadherin-FGFR1 complexes are less abundant at the cell-cell contact sites where N-cadherin is enriched, suggesting dynamic nature of this interaction [97]. Formation of N-cadherin complexes with FGFR1 in breast cancer cells causes decreased internalization and lysosomal degradation of FGFR1, and sustained receptor signaling via MAPKs. Thus, N-cadherin may promote invasiveness of cancer cells not only by regulating cell-cell interactions, but also by affecting FGFR1 levels and activity [98–100]. Silencing of N-cadherin results in the accelerated FGFR1 degradation, whereas overproduction of N-cadherin is accompanied by increased levels of FGFR1. Thus, N-cadherin stabilizes FGFR1 and simultaneously enhances FGF2-induced proliferation and differentiation of epiblast stem cells [101]. Using coimmunoprecipitation, the interaction of N-cadherin with FGFR4 was demonstrated in pancreatic tumor cells and was dependent on neural cell adhesion molecule (N-CAM) [102]. Moreover, FGFR4-388Arg mutant frequently observed in various cancers induces signaling cascades that lead to enhanced N-cadherin expression and modulates epithelial to mesenchymal transition (EMT) [103].

Cadherin-11 is widely expressed in mesenchymal cells like osteoblasts and neurons, and is important for tissue development during embryogenesis [104,105]. It is implicated in migration of cancer cells and in epithelial to mesenchymal transition [106–109]. The formation of complexes between FGFR1 and the cadherin-11/β-catenin adhesion complexes was demonstrated by coimmunoprecipitation (Figure 2a) [110]. Pull-down experiments revealed that the cadherin-11-FGFR1 interaction occurs through their extracellular domains. Cadherin-11 initiates intracellular signaling pathways via FGFR1 and recruits FGFR1 into areas of cell-cell contacts [49]. The cadherin-11-induced FGFR1 signaling stimulates neurite outgrowth [49].

4.2. Nectins

Nectins comprise a group of four plasma membrane proteins (Nectin-1–4) involved in formation of cell-cell contacts that are relevant in the neural development and disorders, and cancer [111]. Nectins contain an extracellular region composed of three immunolglobulin-like (Ig) domains, a single transmembrane helix, and a cytosolic domain (Figure 2b). Nectins from one cell can oligomerize in trans orientation with nectins present on the opposing cell, which results in cell adhesion. Depending on the involvement of accessory proteins nectins can be involved in establishing several types of adhesion complexes [111,112]. Using surface plasmon resonance (SPR) a direct interaction between Ig2–Ig3 domains of FGFRs and Ig3 of nectin-1 was demonstrated (Figure 2b). Binding of Ig3 of nectin-1 to FGFR1 results in receptor activation. Nectin-1 induces neurite outgrowth in hippocampal neurons in FGFR1-dependent manner, indicating that nectin-1 co-clusters with FGFR1 at the cell–cell contacts to stimulate differentiation and development of neurons [113].

4.3. Neuroplastins

Neuroplastins are cell adhesion molecules from immunoglobulin superfamily [114]. Neuroplastin Np55 is expressed in numerous cell types and tissues [115]. Np55 contains two Ig-like domains—Ig2 and Ig3—oriented towards the extracellular space, a single transmembrane span, and a short cytoplasmic tail (Figure 2c) [116]. SPR analysis revealed that Np55 directly interacts with the Ig2–Ig3 region of FGFR1 (Figure 2c). Binding of Np55 to FGFR1 present on the cell surface leads to receptor activation and initiation of downstream signaling. Although FGF2 and Np55 bind to the same region of FGFR1, these proteins elicit different effects on the receptor. Np55-FGFR1 complexes stimulate neurite outgrowth in primary hippocampal neurons, while FGF2-FGFR1 does not, which suggests different mode of intracellular signaling activation by these two FGFR1 ligands. Peptide based on Np55 extracellular domain was able to activate FGFR1 and downstream signaling and displayed antidepressant effects [117].

4.4. N-CAMs

Neural cell adhesion molecules (N-CAMs) are cell surface glycoproteins involved in axonal growth, cell migration, synaptic plasticity, and cell differentiation, and are implicated in various diseases including cancer [118,119]. N-CAMs contain five Ig-like domains and two FN3 domains in their extracellular region. NCAM-140 and NCAM-180 are embedded in the plasma membrane via transmembrane helices and display cytoplasmic tails of different length (Figure 2d). In contrast NCAM-120 utilizes the glycosylphosphatidylinositol (GPI) moiety for attachment to the cell surface [120].

The functional interplay between FGFRs and N-CAMs in neurite outgrowth was initially demonstrated by Williams et al. [94]. Subsequent studies confirmed a direct interaction of N-CAMs and FGFRs in different types of cells, including cancer cells [97,102,121–123]. The FN3 domains are responsible for the N-CAMs interaction with the Ig2–Ig3 region of FGFRs (Figure 2d) [124–126]. N-CAMs bind to FGFR1-FGFR3, but not to FGFR4, and these interactions depend on the receptor splice variants [127]. Binding of N-CAMs to FGFRs results in activation of the receptor and initiation of signaling cascades. The N-CAMs-FGFRs interplay is important for neuronal tissue development, but is also implicated in cancer. The N-CAMs/FGFRs complexes are observed in epithelial ovarian carcinoma, where they stimulate cancer cell migration and invasion [128,129]. The N-CAMs/FGFRs signaling may also modulate EMT [130]. Interestingly, N-CAMs can affect the cellular trafficking of FGFRs. Activation of FGFR1 by FGFs triggers receptor internalization and lysosomal degradation. In contrast, N-CAM-FGFR1 complexes are internalized, but the majority of the receptor is recycled from endosomes to the cell surface [121]. This differential FGFR1 cellular transport determines distinct cell fate depending on stimulation with FGF or N-CAM proteins [73].

4.5. L1-CAMs

L1-CAM is a cell surface glycoprotein that contains six Ig-like domains and five FN3 motifs in its extracellular region, a single TM span, and an intracellular tail that binds several signaling proteins (Figure 2e) [118]. The functional link between FGFR1 and L1-CAM was established by the observation that extracellular region of L1-CAM activates FGFR1, stimulating neurite outgrowth [94]. SPR experiments demonstrated a direct interaction between L1-CAM FN3 domains 1–5 and FGFR1 Ig2–Ig3 domains that was dependent on ATP [131]. Noteworthy, the cross-talk between FGFR1 and L1-CAM plays a role in proliferation and motility of glioma cells. The soluble, extracellular region of L1-CAM is often released by the cells due to the limited proteolysis involving ADAM-10 protease [132]. By binding to FGFR1 the extracellular region of L1-CAM leads to receptor activation, resulting in stimulation of glioma cell proliferation and motility [133]. The multiprotein complex of L1-CAM, FGFR1, and secreted glycoprotein Anosmin-1, which is involved in cell adhesion, motility, and differentiation, were also implicated in neurite branching [134–139].

Neurofascins are L1-CAM group members that control neurite outgrowth and synaptic organization [140]. The interaction between neurofascin (isoform NF166) and FGFR1 was demonstrated by coimmunoprecipitation [141]. Experiments with truncated versions of neurofascin revealed presence of two binding sites for FGFR1: an extracellular and an intracellular. Nevertheless, only the intracellular region of neurofascin is critical for FGFR1-dependent neurite outgrowth [141,142].

4.6. Neurexins

Neurexins and neuroligins are neuronal CAMs that regulate synaptic organization and function [143,144]. Presynaptic neurexins consist of the extracellular region containing from one to six laminin- neurexin-sex hormone binding globulin domains (LNS) and three epidermal growth factor like (EGF-like) domains, O-glycosylation sites, a single transmembrane span, and the cytosolic region recruiting various intracellular cytoskeletal and signaling proteins (Figure 2f) [144]. Postsynaptic neuroligins are composed of the extracellular acetylcholinesterase-like domain, a region enriched in glycosylation sites, a single transmembrane helix and the C-terminal intracellular PDZ domain recognition motif. Neurexins and neuroligins form trans-synaptic tethers that organize structure and function of synapses [144]. SPR experiments revealed a direct interaction between extracellular domain of FGFR1 and ectodomain of neurexin-1β (Figure 2f) [145]. Neurexin-1β binding leads to the activation of FGFR1 and receptor-downstream signaling cascades in a dose-dependent manner [145].

4.7. IgLONs

IgLONs are CAMs from immunoglobulin superfamily composed of three Ig-like domains that are attached to the cell membrane via GPI anchor (Figure 2g) [146]. Neuronal growth regulator 1 (NEGR1) is IgLON member that regulates neuronal maturation [147]. The functional interplay between NEGR1 and FGFRs in neuronal development and disease was initially suggested by Pischedda et al. and Casey et al. [148,149]. This was further confirmed by detection of the interaction between extracellular regions of NEGR1 and FGFR2 (Figure 2g). NEGR1 influences FGFR2 intracellular trafficking, favoring receptor recycling. The prolonged intracellular trafficking of FGFR2 in endosome compartments results in enhanced receptor-dependent signaling. Importantly, it was demonstrated that the coordinated cortical development requires the functional interplay between FGFR2 and NEGR1 [150].

Opioid binding protein cell adhesion molecule (OPCML) is another IgLON member linked with FGFRs. OPCML is a tumor suppressor implicated in various cancers [151–155]. Coimmunoprecipitation revealed that OPCML interacts with FGFR1. Furthermore, pull down experiments with recombinant OPCML and FGFR1 truncations showed that the Ig1–Ig3 region of OPCML directly interacts with the extracellular domain of FGFR1 (Figure 2g). Binding of OPCML to FGFR1 and a few other RTK members results in their downregulation, which is likely a result of their altered intracellular trafficking and decreased recycling [156].

4.8. FLRTs

Fibronectin leucine-rich transmembrane (FLRTs) proteins comprise a group of three cell surface glycoproteins involved in cell adhesion during vascularization and synapse development [157–161]. FLRTs contain the N-terminal extracellular region composed of the leucine-rich repeat domain (LRR) and the FN3 domain. FLTRs are embedded in the cell membrane via a single transmembrane helix and contain a short cytoplasmic tail (Figure 2h) [162]. FLRTs mediate cell-cell contacts mainly through the interaction of LRR domains of FLRTs on neighboring cells or with latrophilin [157,162]. Coimmunoprecipitation, pull-down and BRET experiments revealed that the FN3 domain of FLRT2 and FLRT3 interacts with FGFR2 and FGFR1, respectively (Figure 2h) [162,163]. Assembly of the FLRT-FGFR complexes is mediated by the interaction between intracellular regions of these proteins [164,165]. FGFR1-dependent signaling leads to the tyrosine phosphorylation of the intracellular tail of FLRT1. In addition, formation of the FLRT1-FGFR1 complexes enhances receptor signaling upon stimulation with FGF ligand, which accelerates neurite outgrowth in MAPK-dependent manner [166].

4.9. Integrins

Integrins are adhesion molecules that recognize ligands present in the extracellular matrix and on the cell surface, playing a key role in establishing cell contacts and regulating intracellular signaling [167]. Subunits α (18 isoforms) and β (8 isoforms) assemble into 24 functional integrins that vary in terms of ligand specificity and cellular function (Figure 2i) [168]. Integrin-dependent signaling modulates survival, migration, and differentiation of cells [169]. Dysregulation of integrin adhesion complexes is widely implicated in various cancer types [170]. Coimmunoprecipitation experiments confirmed assembly of the ternary complex containing FGF1, FGFR1 and integrin $\alpha v\beta 3$, with FGF1 acting as a bridging factor (Figure 2i). These multiprotein complexes are important for sustained activation of FGFR1-dependent kinases ERK1/2 [171]. Interestingly, the integrin binding-deficient mutant of FGF-1 (R50E) is capable of binding and activating FGFR1, however it fails to induce cell proliferation and migration, pointing on the functional relevance of integrin $\alpha v\beta 3$ in FGF1 action [172,173]. The integrin binding site within FGF2 was identified as well; however the involvement of FGF2 in bridging FGFR1 and integrin $\alpha v\beta 3$ has still to be determined [174].

Cell-cell contacts are complex signaling platforms that regulate behavior of neighboring cells and thus are strongly implicated in cancer. FGFRs are modulated by a number of different CAMs at the cell-cell interface. The FGFR-CAM interaction involves extracellular domains of these proteins, suggesting formation of complexes in cis and trans orientation. The FGFRs-CAMs interplay may adjust the strength of cell-cell attachment, which is relevant for migration of cancer cells and thus may constitute the target for future anticancer therapies.

5. Novel Activities Acquired by FGFRs upon Binding to Specific Coreceptors

Coreceptors are cell surface molecules that modulate the interaction of primary receptors with ligands. Usually, specific ligands require assembly of the ternary complexes involving ligand, receptor and coreceptor to initiate signal propagation. The perfect examples of FGFRs coreceptors are Klotho proteins that are necessary for endocrine FGFs (FGF19, FGF21, and FGF23) to trigger signaling. Functional FGFR signaling modules involve also specific polysaccharides, heparan sulfate (HS) chains, which stabilize receptor-ligand complexes. In this chapter we focus on coreceptors of FGFRs and their role in modulating FGFRs specificity and activity.

5.1. Heparan Sulfate Proteoglycans

The formation of FGF-FGFR complexes requires presence of HS [175,176]. HS directly binds FGFs and FGFRs stabilizing the ternary complex and facilitating FGFR autophosphorylation [177].

HS chains are covalently attached to the serine residues of a subset of cell surface proteins, forming heparan sulfate proteoglycans (HSPGs). HSPGs are secreted into the extracellular space or are attached to the plasma membrane either via GPI anchor or transmembrane helix [178]. HSPGs participate in FGF signaling by regulating availability of FGFs to FGFRs and by adjusting the FGF-FGFR complex dynamics (Figure 3a) [179].

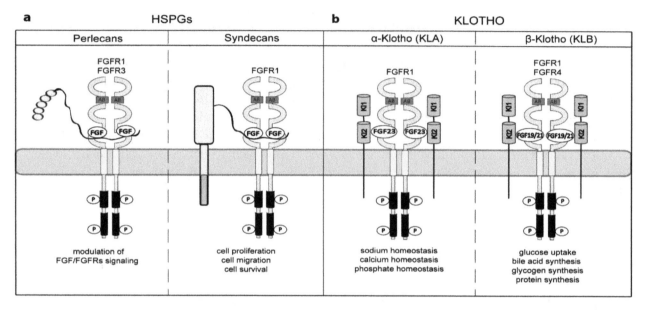

Figure 3. Involvement of coreceptors in the FGFRs signaling. (**a**) Heparan sulfate proteoglycans (HSPGs) provide polysaccharide chains that stabilize FGF-FGFR complexes and regulate availability of ligands. HSPGs are either integral membrane proteins (syndecans) or secreted glycoproteins (perlecans), which form ternary complexes with FGF-FGFR. (**b**) Klotho proteins α (KLA) and β (KLB) are necessary for FGF23 and FGF19/FGF21 signaling, respectively.

Perlecan is high molecular weight, multidomain HSPG ubiquitous in the extracellular space. The HS chains are attached to the N-terminal domain of perlecan [180]. Perlecan interacts with several FGFs, providing their storage in the extracellular matrix, thus adjusting their accessibility to FGFRs [181–184]. In the absence of FGF perlecan is able to bind FGFR3, but perlecan-FGFR1 interaction requires presence of the growth factor [182]. The ternary complexes involving FGFs (FGF20 or FGF18), perlecan, and FGFRs affect FGFRs signaling and resulting cellular response (Figure 3a) [181,185]. Interestingly, perlecan isolated from diverse tissues differentially modulates FGF/FGFRs signaling, highlighting the importance of HS structure for FGFRs [183].

Syndecans are composed of an N-terminal extracellular domain with attached several sugar chains, including HS, a single transmembrane helix and a C-terminal cytosolic tail [186]. The N-terminal domain of syndecans interacts with several proteins, including growth factors, extracellular matrix proteins and chemokines, the transmembrane helix facilitates oligomerization of syndecans, while the intracellular region interacts with numerous signaling and cytoskeletal proteins [187,188]. Syndecans via HS chains interact with FGFs and FGFRs with relatively low affinity, but still facilitating formation of ternary signaling complexes [189–192]. Syndecan-dependent modulation of FGF/FGFR complexes is relevant for cell proliferation, migration and survival (Figure 3a) [193–196]. The cellular trafficking of FGFRs is tightly regulated and constitutes a mechanism for adjustment of signaling pathways and cellular fate [29]. In endothelial cells syndecan-4 initiates the internalization of syndecan-4/FGF2/FGFR1 complexes via micropinocytosis that is independent of clathrin and dynamin, and involves RhoG and Rab4. The altered trafficking of FGFR1 changes kinetics of MAPK signaling important for survival of endothelial cells [197].

Another group of HSPGs that adjust cellular signaling pathways triggered by growth factors are GPI-anchored glypicans [198]. Glypican-1 interacts with FGFs, modulating their activity and accessibility for FGFRs [199–201]. However, in brain endothelial cells and in glioma cells the overexpression of glypican-1 facilitates mitogenic response triggered by FGF2 [202,203].

5.2. Klotho Coreceptors

The FGF family includes a subgroup of endocrine FGFs—FGF19, FGF21, and FGF23—which largely differ from canonical FGFs in their structure and mode of action. Endocrine FGFs circulate throughout the human body regulating numerous metabolic processes [204,205]. In contrast to canonical FGFs, endocrine FGFs display low affinity to FGFRs and cell surface heparans [206–208]. To form functional signaling complexes with FGFRs endocrine FGFs require obligatory coreceptors from Klotho family: α-Klotho (KLA) and β-Klotho (KLB) [209–213]. Klotho proteins are plasma membrane proteins containing two tandem KL1 and KL2 repeats with similarity to family 1 glucosidases in their extracellular region, a single transmembrane helix, and a short cytoplasmic tail [214,215].

KLA was discovered as a protein involved in aging process and is necessary for FGF23 signaling [211,214]. The obligate involvement of KLA in the formation of productive FGF23-FGFR1 signaling complex was enlightened by recent structural studies [216]. KLA interacts directly with FGFR1 and forms a high-affinity binding site for FGF23. FGF23 binds FGFR1 with its N-terminus, while the C-terminal region of FGF23 directly interacts with KLA, forming the KLA-FGF23-FGFR1 signaling complex (Figure 3b) [216]. Interestingly, dimerization of such complexes and receptor activation remain dependent on the binding of heparan sulfate [216]. This ternary complex acts mainly in kidneys, regulating sodium, calcium and phosphate homeostasis and its imbalance leads to various metabolic diseases, like acute and chronic uremia and premature aging [217–225].

KLB is a homologue of KLA that facilitates formation of signaling complexes containing FGFR–FGF19, mainly in hepatocytes, and FGFRs-FGF21 in adipocytes (Figure 3b) [226–229]. The molecular bases of FGFR-FGF19/FGF21-KLB signaling complex assembly largely resemble FGFR1-FGF23-KLA. KLB utilizes both KL1 and KL2 of the extracellular domain for direct binding to FGF19/FGF21 C-terminal domains [230,231]. The KLB-FGF19 complex binds FGFR1 and FGFR4, while KLB-FGF21 can form the ternary complex only with FGFR1 [227]. The dimerized KLB-FGF21-FGFR1 complexes in adipocytes induce catabolic processes, stimulate glucose uptake, and improve insulin sensitivity [232]. Noteworthy, acting as a fasting hormone, FGF21 significantly extends lifespan [233,234]. In hepatocytes the KLB-FGF19-FGFR4 complexes are formed in response to feeding and downregulate synthesis of bile acid [235,236]. Additionally, these complexes contribute to the regulation of blood glucose level by stimulating synthesis of glycogen [237,238]. The dysregulation of FGF19/FGF21 is implicated in metabolic diseases, aging, and cancer [217,239–241].

6. Modulation of FGFRs by Other Cell Surface Proteins

There are plasma membrane proteins that interact with FGFRs but cannot be assigned to the above described categories. One of them is transforming growth factor β receptor III (TGFBRIII), which is also known as betaglycan. It is a coreceptor of TGFBRI and TBFBRII that lacks an intracellular kinase activity [242]. The interaction between TGFBRIII and FGFR1 was demonstrated by coimmunoprecipitation in neuroblastoma. The TGFBRIII-FGFR1 interaction is stimulated by FGF2 and the assembly of ternary complexes enhances FGF2 signaling and promotes neuronal differentiation [243]. In addition, FGF2 binds to the glysocaminoglycan chains (GAG) present on the extracellular region of TGFBRIII, which may regulate availability of the FGF2 to FGFRs on the cell surface [244].

Another FGFRs' interactor is Sef (similar expression to fgf genes), a receptor-like protein composed of an extracellular region containing the FN3 domain, a single transmembrane helix and an intracellular domain with similarity to the interleukin 17 receptor [245]. Besides membrane bound Sef, secreted and cytosolic isoforms of Sef are generated [246]. The expression of Sef is induced by FGF signaling in various cell types [245,247–249]. The interaction of various Sef isoforms with intracellular region of FGFRs was demonstrated with coimmunoprecipitation [246,249–253]. Sef is an inhibitor of FGFR-dependent signaling acting either directly at the level of the receptor and/or on downstream intracellular kinases [254]. FGFR-dependent activation of ERK/MAPK and Akt is blocked by Sef, resulting in inhibition of cell proliferation [250,255]. Sef can also induce apoptosis and affect FGF-induced differentiation in various cell types [255]. Notably, the FGFRs-Sef interplay was implicated in prostate cancer [256,257].

7. Conclusions

The cellular fate is very rarely determined by isolated signaling units. Instead, it is rather a result of extensive cross-communication between numerous diverse ligand/receptor systems. Secreted FGFs and their receptors are well studied signaling molecules. However, a number of recent reports largely changed the view about FGFs/FGFRs as separate signaling modules. FGFs/FGFRs are integrated into the complex cellular signaling at many levels and are subjected to diverse regulatory mechanisms. The cross-talk between FGFRs and other cell surface receptors, adhesion molecules, and coreceptors effectively modulates cellular processes such as proliferation, motility, differentiation, and death. The list of FGFRs binding partners within the plasma membrane is expanding; however it is still far from complete. As FGFRs expose large domains towards the extracellular space and the cytosol, the activity of these receptors might be further modulated by currently unknown secreted and/or intracellular proteins, respectively. Certainly, further studies aiming on the identification of novel FGFRs binding proteins and deciphering the relevance of FGFRs' complexes are required. Moreover, the application of complementary in vitro and in vivo experimental approaches is required for the validation and in-depth characterization of identified interactions. Structural data revealed the molecular mechanism of FGFR tyrosine kinase activation facilitating the design of diverse FGFR small molecule inhibitors that are currently tested as anticancer drugs [258]. Similarly, understanding how FGFRs cooperate with other cell surface receptors may lead to the development of novel inhibitors targeting FGFR-dependent processes.

As FGFRs are embedded in the plasma membrane, the activity and distribution of these receptors can be additionally affected by properties of the cell membrane (membrane composition, organization, curvature, etc.). Additionally, the alternative splicing of FGFRs and partner proteins may constitute another regulatory mechanism of the assembly of multiprotein signaling complexes. Further studies in this direction are unquestionably required. The spatiotemporal regulation of FGFRs constitutes another way to adjust cellular signaling. Some binding partners affect cellular trafficking of FGFRs, influencing selected transport mechanism and subcellular destination of the receptors. This in turn affects the kinetics and specificity of signaling and modulates cellular response. As FGFRs and number of partner proteins are implicated in various diseases including cancer, the deeper understanding of the interplay between FGFRs and other components of the cell membrane may facilitate treatment of life-threatening diseases.

Author Contributions: Ł.O., M.Z. and J.O. designed the manuscript. M.L., A.C., N.P. and M.K. performed literature studies and collected data. N.P. and M.K. prepared the figures. Ł.O. and M.Z. wrote the manuscript. All authors provided conceptual input, edited the text, and approved the final version of the manuscript.

Abbreviations

5-HT1A	5-hydroksytriptamine receptor 1A
8-OH-DPAT	7(Dipropylamino)-5,6,7,8-tetrahydronaphthalen-1-ol
A2AR	adenosine receptor
AJ	adherens junction
AKT	protein kinase B
ATP	adenosine triphosphate
BRET	bioluminescence resonance energy transfer
CAMs	cell adhesion molecules
CB1R	cannabinoid receptor 1
CNS	central nervous system
Dlg-1	disks large homolog 1
EGFs	epidermal growth factors
EGFRs	epidermal growth factor receptors
EMT	epithelial to mesenchymal transition
Eph	ephrin
ERK1/2	extracellular regulated kinases 1/2
FGFs	fibroblast growth factors
FGFRs	fibroblast growth factor receptors
FGFRL1	fibroblast growth factor receptor like 1
FLRTs	fibronectin leucine-rich transmembranes
FN3	fibronectin type III
FRET	Förster Resonance Energy Transfer
FRS2	fibroblast growth factor receptor substrate 2
GAG	glysocaminoglycan
GPCRs	G-protein-coupled receptors
GPI	glycosylphosphatidylinositol
HS	heparan sulfate
HSPGs	heparan sulfate proteoglycans
IGFR	insulin-like growth factor receptor
JM	juxtamembrane
KLA	α-klotho
KLB	β-klotho
L1-CAM	L1 cell adhesion molecule
LNS	laminin, neurexin, sex hormone binding globulin
LPR	leucine-rich repeat domain
mAChR	muscarinic acetylcholine receptor
MAPK	mitogen-activated protein kinase
MOR	mu-opioid receptor
mTOR	mammalian target of rapamycin
N-CAMs	neural cell adhesion molecules
NEGR1	neuronal growth regulator 1
NFs	neurofascins
Np55	neuroplastin 55
OPCML	opioid binding protein cell adhesion molecule
PDGFs	platelet-derived growth factors
PDGFRs	platelet-derived growth factor receptors
PI13K	phosphoinositide 3-kinase
PKC	protein kinase C
PLA	proximity ligation assay
PLCY	phospholipase CY
RTKs	receptor tyrosine kinases
SAM	sterile alpha motif

Sef   similar expression to fgf genes
SPA   solid-phase assay
SPR   surface plasmon resonance
STAT   signal transducer and activator of transcription
TGFs   transforming growth factors
TGFBRs   transforming growth factor receptors
Y2H   yeast two-hybrid

References

1. Ornitz, D.M.; Itoh, N. The Fibroblast Growth Factor signaling pathway. *Wiley Interdiscip. Rev. Dev. Biol.* **2015**, *4*, 215–266. [CrossRef] [PubMed]

2. Helsten, T.; Elkin, S.; Arthur, E.; Tomson, B.N.; Carter, J.; Kurzrock, R. The FGFR Landscape in Cancer: Analysis of 4,853 Tumors by Next-Generation Sequencing. *Clin. Cancer Res.* **2016**, *22*, 259–267. [CrossRef] [PubMed]

3. Hallinan, N.; Finn, S.; Cuffe, S.; Rafee, S.; O'Byrne, K.; Gately, K. Targeting the fibroblast growth factor receptor family in cancer. *Cancer Treat. Rev.* **2016**, *46*, 51–62. [CrossRef] [PubMed]

4. Ornitz, D.M.; Marie, P.J. Fibroblast growth factor signaling in skeletal development and disease. *Genes Dev.* **2015**, *29*, 1463–1486. [CrossRef] [PubMed]

5. Mohammadi, M.; Olsen, S.K.; Ibrahimi, O.A. Structural basis for fibroblast growth factor receptor activation. *Cytokine Growth Factor Rev.* **2005**, *16*, 107–137. [CrossRef] [PubMed]

6. Plotnikov, A.N.; Hubbard, S.R.; Schlessinger, J.; Mohammadi, M. Crystal structures of two FGF-FGFR complexes reveal the determinants of ligand-receptor specificity. *Cell* **2000**, *101*, 413–424. [CrossRef]

7. Olsen, S.K.; Ibrahimi, O.A.; Raucci, A.; Zhang, F.; Eliseenkova, A.V.; Yayon, A.; Basilico, C.; Linhardt, R.J.; Schlessinger, J.; Mohammadi, M. Insights into the molecular basis for fibroblast growth factor receptor autoinhibition and ligand-binding promiscuity. *Proc. Natl. Acad. Sci. USA* **2004**, *101*, 935–940. [CrossRef]

8. Kalinina, J.; Dutta, K.; Ilghari, D.; Beenken, A.; Goetz, R.; Eliseenkova, A.V.; Cowburn, D.; Mohammadi, M. The alternatively spliced acid box region plays a key role in FGF receptor autoinhibition. *Structure* **2012**, *20*, 77–88. [CrossRef]

9. Kiselyov, V.V.; Kochoyan, A.; Poulsen, F.M.; Bock, E.; Berezin, V. Elucidation of the mechanism of the regulatory function of the Ig1 module of the fibroblast growth factor receptor 1. *Protein Sci.* **2006**, *15*, 2318–2322. [CrossRef] [PubMed]

10. Opalinski, L.; Szczepara, M.; Sokolowska-Wedzina, A.; Zakrzewska, M.; Otlewski, J. The autoinhibitory function of D1 domain of FGFR1 goes beyond the inhibition of ligand binding. *Int. J. Biochem. Cell Biol.* **2017**, *89*, 193–198. [CrossRef] [PubMed]

11. Peng, W.C.; Lin, X.; Torres, J. The strong dimerization of the transmembrane domain of the fibroblast growth factor receptor (FGFR) is modulated by C-terminal juxtamembrane residues. *Protein Sci.* **2009**, *18*, 450–459. [CrossRef] [PubMed]

12. Lin, H.Y.; Xu, J.; Ischenko, I.; Ornitz, D.M.; Halegoua, S.; Hayman, M.J. Identification of the cytoplasmic regions of fibroblast growth factor (FGF) receptor 1 which play important roles in induction of neurite outgrowth in PC12 cells by FGF-1. *Mol. Cell Biol.* **1998**, *18*, 3762–3770. [CrossRef]

13. Burgar, H.R.; Burns, H.D.; Elsden, J.L.; Lalioti, M.D.; Heath, J.K. Association of the signaling adaptor FRS2 with fibroblast growth factor receptor 1 (Fgfr1) is mediated by alternative splicing of the juxtamembrane domain. *J. Biol. Chem.* **2002**, *277*, 4018–4023. [CrossRef] [PubMed]

14. Sarabipour, S.; Hristova, K. FGFR3 unliganded dimer stabilization by the juxtamembrane domain. *J. Mol. Biol.* **2015**, *427*, 1705–1714. [CrossRef]

15. Miki, T.; Bottaro, D.P.; Fleming, T.P.; Smith, C.L.; Burgess, W.H.; Chan, A.M.; Aaronson, S.A. Determination of ligand-binding specificity by alternative splicing: Two distinct growth factor receptors encoded by a single gene. *Proc. Natl. Acad. Sci. USA* **1992**, *89*, 246–250. [CrossRef] [PubMed]

16. Chellaiah, A.T.; McEwen, D.G.; Werner, S.; Xu, J.; Ornitz, D.M. Fibroblast growth factor receptor (FGFR) 3. Alternative splicing in immunoglobulin-like domain III creates a receptor highly specific for acidic FGF/FGF-1. *J. Biol. Chem.* **1994**, *269*, 11620–11627. [PubMed]

17. Gong, S.G. Isoforms of receptors of fibroblast growth factors. *J. Cell. Physiol.* **2014**, *229*, 1887–1895. [CrossRef] [PubMed]
18. Wiedemann, M.; Trueb, B. Characterization of a novel protein (FGFRL1) from human cartilage related to FGF receptors. *Genomics* **2000**, *69*, 275–279. [CrossRef] [PubMed]
19. Trueb, B.; Zhuang, L.; Taeschler, S.; Wiedemann, M. Characterization of FGFRL1, a novel fibroblast growth factor (FGF) receptor preferentially expressed in skeletal tissues. *J. Biol. Chem.* **2003**, *278*, 33857–33865. [CrossRef]
20. Goetz, R.; Mohammadi, M. Exploring mechanisms of FGF signalling through the lens of structural biology. *Nat. Rev. Mol. Cell Biol.* **2013**, *14*, 166–180. [CrossRef]
21. Furdui, C.M.; Lew, E.D.; Schlessinger, J.; Anderson, K.S. Autophosphorylation of FGFR1 kinase is mediated by a sequential and precisely ordered reaction. *Mol. Cell* **2006**, *21*, 711–717. [CrossRef]
22. Zinkle, A.; Mohammadi, M. A threshold model for receptor tyrosine kinase signaling specificity and cell fate determination. *F1000Researcharch* **2018**, *7*. [CrossRef]
23. Sarabipour, S.; Hristova, K. Mechanism of FGF receptor dimerization and activation. *Nat. Commun.* **2016**, *7*, 10262. [CrossRef]
24. Duchesne, L.; Tissot, B.; Rudd, T.R.; Dell, A.; Fernig, D.G. N-glycosylation of fibroblast growth factor receptor 1 regulates ligand and heparan sulfate co-receptor binding. *J. Biol. Chem.* **2006**, *281*, 27178–27189. [CrossRef] [PubMed]
25. Polanska, U.M.; Duchesne, L.; Harries, J.C.; Fernig, D.G.; Kinnunen, T.K. N-Glycosylation regulates fibroblast growth factor receptor/EGL-15 activity in Caenorhabditis elegans in vivo. *J. Biol. Chem.* **2009**, *284*, 33030–33039. [CrossRef] [PubMed]
26. Brooks, A.N.; Kilgour, E.; Smith, P.D. Molecular pathways: Fibroblast growth factor signaling: A new therapeutic opportunity in cancer. *Clin. Cancer Res.* **2012**, *18*, 1855–1862. [CrossRef] [PubMed]
27. Haugsten, E.M.; Malecki, J.; Bjorklund, S.M.; Olsnes, S.; Wesche, J. Ubiquitination of fibroblast growth factor receptor 1 is required for its intracellular sorting but not for its endocytosis. *Mol. Biol. Cell* **2008**, *19*, 3390–3403. [CrossRef] [PubMed]
28. Zakrzewska, M.; Haugsten, E.M.; Nadratowska-Wesolowska, B.; Oppelt, A.; Hausott, B.; Jin, Y.; Otlewski, J.; Wesche, J.; Wiedlocha, A. ERK-mediated phosphorylation of fibroblast growth factor receptor 1 on Ser777 inhibits signaling. *Sci. Signal.* **2013**, *6*, ra11. [CrossRef]
29. Porebska, N.; Latko, M.; Kucinska, M.; Zakrzewska, M.; Otlewski, J.; Opalinski, L. Targeting Cellular Trafficking of Fibroblast Growth Factor Receptors as a Strategy for Selective Cancer Treatment. *J. Clin. Med.* **2018**, *8*, 7. [CrossRef] [PubMed]
30. Hitosugi, T.; Fan, J.; Chung, T.W.; Lythgoe, K.; Wang, X.; Xie, J.; Ge, Q.; Gu, T.L.; Polakiewicz, R.D.; Roesel, J.L.; et al. Tyrosine phosphorylation of mitochondrial pyruvate dehydrogenase kinase 1 is important for cancer metabolism. *Mol. Cell* **2011**, *44*, 864–877. [CrossRef] [PubMed]
31. Stachowiak, M.K.; Maher, P.A.; Stachowiak, E.K. Integrative nuclear signaling in cell development–a role for FGF receptor-1. *DNA Cell Biol.* **2007**, *26*, 811–826. [CrossRef] [PubMed]
32. Itoh, N.; Ohta, H.; Konishi, M. Endocrine FGFs: Evolution, Physiology, Pathophysiology, and Pharmacotherapy. *Front. Endocrinol. (Lausanne)* **2015**, *6*, 154. [CrossRef] [PubMed]
33. Vecchione, A.; Cooper, H.J.; Trim, K.J.; Akbarzadeh, S.; Heath, J.K.; Wheldon, L.M. Protein partners in the life history of activated fibroblast growth factor receptors. *Proteomics* **2007**, *7*, 4565–4578. [CrossRef] [PubMed]
34. Balek, L.; Nemec, P.; Konik, P.; Kunova Bosakova, M.; Varecha, M.; Gudernova, I.; Medalova, J.; Krakow, D.; Krejci, P. Proteomic analyses of signalling complexes associated with receptor tyrosine kinase identify novel members of fibroblast growth factor receptor 3 interactome. *Cell Signal.* **2018**, *42*, 144–154. [CrossRef] [PubMed]
35. Kostas, M.; Haugsten, E.M.; Zhen, Y.; Sorensen, V.; Szybowska, P.; Fiorito, E.; Lorenz, S.; Jones, N.; de Souza, G.A.; Wiedlocha, A.; et al. Protein Tyrosine Phosphatase Receptor Type G (PTPRG) Controls Fibroblast Growth Factor Receptor (FGFR) 1 Activity and Influences Sensitivity to FGFR Kinase Inhibitors. *Mol. Cell Proteom.* **2018**, *17*, 850–870. [CrossRef] [PubMed]
36. Calebiro, D.; Koszegi, Z. The subcellular dynamics of GPCR signaling. *Mol. Cell Endocrinol.* **2019**. [CrossRef] [PubMed]
37. Calebiro, D.; Jobin, M.L. Hot spots for GPCR signaling: Lessons from single-molecule microscopy. *Curr. Opin. Cell Biol.* **2018**, *57*, 57–63. [CrossRef]

38. Milligan, G.; Ward, R.J.; Marsango, S. GPCR homo-oligomerization. *Curr. Opin. Cell. Biol.* **2018**, *57*, 40–47. [CrossRef] [PubMed]

39. Thal, D.M.; Glukhova, A.; Sexton, P.M.; Christopoulos, A. Structural insights into G-protein-coupled receptor allostery. *Nature* **2018**, *559*, 45–53. [CrossRef] [PubMed]

40. Mahoney, J.P.; Sunahara, R.K. Mechanistic insights into GPCR-G protein interactions. *Curr. Opin. Struct. Biol.* **2016**, *41*, 247–254. [CrossRef] [PubMed]

41. Husted, A.S.; Trauelsen, M.; Rudenko, O.; Hjorth, S.A.; Schwartz, T.W. GPCR-Mediated Signaling of Metabolites. *Cell Metab.* **2017**, *25*, 777–796. [CrossRef] [PubMed]

42. Di Liberto, V.; Mudo, G.; Belluardo, N. Crosstalk between receptor tyrosine kinases (RTKs) and G protein-coupled receptors (GPCR) in the brain: Focus on heteroreceptor complexes and related functional neurotrophic effects. *Neuropharmacology* **2018**. [CrossRef] [PubMed]

43. Leung, C.C.Y.; Wong, Y.H. Role of G Protein-Coupled Receptors in the Regulation of Structural Plasticity and Cognitive Function. *Molecules* **2017**, *22*, 1239. [CrossRef] [PubMed]

44. Belluardo, N.; Wu, G.; Mudo, G.; Hansson, A.C.; Pettersson, R.; Fuxe, K. Comparative localization of fibroblast growth factor receptor-1, -2, and -3 mRNAs in the rat brain: In situ hybridization analysis. *J. Comp. Neurol.* **1997**, *379*, 226–246. [CrossRef]

45. Gonzalez, A.M.; Berry, M.; Maher, P.A.; Logan, A.; Baird, A. A comprehensive analysis of the distribution of FGF-2 and FGFR1 in the rat brain. *Brain Res.* **1995**, *701*, 201–226. [CrossRef]

46. Ford-Perriss, M.; Abud, H.; Murphy, M. Fibroblast growth factors in the developing central nervous system. *Clin. Exp. Pharm. Physiol.* **2001**, *28*, 493–503. [CrossRef]

47. Choubey, L.; Collette, J.C.; Smith, K.M. Quantitative assessment of fibroblast growth factor receptor 1 expression in neurons and glia. *PeerJ.* **2017**, *5*, e3173. [CrossRef] [PubMed]

48. Itoh, N.; Yazaki, N.; Tagashira, S.; Miyake, A.; Ozaki, K.; Minami, M.; Satoh, M.; Ohta, M.; Kawasaki, T. Rat FGF receptor-4 mRNA in the brain is expressed preferentially in the medial habenular nucleus. *Brain Res. Mol. Brain Res.* **1994**, *21*, 344–348. [CrossRef]

49. Turner, C.A.; Akil, H.; Watson, S.J.; Evans, S.J. The fibroblast growth factor system and mood disorders. *Biol. Psychiatry* **2006**, *59*, 1128–1135. [CrossRef]

50. Flajolet, M.; Wang, Z.; Futter, M.; Shen, W.; Nuangchamnong, N.; Bendor, J.; Wallach, I.; Nairn, A.C.; Surmeier, D.J.; Greengard, P. FGF acts as a co-transmitter through adenosine A(2A) receptor to regulate synaptic plasticity. *Nat. Neurosci.* **2008**, *11*, 1402–1409. [CrossRef]

51. Howlett, A.C.; Barth, F.; Bonner, T.I.; Cabral, G.; Casellas, P.; Devane, W.A.; Felder, C.C.; Herkenham, M.; Mackie, K.; Martin, B.R.; et al. International Union of Pharmacology. XXVII. Classification of cannabinoid receptors. *Pharm. Rev.* **2002**, *54*, 161–202. [CrossRef] [PubMed]

52. Asimaki, O.; Leondaritis, G.; Lois, G.; Sakellaridis, N.; Mangoura, D. Cannabinoid 1 receptor-dependent transactivation of fibroblast growth factor receptor 1 emanates from lipid rafts and amplifies extracellular signal-regulated kinase 1/2 activation in embryonic cortical neurons. *J. Neurochem.* **2011**, *116*, 866–873. [CrossRef] [PubMed]

53. Di Liberto, V.; Borroto-Escuela, D.O.; Frinchi, M.; Verdi, V.; Fuxe, K.; Belluardo, N.; Mudo, G. Existence of muscarinic acetylcholine receptor (mAChR) and fibroblast growth factor receptor (FGFR) heteroreceptor complexes and their enhancement of neurite outgrowth in neural hippocampal cultures. *Biochim. Biophys. Acta Gen. Subj.* **2017**, *1861*, 235–245. [CrossRef]

54. Borroto-Escuela, D.O.; Narvaez, M.; Perez-Alea, M.; Tarakanov, A.O.; Jimenez-Beristain, A.; Mudo, G.; Agnati, L.F.; Ciruela, F.; Belluardo, N.; Fuxe, K. Evidence for the existence of FGFR1-5-HT1A heteroreceptor complexes in the midbrain raphe 5-HT system. *Biochem. Biophys. Res. Commun.* **2015**, *456*, 489–493. [CrossRef] [PubMed]

55. Borroto-Escuela, D.O.; Romero-Fernandez, W.; Mudo, G.; Perez-Alea, M.; Ciruela, F.; Tarakanov, A.O.; Narvaez, M.; Di Liberto, V.; Agnati, L.F.; Belluardo, N.; et al. Fibroblast growth factor receptor 1-5-hydroxytryptamine 1A heteroreceptor complexes and their enhancement of hippocampal plasticity. *Biol. Psychiatry* **2012**, *71*, 84–91. [CrossRef]

56. Borroto-Escuela, D.O.; Perez-Alea, M.; Narvaez, M.; Tarakanov, A.O.; Mudo, G.; Jimenez-Beristain, A.; Agnati, L.F.; Ciruela, F.; Belluardo, N.; Fuxe, K. Enhancement of the FGFR1 signaling in the FGFR1-5-HT1A heteroreceptor complex in midbrain raphe 5-HT neuron systems. Relevance for neuroplasticity and depression. *Biochem. Biophys. Res. Commun.* **2015**, *463*, 180–186. [CrossRef] [PubMed]

57. Borroto-Escuela, D.O.; DuPont, C.M.; Li, X.; Savelli, D.; Lattanzi, D.; Srivastava, I.; Narvaez, M.; Di Palma, M.; Barbieri, E.; Andrade-Talavera, Y.; et al. Disturbances in the FGFR1-5-HT1A Heteroreceptor Complexes in the Raphe-Hippocampal 5-HT System Develop in a Genetic Rat Model of Depression. *Front. Cell Neurosci.* **2017**, *11*, 309. [CrossRef] [PubMed]

58. Borroto-Escuela, D.O.; Tarakanov, A.O.; Fuxe, K. FGFR1-5-HT1A Heteroreceptor Complexes: Implications for Understanding and Treating Major Depression. *Trends Neurosci.* **2016**, *39*, 5–15. [CrossRef] [PubMed]

59. Borroto-Escuela, D.O.; Carlsson, J.; Ambrogini, P.; Narvaez, M.; Wydra, K.; Tarakanov, A.O.; Li, X.; Millon, C.; Ferraro, L.; Cuppini, R.; et al. Understanding the Role of GPCR Heteroreceptor Complexes in Modulating the Brain Networks in Health and Disease. *Front. Cell Neurosci.* **2017**, *11*, 37. [CrossRef] [PubMed]

60. Belcheva, M.M.; Haas, P.D.; Tan, Y.; Heaton, V.M.; Coscia, C.J. The fibroblast growth factor receptor is at the site of convergence between mu-opioid receptor and growth factor signaling pathways in rat C6 glioma cells. *J. Pharm. Exp.* **2002**, *303*, 909–918. [CrossRef] [PubMed]

61. Volinsky, N.; Kholodenko, B.N. Complexity of receptor tyrosine kinase signal processing. *Cold Spring Harb. Perspect. Biol.* **2013**, *5*, a009043. [CrossRef] [PubMed]

62. Saha, N.; Robev, D.; Mason, E.O.; Himanen, J.P.; Nikolov, D.B. Therapeutic potential of targeting the Eph/ephrin signaling complex. *Int. J. Biochem. Cell Biol.* **2018**, *105*, 123–133. [CrossRef]

63. Pasquale, E.B. Eph receptor signalling casts a wide net on cell behaviour. *Nat. Rev. Mol. Cell Biol.* **2005**, *6*, 462–475. [CrossRef]

64. Lisabeth, E.M.; Falivelli, G.; Pasquale, E.B. Eph receptor signaling and ephrins. *Cold Spring Harb. Perspect. Biol.* **2013**, *5*. [CrossRef] [PubMed]

65. Shiuan, E.; Chen, J. Eph Receptor Tyrosine Kinases in Tumor Immunity. *Cancer Res.* **2016**, *76*, 6452–6457. [CrossRef] [PubMed]

66. Himanen, J.P.; Rajashankar, K.R.; Lackmann, M.; Cowan, C.A.; Henkemeyer, M.; Nikolov, D.B. Crystal structure of an Eph receptor-ephrin complex. *Nature* **2001**, *414*, 933–938. [CrossRef] [PubMed]

67. Kalo, M.S.; Pasquale, E.B. Signal transfer by Eph receptors. *Cell Tissue Res.* **1999**, *298*, 1–9. [CrossRef]

68. Janes, P.W.; Nievergall, E.; Lackmann, M. Concepts and consequences of Eph receptor clustering. *Semin. Cell Dev. Biol.* **2012**, *23*, 43–50. [CrossRef]

69. Himanen, J.P.; Yermekbayeva, L.; Janes, P.W.; Walker, J.R.; Xu, K.; Atapattu, L.; Rajashankar, K.R.; Mensinga, A.; Lackmann, M.; Nikolov, D.B.; et al. Architecture of Eph receptor clusters. *Proc. Natl. Acad. Sci. USA* **2010**, *107*, 10860–10865. [CrossRef] [PubMed]

70. Schmucker, D.; Zipursky, S.L. Signaling downstream of Eph receptors and ephrin ligands. *Cell* **2001**, *105*, 701–704. [CrossRef]

71. Yokote, H.; Fujita, K.; Jing, X.; Sawada, T.; Liang, S.; Yao, L.; Yan, X.; Zhang, Y.; Schlessinger, J.; Sakaguchi, K. Trans-activation of EphA4 and FGF receptors mediated by direct interactions between their cytoplasmic domains. *Proc. Natl. Acad. Sci. USA* **2005**, *102*, 18866–18871. [CrossRef] [PubMed]

72. Fukai, J.; Yokote, H.; Yamanaka, R.; Arao, T.; Nishio, K.; Itakura, T. EphA4 promotes cell proliferation and migration through a novel EphA4-FGFR1 signaling pathway in the human glioma U251 cell line. *Mol. Cancer* **2008**, *7*, 2768–2778. [CrossRef] [PubMed]

73. Gotoh, N. Regulation of growth factor signaling by FRS2 family docking/scaffold adaptor proteins. *Cancer Sci.* **2008**, *99*, 1319–1325. [CrossRef] [PubMed]

74. Sawada, T.; Jing, X.; Zhang, Y.; Shimada, E.; Yokote, H.; Miyajima, M.; Sakaguchi, K. Ternary complex formation of EphA4, FGFR and FRS2alpha plays an important role in the proliferation of embryonic neural stem/progenitor cells. *Genes Cells* **2010**, *15*, 297–311. [CrossRef] [PubMed]

75. Sawada, T.; Arai, D.; Jing, X.; Furushima, K.; Chen, Q.; Kawakami, K.; Yokote, H.; Miyajima, M.; Sakaguchi, K. Trans-Activation between EphA and FGFR Regulates Self-Renewal and Differentiation of Mouse Embryonic Neural Stem/Progenitor Cells via Differential Activation of FRS2alpha. *PLoS ONE* **2015**, *10*, e0128826. [CrossRef]

76. Zhang, Y.; Sawada, T.; Jing, X.; Yokote, H.; Yan, X.; Sakaguchi, K. Regulation of ephexin1, a guanine nucleotide exchange factor of Rho family GTPases, by fibroblast growth factor receptor-mediated tyrosine phosphorylation. *J. Biol. Chem.* **2007**, *282*, 31103–31112. [CrossRef] [PubMed]

77. Lee, S.; Shatadal, S.; Griep, A.E. Dlg-1 Interacts With and Regulates the Activities of Fibroblast Growth Factor Receptors and EphA2 in the Mouse Lens. *Invest. Ophthalmol. Vis. Sci.* **2016**, *57*, 707–718. [CrossRef] [PubMed]

78. Lee, S.; Griep, A.E. Loss of Dlg-1 in the mouse lens impairs fibroblast growth factor receptor signaling. *PLoS ONE* **2014**, *9*, e97470. [CrossRef] [PubMed]

79. Cao, Y. Multifarious functions of PDGFs and PDGFRs in tumor growth and metastasis. *Trends Mol. Med.* **2013**, *19*, 460–473. [CrossRef]

80. Papadopoulos, N.; Lennartsson, J. The PDGF/PDGFR pathway as a drug target. *Mol. Asp. Med.* **2018**, *62*, 75–88. [CrossRef] [PubMed]

81. Shim, A.H.; Liu, H.; Focia, P.J.; Chen, X.; Lin, P.C.; He, X. Structures of a platelet-derived growth factor/propeptide complex and a platelet-derived growth factor/receptor complex. *Proc. Natl. Acad. Sci. USA* **2010**, *107*, 11307–11312. [CrossRef] [PubMed]

82. Miyazawa, K.; Backstrom, G.; Leppanen, O.; Persson, C.; Wernstedt, C.; Hellman, U.; Heldin, C.H.; Ostman, A. Role of immunoglobulin-like domains 2-4 of the platelet-derived growth factor alpha-receptor in ligand-receptor complex assembly. *J. Biol. Chem.* **1998**, *273*, 25495–25502. [CrossRef]

83. Faraone, D.; Aguzzi, M.S.; Ragone, G.; Russo, K.; Capogrossi, M.C.; Facchiano, A. Heterodimerization of FGF-receptor 1 and PDGF-receptor-alpha: A novel mechanism underlying the inhibitory effect of PDGF-BB on FGF-2 in human cells. *Blood* **2006**, *107*, 1896–1902. [CrossRef] [PubMed]

84. Chen, P.Y.; Simons, M.; Friesel, R. FRS2 via fibroblast growth factor receptor 1 is required for platelet-derived growth factor receptor beta-mediated regulation of vascular smooth muscle marker gene expression. *J. Biol. Chem.* **2009**, *284*, 15980–15992. [CrossRef] [PubMed]

85. Russo, K.; Ragone, R.; Facchiano, A.M.; Capogrossi, M.C.; Facchiano, A. Platelet-derived growth factor-BB and basic fibroblast growth factor directly interact *in vitro* with high affinity. *J. Biol. Chem.* **2002**, *277*, 1284–1291. [CrossRef] [PubMed]

86. De Marchis, F.; Ribatti, D.; Giampietri, C.; Lentini, A.; Faraone, D.; Scoccianti, M.; Capogrossi, M.C.; Facchiano, A. Platelet-derived growth factor inhibits basic fibroblast growth factor angiogenic properties in vitro and in vivo through its alpha receptor. *Blood* **2002**, *99*, 2045–2053. [CrossRef]

87. Facchiano, A.; De Marchis, F.; Turchetti, E.; Facchiano, F.; Guglielmi, M.; Denaro, A.; Palumbo, R.; Scoccianti, M.; Capogrossi, M.C. The chemotactic and mitogenic effects of platelet-derived growth factor-BB on rat aorta smooth muscle cells are inhibited by basic fibroblast growth factor. *J. Cell Sci.* **2000**, *113 Pt 16*, 2855–2863.

88. Kono, S.A.; Heasley, L.E.; Doebele, R.C.; Camidge, D.R. Adding to the mix: Fibroblast growth factor and platelet-derived growth factor receptor pathways as targets in non-small cell lung cancer. *Curr. Cancer Drug Targets* **2012**, *12*, 107–123. [CrossRef] [PubMed]

89. Takeichi, M. Morphogenetic roles of classic cadherins. *Curr. Opin. Cell Biol.* **1995**, *7*, 619–627. [CrossRef]

90. Kourtidis, A.; Lu, R.; Pence, L.J.; Anastasiadis, P.Z. A central role for cadherin signaling in cancer. *Exp. Cell Res.* **2017**, *358*, 78–85. [CrossRef]

91. Gloushankova, N.A.; Rubtsova, S.N.; Zhitnyak, I.Y. Cadherin-mediated cell-cell interactions in normal and cancer cells. *Tissue Barriers* **2017**, *5*, e1356900. [CrossRef] [PubMed]

92. Fontenete, S.; Pena-Jimenez, D.; Perez-Moreno, M. Heterocellular cadherin connections: Coordinating adhesive cues in homeostasis and cancer. *F1000Research* **2017**, *6*, 1010. [CrossRef]

93. Nguyen, T.; Mege, R.M. N-Cadherin and Fibroblast Growth Factor Receptors crosstalk in the control of developmental and cancer cell migrations. *Eur. J. Cell Biol.* **2016**, *95*, 415–426. [CrossRef]

94. Williams, E.J.; Furness, J.; Walsh, F.S.; Doherty, P. Activation of the FGF receptor underlies neurite outgrowth stimulated by L1, N-CAM, and N-cadherin. *Neuron* **1994**, *13*, 583–594. [CrossRef]

95. Saffell, J.L.; Williams, E.J.; Mason, I.J.; Walsh, F.S.; Doherty, P. Expression of a dominant negative FGF receptor inhibits axonal growth and FGF receptor phosphorylation stimulated by CAMs. *Neuron* **1997**, *18*, 231–242. [CrossRef]

96. Ronn, L.C.; Doherty, P.; Holm, A.; Berezin, V.; Bock, E. Neurite outgrowth induced by a synthetic peptide ligand of neural cell adhesion molecule requires fibroblast growth factor receptor activation. *J. Neurochem.* **2000**, *75*, 665–671. [CrossRef] [PubMed]

97. Sanchez-Heras, E.; Howell, F.V.; Williams, G.; Doherty, P. The fibroblast growth factor receptor acid box is essential for interactions with N-cadherin and all of the major isoforms of neural cell adhesion molecule. *J. Biol. Chem.* **2006**, *281*, 35208–35216. [CrossRef]

98. Suyama, K.; Shapiro, I.; Guttman, M.; Hazan, R.B. A signaling pathway leading to metastasis is controlled by N-cadherin and the FGF receptor. *Cancer Cell* **2002**, *2*, 301–314. [CrossRef]

99. Hulit, J.; Suyama, K.; Chung, S.; Keren, R.; Agiostratidou, G.; Shan, W.; Dong, X.; Williams, T.M.; Lisanti, M.P.; Knudsen, K.; et al. N-cadherin signaling potentiates mammary tumor metastasis via enhanced extracellular signal-regulated kinase activation. *Cancer Res.* **2007**, *67*, 3106–3116. [CrossRef]

100. Qian, X.; Anzovino, A.; Kim, S.; Suyama, K.; Yao, J.; Hulit, J.; Agiostratidou, G.; Chandiramani, N.; McDaid, H.M.; Nagi, C.; et al. N-cadherin/FGFR promotes metastasis through epithelial-to-mesenchymal transition and stem/progenitor cell-like properties. *Oncogene* **2014**, *33*, 3411–3421. [CrossRef]

101. Takehara, T.; Teramura, T.; Onodera, Y.; Frampton, J.; Fukuda, K. Cdh2 stabilizes FGFR1 and contributes to primed-state pluripotency in mouse epiblast stem cells. *Sci. Rep.* **2015**, *5*, 14722. [CrossRef] [PubMed]

102. Cavallaro, U.; Niedermeyer, J.; Fuxa, M.; Christofori, G. N-CAM modulates tumour-cell adhesion to matrix by inducing FGF-receptor signalling. *Nat. Cell Biol.* **2001**, *3*, 650–657. [CrossRef]

103. Quintanal-Villalonga, A.; Ojeda-Marquez, L.; Marrugal, A.; Yague, P.; Ponce-Aix, S.; Salinas, A.; Carnero, A.; Ferrer, I.; Molina-Pinelo, S.; Paz-Ares, L. The FGFR4-388arg Variant Promotes Lung Cancer Progression by N-Cadherin Induction. *Sci. Rep.* **2018**, *8*, 2394. [CrossRef]

104. Kimura, Y.; Matsunami, H.; Inoue, T.; Shimamura, K.; Uchida, N.; Ueno, T.; Miyazaki, T.; Takeichi, M. Cadherin-11 expressed in association with mesenchymal morphogenesis in the head, somite, and limb bud of early mouse embryos. *Dev. Biol.* **1995**, *169*, 347–358. [CrossRef]

105. Sfikakis, P.P.; Vlachogiannis, N.I.; Christopoulos, P.F. Cadherin-11 as a therapeutic target in chronic, inflammatory rheumatic diseases. *Clin. Immunol.* **2017**, *176*, 107–113. [CrossRef]

106. Birtolo, C.; Pham, H.; Morvaridi, S.; Chheda, C.; Go, V.L.; Ptasznik, A.; Edderkaoui, M.; Weisman, M.H.; Noss, E.; Brenner, M.B.; et al. Cadherin-11 Is a Cell Surface Marker Up-Regulated in Activated Pancreatic Stellate Cells and Is Involved in Pancreatic Cancer Cell Migration. *Am. J. Pathol.* **2017**, *187*, 146–155. [CrossRef]

107. Ortiz, A.; Lee, Y.C.; Yu, G.; Liu, H.C.; Lin, S.C.; Bilen, M.A.; Cho, H.; Yu-Lee, L.Y.; Lin, S.H. Angiomotin is a novel component of cadherin-11/beta-catenin/p120 complex and is critical for cadherin-11-mediated cell migration. *FASEB J.* **2015**, *29*, 1080–1091. [CrossRef] [PubMed]

108. Kim, N.H.; Choi, S.H.; Lee, T.R.; Lee, C.H.; Lee, A.Y. Cadherin 11, a miR-675 target, induces N-cadherin expression and epithelial-mesenchymal transition in melasma. *J. Invest. Derm.* **2014**, *134*, 2967–2976. [CrossRef] [PubMed]

109. Chu, K.; Cheng, C.J.; Ye, X.; Lee, Y.C.; Zurita, A.J.; Chen, D.T.; Yu-Lee, L.Y.; Zhang, S.; Yeh, E.T.; Hu, M.C.; et al. Cadherin-11 promotes the metastasis of prostate cancer cells to bone. *Mol. Cancer Res.* **2008**, *6*, 1259–1267. [CrossRef] [PubMed]

110. Boscher, C.; Mege, R.M. Cadherin-11 interacts with the FGF receptor and induces neurite outgrowth through associated downstream signalling. *Cell Signal.* **2008**, *20*, 1061–1072. [CrossRef] [PubMed]

111. Mizutani, K.; Takai, Y. Nectin spot: A novel type of nectin-mediated cell adhesion apparatus. *Biochem. J.* **2016**, *473*, 2691–2715. [CrossRef] [PubMed]

112. Huang, K.; Lui, W.Y. Nectins and nectin-like molecules (Necls): Recent findings and their role and regulation in spermatogenesis. *Semin. Cell Dev. Biol.* **2016**, *59*, 54–61. [CrossRef]

113. Bojesen, K.B.; Clausen, O.; Rohde, K.; Christensen, C.; Zhang, L.; Li, S.; Kohler, L.; Nielbo, S.; Nielsen, J.; Gjorlund, M.D.; et al. Nectin-1 binds and signals through the fibroblast growth factor receptor. *J. Biol. Chem.* **2012**, *287*, 37420–37433. [CrossRef]

114. Owczarek, S.; Berezin, V. Neuroplastin: Cell adhesion molecule and signaling receptor. *Int. J. Biochem. Cell Biol.* **2012**, *44*, 1–5. [CrossRef]

115. Langnaese, K.; Mummery, R.; Gundelfinger, E.D.; Beesley, P.W. Immunoglobulin superfamily members gp65 and gp55: Tissue distribution of glycoforms. *FEBS Lett.* **1998**, *429*, 284–288. [CrossRef]

116. Langnaese, K.; Beesley, P.W.; Gundelfinger, E.D. Synaptic membrane glycoproteins gp65 and gp55 are new members of the immunoglobulin superfamily. *J. Biol. Chem.* **1997**, *272*, 821–827. [CrossRef] [PubMed]

117. Owczarek, S.; Kiryushko, D.; Larsen, M.H.; Kastrup, J.S.; Gajhede, M.; Sandi, C.; Berezin, V.; Bock, E.; Soroka, V. Neuroplastin-55 binds to and signals through the fibroblast growth factor receptor. *FASEB J.* **2010**, *24*, 1139–1150. [CrossRef] [PubMed]

118. Colombo, F.; Meldolesi, J. L1-CAM and N-CAM: From Adhesion Proteins to Pharmacological Targets. *Trends Pharm. Sci.* **2015**, *36*, 769–781. [CrossRef]

119. Sytnyk, V.; Leshchyns'ka, I.; Schachner, M. Neural Cell Adhesion Molecules of the Immunoglobulin Superfamily Regulate Synapse Formation, Maintenance, and Function. *Trends Neurosci.* **2017**, *40*, 295–308. [CrossRef] [PubMed]

120. Aonurm-Helm, A.; Jaako, K.; Jurgenson, M.; Zharkovsky, A. Pharmacological approach for targeting dysfunctional brain plasticity: Focus on neural cell adhesion molecule (NCAM). *Pharm. Res.* **2016**, *113*, 731–738. [CrossRef] [PubMed]

121. Francavilla, C.; Cattaneo, P.; Berezin, V.; Bock, E.; Ami, D.; de Marco, A.; Christofori, G.; Cavallaro, U. The binding of NCAM to FGFR1 induces a specific cellular response mediated by receptor trafficking. *J. Cell Biol.* **2009**, *187*, 1101–1116. [CrossRef]

122. Christensen, C.; Lauridsen, J.B.; Berezin, V.; Bock, E.; Kiselyov, V.V. The neural cell adhesion molecule binds to fibroblast growth factor receptor 2. *FEBS Lett.* **2006**, *580*, 3386–3390. [CrossRef] [PubMed]

123. Francavilla, C.; Loeffler, S.; Piccini, D.; Kren, A.; Christofori, G.; Cavallaro, U. Neural cell adhesion molecule regulates the cellular response to fibroblast growth factor. *J. Cell Sci.* **2007**, *120*, 4388–4394. [CrossRef] [PubMed]

124. Hansen, S.M.; Li, S.; Bock, E.; Berezin, V. Synthetic NCAM-derived ligands of the fibroblast growth factor receptor. *Adv. Exp. Med. Biol.* **2010**, *663*, 355–372. [CrossRef] [PubMed]

125. Abe, K.; Ohuchi, H.; Tanabe, H.; Imanaka, K.; Asano, H.; Kato, M.; Yokote, Y.; Kyo, S. Aortic root remodeling and coronary artery bypass grafting for acute type A aortic dissection involving the left main coronary artery; report of a case. *Kyobu Geka* **2005**, *58*, 897–901.

126. Kiselyov, V.V.; Soroka, V.; Berezin, V.; Bock, E. Structural biology of NCAM homophilic binding and activation of FGFR. *J. Neurochem.* **2005**, *94*, 1169–1179. [CrossRef]

127. Christensen, C.; Berezin, V.; Bock, E. Neural cell adhesion molecule differentially interacts with isoforms of the fibroblast growth factor receptor. *Neuroreport* **2011**, *22*, 727–732. [CrossRef]

128. Zecchini, S.; Bombardelli, L.; Decio, A.; Bianchi, M.; Mazzarol, G.; Sanguineti, F.; Aletti, G.; Maddaluno, L.; Berezin, V.; Bock, E.; et al. The adhesion molecule NCAM promotes ovarian cancer progression via FGFR signalling. *EMBO Mol. Med.* **2011**, *3*, 480–494. [CrossRef]

129. Colombo, N.; Cavallaro, U. The interplay between NCAM and FGFR signalling underlies ovarian cancer progression. *Ecancermedicalscience* **2011**, *5*, 226. [CrossRef]

130. Zivotic, M.; Tampe, B.; Muller, G.; Muller, C.; Lipkovski, A.; Xu, X.; Nyamsuren, G.; Zeisberg, M.; Markovic-Lipkovski, J. Modulation of NCAM/FGFR1 signaling suppresses EMT program in human proximal tubular epithelial cells. *PLoS ONE* **2018**, *13*, e0206786. [CrossRef] [PubMed]

131. Kulahin, N.; Li, S.; Hinsby, A.; Kiselyov, V.; Berezin, V.; Bock, E. Fibronectin type III (FN3) modules of the neuronal cell adhesion molecule L1 interact directly with the fibroblast growth factor (FGF) receptor. *Mol. Cell Neurosci.* **2008**, *37*, 528–536. [CrossRef]

132. Riedle, S.; Kiefel, H.; Gast, D.; Bondong, S.; Wolterink, S.; Gutwein, P.; Altevogt, P. Nuclear translocation and signalling of L1-CAM in human carcinoma cells requires ADAM10 and presenilin/gamma-secretase activity. *Biochem. J.* **2009**, *420*, 391–402. [CrossRef] [PubMed]

133. Mohanan, V.; Temburni, M.K.; Kappes, J.C.; Galileo, D.S. L1CAM stimulates glioma cell motility and proliferation through the fibroblast growth factor receptor. *Clin. Exp. Metastasis* **2013**, *30*, 507–520. [CrossRef] [PubMed]

134. Gonzalez-Martinez, D.; Kim, S.H.; Hu, Y.; Guimond, S.; Schofield, J.; Winyard, P.; Vannelli, G.B.; Turnbull, J.; Bouloux, P.M. Anosmin-1 modulates fibroblast growth factor receptor 1 signaling in human gonadotropin-releasing hormone olfactory neuroblasts through a heparan sulfate-dependent mechanism. *J. Neurosci.* **2004**, *24*, 10384–10392. [CrossRef]

135. Bribian, A.; Barallobre, M.J.; Soussi-Yanicostas, N.; de Castro, F. Anosmin-1 modulates the FGF-2-dependent migration of oligodendrocyte precursors in the developing optic nerve. *Mol. Cell Neurosci.* **2006**, *33*, 2–14. [CrossRef] [PubMed]

136. Garcia-Gonzalez, D.; Clemente, D.; Coelho, M.; Esteban, P.F.; Soussi-Yanicostas, N.; de Castro, F. Dynamic roles of FGF-2 and Anosmin-1 in the migration of neuronal precursors from the subventricular zone during pre- and postnatal development. *Exp. Neurol.* **2010**, *222*, 285–295. [CrossRef] [PubMed]

137. Murcia-Belmonte, V.; Esteban, P.F.; Garcia-Gonzalez, D.; De Castro, F. Biochemical dissection of Anosmin-1 interaction with FGFR1 and components of the extracellular matrix. *J. Neurochem.* **2010**, *115*, 1256–1265. [CrossRef]

138. Hu, Y.; Guimond, S.E.; Travers, P.; Cadman, S.; Hohenester, E.; Turnbull, J.E.; Kim, S.H.; Bouloux, P.M. Novel mechanisms of fibroblast growth factor receptor 1 regulation by extracellular matrix protein anosmin-1. *J. Biol. Chem.* **2009**, *284*, 29905–29920. [CrossRef] [PubMed]

139. Diaz-Balzac, C.A.; Lazaro-Pena, M.I.; Ramos-Ortiz, G.A.; Bulow, H.E. The Adhesion Molecule KAL-1/anosmin-1 Regulates Neurite Branching through a SAX-7/L1CAM-EGL-15/FGFR Receptor Complex. *Cell Rep.* **2015**, *11*, 1377–1384. [CrossRef] [PubMed]

140. Kriebel, M.; Wuchter, J.; Trinks, S.; Volkmer, H. Neurofascin: A switch between neuronal plasticity and stability. *Int. J. Biochem. Cell Biol.* **2012**, *44*, 694–697. [CrossRef]

141. Kirschbaum, K.; Kriebel, M.; Kranz, E.U.; Potz, O.; Volkmer, H. Analysis of non-canonical fibroblast growth factor receptor 1 (FGFR1) interaction reveals regulatory and activating domains of neurofascin. *J. Biol. Chem.* **2009**, *284*, 28533–28542. [CrossRef] [PubMed]

142. Pruss, T.; Kranz, E.U.; Niere, M.; Volkmer, H. A regulated switch of chick neurofascin isoforms modulates ligand recognition and neurite extension. *Mol. Cell Neurosci.* **2006**, *31*, 354–365. [CrossRef] [PubMed]

143. Sudhof, T.C. Neuroligins and neurexins link synaptic function to cognitive disease. *Nature* **2008**, *455*, 903–911. [CrossRef] [PubMed]

144. Rudenko, G. Neurexins - versatile molecular platforms in the synaptic cleft. *Curr. Opin. Struct. Biol.* **2019**, *54*, 112–121. [CrossRef] [PubMed]

145. Gjorlund, M.D.; Nielsen, J.; Pankratova, S.; Li, S.; Korshunova, I.; Bock, E.; Berezin, V. Neuroligin-1 induces neurite outgrowth through interaction with neurexin-1beta and activation of fibroblast growth factor receptor-1. *FASEB J.* **2012**, *26*, 4174–4186. [CrossRef]

146. Kubick, N.; Brosamle, D.; Mickael, M.E. Molecular Evolution and Functional Divergence of the IgLON Family. *Evol. Bioinform. Online* **2018**, *14*, 1176934318775081. [CrossRef] [PubMed]

147. Funatsu, N.; Miyata, S.; Kumanogoh, H.; Shigeta, M.; Hamada, K.; Endo, Y.; Sokawa, Y.; Maekawa, S. Characterization of a novel rat brain glycosylphosphatidylinositol-anchored protein (Kilon), a member of the IgLON cell adhesion molecule family. *J. Biol. Chem.* **1999**, *274*, 8224–8230. [CrossRef] [PubMed]

148. Pischedda, F.; Piccoli, G. The IgLON Family Member Negr1 Promotes Neuronal Arborization Acting as Soluble Factor via FGFR2. *Front. Mol. Neurosci.* **2015**, *8*, 89. [CrossRef]

149. Casey, J.P.; Magalhaes, T.; Conroy, J.M.; Regan, R.; Shah, N.; Anney, R.; Shields, D.C.; Abrahams, B.S.; Almeida, J.; Bacchelli, E.; et al. A novel approach of homozygous haplotype sharing identifies candidate genes in autism spectrum disorder. *Hum. Genet.* **2012**, *131*, 565–579. [CrossRef] [PubMed]

150. Szczurkowska, J.; Pischedda, F.; Pinto, B.; Manago, F.; Haas, C.A.; Summa, M.; Bertorelli, R.; Papaleo, F.; Schafer, M.K.; Piccoli, G.; et al. NEGR1 and FGFR2 cooperatively regulate cortical development and core behaviours related to autism disorders in mice. *Brain* **2018**, *141*, 2772–2794. [CrossRef]

151. Sellar, G.C.; Watt, K.P.; Rabiasz, G.J.; Stronach, E.A.; Li, L.; Miller, E.P.; Massie, C.E.; Miller, J.; Contreras-Moreira, B.; Scott, D.; et al. OPCML at 11q25 is epigenetically inactivated and has tumor-suppressor function in epithelial ovarian cancer. *Nat. Genet.* **2003**, *34*, 337–343. [CrossRef] [PubMed]

152. Chen, H.; Ye, F.; Zhang, J.; Lu, W.; Cheng, Q.; Xie, X. Loss of OPCML expression and the correlation with CpG island methylation and LOH in ovarian serous carcinoma. *Eur. J. Gynaecol. Oncol.* **2007**, *28*, 464–467.

153. Reed, J.E.; Dunn, J.R.; du Plessis, D.G.; Shaw, E.J.; Reeves, P.; Gee, A.L.; Warnke, P.C.; Sellar, G.C.; Moss, D.J.; Walker, C. Expression of cellular adhesion molecule 'OPCML' is down-regulated in gliomas and other brain tumours. *Neuropathol. Appl. Neurobiol.* **2007**, *33*, 77–85. [CrossRef]

154. Cui, Y.; Ying, Y.; van Hasselt, A.; Ng, K.M.; Yu, J.; Zhang, Q.; Jin, J.; Liu, D.; Rhim, J.S.; Rha, S.Y.; et al. OPCML is a broad tumor suppressor for multiple carcinomas and lymphomas with frequently epigenetic inactivation. *PLoS ONE* **2008**, *3*, e2990. [CrossRef]

155. Zhang, N.; Xu, J.; Wang, Y.; Heng, X.; Yang, L.; Xing, X. Loss of opioid binding protein/cell adhesion molecule-like gene expression in gastric cancer. *Oncol. Lett.* **2018**, *15*, 9973–9977. [CrossRef] [PubMed]

156. McKie, A.B.; Vaughan, S.; Zanini, E.; Okon, I.S.; Louis, L.; de Sousa, C.; Greene, M.I.; Wang, Q.; Agarwal, R.; Shaposhnikov, D.; et al. The OPCML tumor suppressor functions as a cell surface repressor-adaptor, negatively regulating receptor tyrosine kinases in epithelial ovarian cancer. *Cancer Discov.* **2012**, *2*, 156–171. [CrossRef]

157. O'Sullivan, M.L.; de Wit, J.; Savas, J.N.; Comoletti, D.; Otto-Hitt, S.; Yates, J.R., 3rd; Ghosh, A. FLRT proteins are endogenous latrophilin ligands and regulate excitatory synapse development. *Neuron* **2012**, *73*, 903–910. [CrossRef] [PubMed]

158. Sando, R.; Jiang, X.; Sudhof, T.C. Latrophilin GPCRs direct synapse specificity by coincident binding of FLRTs and teneurins. *Science* **2019**, *363*. [CrossRef] [PubMed]

159. Del Toro, D.; Ruff, T.; Cederfjall, E.; Villalba, A.; Seyit-Bremer, G.; Borrell, V.; Klein, R. Regulation of Cerebral Cortex Folding by Controlling Neuronal Migration via FLRT Adhesion Molecules. *Cell* **2017**, *169*, 621–635.e616. [CrossRef] [PubMed]

160. Jackson, V.A.; Mehmood, S.; Chavent, M.; Roversi, P.; Carrasquero, M.; Del Toro, D.; Seyit-Bremer, G.; Ranaivoson, F.M.; Comoletti, D.; Sansom, M.S.; et al. Super-complexes of adhesion GPCRs and neural guidance receptors. *Nat. Commun.* **2016**, *7*, 11184. [CrossRef] [PubMed]

161. Lacy, S.E.; Bonnemann, C.G.; Buzney, E.A.; Kunkel, L.M. Identification of FLRT1, FLRT2, and FLRT3: A novel family of transmembrane leucine-rich repeat proteins. *Genomics* **1999**, *62*, 417–426. [CrossRef] [PubMed]

162. Karaulanov, E.E.; Bottcher, R.T.; Niehrs, C. A role for fibronectin-leucine-rich transmembrane cell-surface proteins in homotypic cell adhesion. *EMBO Rep.* **2006**, *7*, 283–290. [CrossRef] [PubMed]

163. Bottcher, R.T.; Pollet, N.; Delius, H.; Niehrs, C. The transmembrane protein XFLRT3 forms a complex with FGF receptors and promotes FGF signalling. *Nat. Cell Biol.* **2004**, *6*, 38–44. [CrossRef] [PubMed]

164. Wei, K.; Xu, Y.; Tse, H.; Manolson, M.F.; Gong, S.G. Mouse FLRT2 interacts with the extracellular and intracellular regions of FGFR2. *J. Dent. Res.* **2011**, *90*, 1234–1239. [CrossRef] [PubMed]

165. Haines, B.P.; Wheldon, L.M.; Summerbell, D.; Heath, J.K.; Rigby, P.W. Regulated expression of FLRT genes implies a functional role in the regulation of FGF signalling during mouse development. *Dev. Biol.* **2006**, *297*, 14–25. [CrossRef] [PubMed]

166. Wheldon, L.M.; Haines, B.P.; Rajappa, R.; Mason, I.; Rigby, P.W.; Heath, J.K. Critical role of FLRT1 phosphorylation in the interdependent regulation of FLRT1 function and FGF receptor signalling. *PLoS ONE* **2010**, *5*, e10264. [CrossRef] [PubMed]

167. Humphries, J.D.; Chastney, M.R.; Askari, J.A.; Humphries, M.J. Signal transduction via integrin adhesion complexes. *Curr. Opin. Cell Biol.* **2019**, *56*, 14–21. [CrossRef]

168. Barczyk, M.; Carracedo, S.; Gullberg, D. Integrins. *Cell Tissue Res.* **2010**, *339*, 269–280. [CrossRef] [PubMed]

169. Harburger, D.S.; Calderwood, D.A. Integrin signalling at a glance. *J. Cell Sci.* **2009**, *122*, 159–163. [CrossRef]

170. Hamidi, H.; Ivaska, J. Every step of the way: Integrins in cancer progression and metastasis. *Nat. Rev. Cancer* **2018**, *18*, 533–548. [CrossRef]

171. Yamaji, S.; Saegusa, J.; Ieguchi, K.; Fujita, M.; Mori, S.; Takada, Y.K.; Takada, Y. A novel fibroblast growth factor-1 (FGF1) mutant that acts as an FGF antagonist. *PLoS ONE* **2010**, *5*, e10273. [CrossRef] [PubMed]

172. Mori, S.; Wu, C.Y.; Yamaji, S.; Saegusa, J.; Shi, B.; Ma, Z.; Kuwabara, Y.; Lam, K.S.; Isseroff, R.R.; Takada, Y.K.; et al. Direct binding of integrin alphavbeta3 to FGF1 plays a role in FGF1 signaling. *J. Biol. Chem.* **2008**, *283*, 18066–18075. [CrossRef]

173. Mori, S.; Tran, V.; Nishikawa, K.; Kaneda, T.; Hamada, Y.; Kawaguchi, N.; Fujita, M.; Saegusa, J.; Takada, Y.K.; Matsuura, N.; et al. A dominant-negative FGF1 mutant (the R50E mutant) suppresses tumorigenesis and angiogenesis. *PLoS ONE* **2013**, *8*, e57927. [CrossRef]

174. Mori, S.; Hatori, N.; Kawaguchi, N.; Hamada, Y.; Shih, T.C.; Wu, C.Y.; Lam, K.S.; Matsuura, N.; Yamamoto, H.; Takada, Y.K.; et al. The integrin-binding defective FGF2 mutants potently suppress FGF2 signalling and angiogenesis. *Biosci. Rep.* **2017**, *37*. [CrossRef] [PubMed]

175. Yayon, A.; Klagsbrun, M.; Esko, J.D.; Leder, P.; Ornitz, D.M. Cell surface, heparin-like molecules are required for binding of basic fibroblast growth factor to its high affinity receptor. *Cell* **1991**, *64*, 841–848. [CrossRef]

176. Rapraeger, A.C.; Krufka, A.; Olwin, B.B. Requirement of heparan sulfate for bFGF-mediated fibroblast growth and myoblast differentiation. *Science* **1991**, *252*, 1705–1708. [CrossRef]

177. Pellegrini, L.; Burke, D.F.; von Delft, F.; Mulloy, B.; Blundell, T.L. Crystal structure of fibroblast growth factor receptor ectodomain bound to ligand and heparin. *Nature* **2000**, *407*, 1029–1034. [CrossRef]

178. Bishop, J.R.; Schuksz, M.; Esko, J.D. Heparan sulphate proteoglycans fine-tune mammalian physiology. *Nature* **2007**, *446*, 1030–1037. [CrossRef] [PubMed]

179. Matsuo, I.; Kimura-Yoshida, C. Extracellular modulation of Fibroblast Growth Factor signaling through heparan sulfate proteoglycans in mammalian development. *Curr. Opin. Genet. Dev.* **2013**, *23*, 399–407. [CrossRef] [PubMed]

180. Lord, M.S.; Tang, F.; Rnjak-Kovacina, J.; Smith, J.G.W.; Melrose, J.; Whitelock, J.M. The multifaceted roles of perlecan in fibrosis. *Matrix Biol.* **2018**, *68–69*, 150–166. [CrossRef]

181. Chuang, C.Y.; Lord, M.S.; Melrose, J.; Rees, M.D.; Knox, S.M.; Freeman, C.; Iozzo, R.V.; Whitelock, J.M. Heparan sulfate-dependent signaling of fibroblast growth factor 18 by chondrocyte-derived perlecan. *Biochemistry* **2010**, *49*, 5524–5532. [CrossRef] [PubMed]

182. Knox, S.; Merry, C.; Stringer, S.; Melrose, J.; Whitelock, J. Not all perlecans are created equal: Interactions with fibroblast growth factor (FGF) 2 and FGF receptors. *J. Biol. Chem.* **2002**, *277*, 14657–14665. [CrossRef] [PubMed]

183. Smith, S.M.; West, L.A.; Govindraj, P.; Zhang, X.; Ornitz, D.M.; Hassell, J.R. Heparan and chondroitin sulfate on growth plate perlecan mediate binding and delivery of FGF-2 to FGF receptors. *Matrix Biol.* **2007**, *26*, 175–184. [CrossRef] [PubMed]

184. Smith, S.M.; West, L.A.; Hassell, J.R. The core protein of growth plate perlecan binds FGF-18 and alters its mitogenic effect on chondrocytes. *Arch. Biochem. Biophys.* **2007**, *468*, 244–251. [CrossRef] [PubMed]

185. Aviezer, D.; Hecht, D.; Safran, M.; Eisinger, M.; David, G.; Yayon, A. Perlecan, basal lamina proteoglycan, promotes basic fibroblast growth factor-receptor binding, mitogenesis, and angiogenesis. *Cell* **1994**, *79*, 1005–1013. [CrossRef]

186. Theocharis, A.D.; Karamanos, N.K. Proteoglycans remodeling in cancer: Underlying molecular mechanisms. *Matrix Biol.* **2019**, *75–76*, 220–259. [CrossRef]

187. Afratis, N.A.; Nikitovic, D.; Multhaupt, H.A.; Theocharis, A.D.; Couchman, J.R.; Karamanos, N.K. Syndecans—Key regulators of cell signaling and biological functions. *FEBS J.* **2017**, *284*, 27–41. [CrossRef] [PubMed]

188. Chung, H.; Multhaupt, H.A.; Oh, E.S.; Couchman, J.R. Minireview: Syndecans and their crucial roles during tissue regeneration. *FEBS Lett.* **2016**, *590*, 2408–2417. [CrossRef]

189. Bernfield, M.; Sanderson, R.D. Syndecan, a developmentally regulated cell surface proteoglycan that binds extracellular matrix and growth factors. *Philos. Trans. R. Soc. Lond. B Biol. Sci.* **1990**, *327*, 171–186. [CrossRef] [PubMed]

190. Olwin, B.B.; Rapraeger, A. Repression of myogenic differentiation by aFGF, bFGF, and K-FGF is dependent on cellular heparan sulfate. *J. Cell Biol.* **1992**, *118*, 631–639. [CrossRef]

191. Clasper, S.; Vekemans, S.; Fiore, M.; Plebanski, M.; Wordsworth, P.; David, G.; Jackson, D.G. Inducible expression of the cell surface heparan sulfate proteoglycan syndecan-2 (fibroglycan) on human activated macrophages can regulate fibroblast growth factor action. *J. Biol. Chem.* **1999**, *274*, 24113–24123. [CrossRef] [PubMed]

192. Wu, X.; Kan, M.; Wang, F.; Jin, C.; Yu, C.; McKeehan, W.L. A rare premalignant prostate tumor epithelial cell syndecan-1 forms a fibroblast growth factor-binding complex with progression-promoting ectopic fibroblast growth factor receptor 1. *Cancer Res.* **2001**, *61*, 5295–5302. [PubMed]

193. Iwabuchi, T.; Goetinck, P.F. Syndecan-4 dependent FGF stimulation of mouse vibrissae growth. *Mech. Dev.* **2006**, *123*, 831–841. [CrossRef] [PubMed]

194. Filla, M.S.; Dam, P.; Rapraeger, A.C. The cell surface proteoglycan syndecan-1 mediates fibroblast growth factor-2 binding and activity. *J. Cell Physiol.* **1998**, *174*, 310–321. [CrossRef]

195. Jang, E.; Albadawi, H.; Watkins, M.T.; Edelman, E.R.; Baker, A.B. Syndecan-4 proteoliposomes enhance fibroblast growth factor-2 (FGF-2)-induced proliferation, migration, and neovascularization of ischemic muscle. *Proc. Natl. Acad. Sci. USA* **2012**, *109*, 1679–1684. [CrossRef] [PubMed]

196. Murakami, M.; Nguyen, L.T.; Zhuang, Z.W.; Moodie, K.L.; Carmeliet, P.; Stan, R.V.; Simons, M. The FGF system has a key role in regulating vascular integrity. *J. Clin. Invest.* **2008**, *118*, 3355–3366. [CrossRef] [PubMed]

197. Elfenbein, A.; Lanahan, A.; Zhou, T.X.; Yamasaki, A.; Tkachenko, E.; Matsuda, M.; Simons, M. Syndecan 4 regulates FGFR1 signaling in endothelial cells by directing macropinocytosis. *Sci. Signal.* **2012**, *5*, ra36. [CrossRef] [PubMed]

198. Fico, A.; Maina, F.; Dono, R. Fine-tuning of cell signaling by glypicans. *Cell Mol. Life Sci.* **2011**, *68*, 923–929. [CrossRef]

199. Galli, A.; Roure, A.; Zeller, R.; Dono, R. Glypican 4 modulates FGF signalling and regulates dorsoventral forebrain patterning in Xenopus embryos. *Development* **2003**, *130*, 4919–4929. [CrossRef]

200. Gutierrez, J.; Brandan, E. A novel mechanism of sequestering fibroblast growth factor 2 by glypican in lipid rafts, allowing skeletal muscle differentiation. *Mol. Cell Biol.* **2010**, *30*, 1634–1649. [CrossRef] [PubMed]

201. Berman, B.; Ostrovsky, O.; Shlissel, M.; Lang, T.; Regan, D.; Vlodavsky, I.; Ishai-Michaeli, R.; Ron, D. Similarities and differences between the effects of heparin and glypican-1 on the bioactivity of acidic fibroblast growth factor and the keratinocyte growth factor. *J. Biol. Chem.* **1999**, *274*, 36132–36138. [CrossRef] [PubMed]

202. Qiao, D.; Meyer, K.; Mundhenke, C.; Drew, S.A.; Friedl, A. Heparan sulfate proteoglycans as regulators of fibroblast growth factor-2 signaling in brain endothelial cells. Specific role for glypican-1 in glioma angiogenesis. *J. Biol. Chem.* **2003**, *278*, 16045–16053. [CrossRef] [PubMed]

203. Su, G.; Meyer, K.; Nandini, C.D.; Qiao, D.; Salamat, S.; Friedl, A. Glypican-1 is frequently overexpressed in human gliomas and enhances FGF-2 signaling in glioma cells. *Am. J. Pathol.* **2006**, *168*, 2014–2026. [CrossRef]

204. Itoh, N.; Nakayama, Y.; Konishi, M. Roles of FGFs As Paracrine or Endocrine Signals in Liver Development, Health, and Disease. *Front. Cell Dev. Biol.* **2016**, *4*, 30. [CrossRef] [PubMed]

205. Itoh, N. Hormone-like (endocrine) Fgfs: Their evolutionary history and roles in development, metabolism, and disease. *Cell Tissue Res.* **2010**, *342*, 1–11. [CrossRef] [PubMed]

206. Yu, X.; Ibrahimi, O.A.; Goetz, R.; Zhang, F.; Davis, S.I.; Garringer, H.J.; Linhardt, R.J.; Ornitz, D.M.; Mohammadi, M.; White, K.E. Analysis of the biochemical mechanisms for the endocrine actions of fibroblast growth factor-23. *Endocrinology* **2005**, *146*, 4647–4656. [CrossRef]

207. Yie, J.; Wang, W.; Deng, L.; Tam, L.T.; Stevens, J.; Chen, M.M.; Li, Y.; Xu, J.; Lindberg, R.; Hecht, R.; et al. Understanding the physical interactions in the FGF21/FGFR/beta-Klotho complex: Structural requirements and implications in FGF21 signaling. *Chem. Biol. Drug Des.* **2012**, *79*, 398–410. [CrossRef] [PubMed]

208. Goetz, R.; Ohnishi, M.; Kir, S.; Kurosu, H.; Wang, L.; Pastor, J.; Ma, J.; Gai, W.; Kuro-o, M.; Razzaque, M.S.; et al. Conversion of a paracrine fibroblast growth factor into an endocrine fibroblast growth factor. *J. Biol. Chem.* **2012**, *287*, 29134–29146. [CrossRef] [PubMed]

209. Adams, A.C.; Cheng, C.C.; Coskun, T.; Kharitonenkov, A. FGF21 requires betaklotho to act in vivo. *PLoS ONE* **2012**, *7*, e49977. [CrossRef] [PubMed]

210. Ding, X.; Boney-Montoya, J.; Owen, B.M.; Bookout, A.L.; Coate, K.C.; Mangelsdorf, D.J.; Kliewer, S.A. betaKlotho is required for fibroblast growth factor 21 effects on growth and metabolism. *Cell Metab.* **2012**, *16*, 387–393. [CrossRef] [PubMed]

211. Kurosu, H.; Ogawa, Y.; Miyoshi, M.; Yamamoto, M.; Nandi, A.; Rosenblatt, K.P.; Baum, M.G.; Schiavi, S.; Hu, M.C.; Moe, O.W.; et al. Regulation of fibroblast growth factor-23 signaling by klotho. *J. Biol. Chem.* **2006**, *281*, 6120–6123. [CrossRef]

212. Urakawa, I.; Yamazaki, Y.; Shimada, T.; Iijima, K.; Hasegawa, H.; Okawa, K.; Fujita, T.; Fukumoto, S.; Yamashita, T. Klotho converts canonical FGF receptor into a specific receptor for FGF23. *Nature* **2006**, *444*, 770–774. [CrossRef] [PubMed]

213. Kharitonenkov, A.; Dunbar, J.D.; Bina, H.A.; Bright, S.; Moyers, J.S.; Zhang, C.; Ding, L.; Micanovic, R.; Mehrbod, S.F.; Knierman, M.D.; et al. FGF-21/FGF-21 receptor interaction and activation is determined by betaKlotho. *J. Cell Physiol.* **2008**, *215*, 1–7. [CrossRef] [PubMed]

214. Kuro-o, M.; Matsumura, Y.; Aizawa, H.; Kawaguchi, H.; Suga, T.; Utsugi, T.; Ohyama, Y.; Kurabayashi, M.; Kaname, T.; Kume, E.; et al. Mutation of the mouse klotho gene leads to a syndrome resembling ageing. *Nature* **1997**, *390*, 45–51. [CrossRef] [PubMed]

215. Matsumura, Y.; Aizawa, H.; Shiraki-Iida, T.; Nagai, R.; Kuro-o, M.; Nabeshima, Y. Identification of the human klotho gene and its two transcripts encoding membrane and secreted klotho protein. *Biochem. Biophys. Res. Commun.* **1998**, *242*, 626–630. [CrossRef]

216. Chen, G.; Liu, Y.; Goetz, R.; Fu, L.; Jayaraman, S.; Hu, M.C.; Moe, O.W.; Liang, G.; Li, X.; Mohammadi, M. alpha-Klotho is a non-enzymatic molecular scaffold for FGF23 hormone signalling. *Nature* **2018**, *553*, 461–466. [CrossRef] [PubMed]

217. Luo, Y.; Ye, S.; Li, X.; Lu, W. Emerging Structure-Function Paradigm of Endocrine FGFs in Metabolic Diseases. *Trends Pharm. Sci.* **2019**, *40*, 142–153. [CrossRef] [PubMed]

218. Takashi, Y.; Fukumoto, S. FGF23 beyond Phosphotropic Hormone. *Trends Endocrinol. Metab.* **2018**, *29*, 755–767. [CrossRef] [PubMed]

219. Farrow, E.G.; Davis, S.I.; Summers, L.J.; White, K.E. Initial FGF23-mediated signaling occurs in the distal convoluted tubule. *J. Am. Soc. Nephrol.* **2009**, *20*, 955–960. [CrossRef]

220. Andrukhova, O.; Smorodchenko, A.; Egerbacher, M.; Streicher, C.; Zeitz, U.; Goetz, R.; Shalhoub, V.; Mohammadi, M.; Pohl, E.E.; Lanske, B.; et al. FGF23 promotes renal calcium reabsorption through the TRPV5 channel. *EMBO J.* **2014**, *33*, 229–246. [CrossRef] [PubMed]

221. Goetz, R.; Nakada, Y.; Hu, M.C.; Kurosu, H.; Wang, L.; Nakatani, T.; Shi, M.; Eliseenkova, A.V.; Razzaque, M.S.; Moe, O.W.; et al. Isolated C-terminal tail of FGF23 alleviates hypophosphatemia by inhibiting FGF23-FGFR-Klotho complex formation. *Proc. Natl. Acad. Sci. USA* **2010**, *107*, 407–412. [CrossRef] [PubMed]

222. Andrukhova, O.; Zeitz, U.; Goetz, R.; Mohammadi, M.; Lanske, B.; Erben, R.G. FGF23 acts directly on renal proximal tubules to induce phosphaturia through activation of the ERK1/2-SGK1 signaling pathway. *Bone* **2012**, *51*, 621–628. [CrossRef] [PubMed]

223. Martin, A.; David, V.; Quarles, L.D. Regulation and function of the FGF23/Klotho endocrine pathways. *Physiol. Rev.* **2012**, *92*, 131–155. [CrossRef] [PubMed]

224. Villanueva, L.S.; Gonzalez, S.G.; Tomero, J.A.S.; Aguilera, A.; Junco, E.O. Bone mineral disorder in chronic kidney disease: Klotho and FGF23; cardiovascular implications. *Nefrologia* **2016**, *36*, 333–464. [CrossRef]

225. Lu, X.; Hu, M.C. Klotho/FGF23 Axis in Chronic Kidney Disease and Cardiovascular Disease. *Kidney Dis.* **2017**, *3*, 15–23. [CrossRef]

226. Shiohama, A.; Sasaki, T.; Noda, S.; Minoshima, S.; Shimizu, N. Molecular cloning and expression analysis of a novel gene DGCR8 located in the DiGeorge syndrome chromosomal region. *Biochem. Biophys. Res. Commun.* **2003**, *304*, 184–190. [CrossRef]

227. Kurosu, H.; Choi, M.; Ogawa, Y.; Dickson, A.S.; Goetz, R.; Eliseenkova, A.V.; Mohammadi, M.; Rosenblatt, K.P.; Kliewer, S.A.; Kuro-o, M. Tissue-specific expression of betaKlotho and fibroblast growth factor (FGF) receptor isoforms determines metabolic activity of FGF19 and FGF21. *J. Biol. Chem.* **2007**, *282*, 26687–26695. [CrossRef]

228. Ogawa, Y.; Kurosu, H.; Yamamoto, M.; Nandi, A.; Rosenblatt, K.P.; Goetz, R.; Eliseenkova, A.V.; Mohammadi, M.; Kuro-o, M. BetaKlotho is required for metabolic activity of fibroblast growth factor 21. *Proc. Natl. Acad. Sci. USA* **2007**, *104*, 7432–7437. [CrossRef]

229. Adams, A.C.; Yang, C.; Coskun, T.; Cheng, C.C.; Gimeno, R.E.; Luo, Y.; Kharitonenkov, A. The breadth of FGF21's metabolic actions are governed by FGFR1 in adipose tissue. *Mol. Metab.* **2012**, *2*, 31–37. [CrossRef] [PubMed]

230. Lee, S.; Choi, J.; Mohanty, J.; Sousa, L.P.; Tome, F.; Pardon, E.; Steyaert, J.; Lemmon, M.A.; Lax, I.; Schlessinger, J. Structures of beta-klotho reveal a 'zip code'-like mechanism for endocrine FGF signalling. *Nature* **2018**, *553*, 501–505. [CrossRef] [PubMed]

231. Shi, S.Y.; Lu, Y.W.; Richardson, J.; Min, X.; Weiszmann, J.; Richards, W.G.; Wang, Z.; Zhang, Z.; Zhang, J.; Li, Y. A systematic dissection of sequence elements determining beta-Klotho and FGF interaction and signaling. *Sci. Rep.* **2018**, *8*, 11045. [CrossRef] [PubMed]

232. Kharitonenkov, A.; Shiyanova, T.L.; Koester, A.; Ford, A.M.; Micanovic, R.; Galbreath, E.J.; Sandusky, G.E.; Hammond, L.J.; Moyers, J.S.; Owens, R.A.; et al. FGF-21 as a novel metabolic regulator. *J. Clin. Investig.* **2005**, *115*, 1627–1635. [CrossRef] [PubMed]

233. Zhang, Y.; Xie, Y.; Berglund, E.D.; Coate, K.C.; He, T.T.; Katafuchi, T.; Xiao, G.; Potthoff, M.J.; Wei, W.; Wan, Y.; et al. The starvation hormone, fibroblast growth factor-21, extends lifespan in mice. *eLife* **2012**, *1*, e00065. [CrossRef] [PubMed]

234. Salminen, A.; Kaarniranta, K.; Kauppinen, A. Regulation of longevity by FGF21: Interaction between energy metabolism and stress responses. *Ageing Res. Rev.* **2017**, *37*, 79–93. [CrossRef] [PubMed]

235. Inagaki, T.; Choi, M.; Moschetta, A.; Peng, L.; Cummins, C.L.; McDonald, J.G.; Luo, G.; Jones, S.A.; Goodwin, B.; Richardson, J.A.; et al. Fibroblast growth factor 15 functions as an enterohepatic signal to regulate bile acid homeostasis. *Cell Metab.* **2005**, *2*, 217–225. [CrossRef] [PubMed]

236. Yu, C.; Wang, F.; Kan, M.; Jin, C.; Jones, R.B.; Weinstein, M.; Deng, C.X.; McKeehan, W.L. Elevated cholesterol metabolism and bile acid synthesis in mice lacking membrane tyrosine kinase receptor FGFR4. *J. Biol. Chem.* **2000**, *275*, 15482–15489. [CrossRef] [PubMed]

237. Fu, L.; John, L.M.; Adams, S.H.; Yu, X.X.; Tomlinson, E.; Renz, M.; Williams, P.M.; Soriano, R.; Corpuz, R.; Moffat, B.; et al. Fibroblast growth factor 19 increases metabolic rate and reverses dietary and leptin-deficient diabetes. *Endocrinology* **2004**, *145*, 2594–2603. [CrossRef] [PubMed]

238. Kir, S.; Beddow, S.A.; Samuel, V.T.; Miller, P.; Previs, S.F.; Suino-Powell, K.; Xu, H.E.; Shulman, G.I.; Kliewer, S.A.; Mangelsdorf, D.J. FGF19 as a postprandial, insulin-independent activator of hepatic protein and glycogen synthesis. *Science* **2011**, *331*, 1621–1624. [CrossRef] [PubMed]

239. Kuro, O.M. The Klotho proteins in health and disease. *Nat. Rev. Nephrol.* **2019**, *15*, 27–44. [CrossRef]

240. Babaknejad, N.; Nayeri, H.; Hemmati, R.; Bahrami, S.; Esmaillzadeh, A. An Overview of FGF19 and FGF21: The Therapeutic Role in the Treatment of the Metabolic Disorders and Obesity. *Horm. Metab. Res.* **2018**, *50*, 441–452. [CrossRef] [PubMed]

241. Alvarez-Sola, G.; Uriarte, I.; Latasa, M.U.; Urtasun, R.; Barcena-Varela, M.; Elizalde, M.; Jimenez, M.; Rodriguez-Ortigosa, C.M.; Corrales, F.J.; Fernandez-Barrena, M.G.; et al. Fibroblast Growth Factor 15/19 in Hepatocarcinogenesis. *Dig. Dis.* **2017**, *35*, 158–165. [CrossRef] [PubMed]

242. Lopez-Casillas, F.; Cheifetz, S.; Doody, J.; Andres, J.L.; Lane, W.S.; Massague, J. Structure and expression of the membrane proteoglycan betaglycan, a component of the TGF-beta receptor system. *Cell* **1991**, *67*, 785–795. [CrossRef]

243. Knelson, E.H.; Gaviglio, A.L.; Tewari, A.K.; Armstrong, M.B.; Mythreye, K.; Blobe, G.C. Type III TGF-beta receptor promotes FGF2-mediated neuronal differentiation in neuroblastoma. *J. Clin. Investig.* **2013**, *123*, 4786–4798. [CrossRef] [PubMed]

244. Andres, J.L.; DeFalcis, D.; Noda, M.; Massague, J. Binding of two growth factor families to separate domains of the proteoglycan betaglycan. *J. Biol. Chem.* **1992**, *267*, 5927–5930. [PubMed]

245. Furthauer, M.; Lin, W.; Ang, S.L.; Thisse, B.; Thisse, C. Sef is a feedback-induced antagonist of Ras/MAPK-mediated FGF signalling. *Nat. Cell Biol.* **2002**, *4*, 170–174. [CrossRef] [PubMed]

246. Preger, E.; Ziv, I.; Shabtay, A.; Sher, I.; Tsang, M.; Dawid, I.B.; Altuvia, Y.; Ron, D. Alternative splicing generates an isoform of the human Sef gene with altered subcellular localization and specificity. *Proc. Natl. Acad. Sci. USA* **2004**, *101*, 1229–1234. [CrossRef] [PubMed]

247. Harduf, H.; Halperin, E.; Reshef, R.; Ron, D. Sef is synexpressed with FGFs during chick embryogenesis and its expression is differentially regulated by FGFs in the developing limb. *Dev. Dyn.* **2005**, *233*, 301–312. [CrossRef]

248. Lin, W.; Furthauer, M.; Thisse, B.; Thisse, C.; Jing, N.; Ang, S.L. Cloning of the mouse Sef gene and comparative analysis of its expression with Fgf8 and Spry2 during embryogenesis. *Mech. Dev.* **2002**, *113*, 163–168. [CrossRef]

249. Yang, R.B.; Ng, C.K.; Wasserman, S.M.; Komuves, L.G.; Gerritsen, M.E.; Topper, J.N. A novel interleukin-17 receptor-like protein identified in human umbilical vein endothelial cells antagonizes basic fibroblast growth factor-induced signaling. *J. Biol. Chem.* **2003**, *278*, 33232–33238. [CrossRef] [PubMed]

250. Kovalenko, D.; Yang, X.; Nadeau, R.J.; Harkins, L.K.; Friesel, R. Sef inhibits fibroblast growth factor signaling by inhibiting FGFR1 tyrosine phosphorylation and subsequent ERK activation. *J. Biol. Chem.* **2003**, *278*, 14087–14091. [CrossRef]

251. Xiong, S.; Zhao, Q.; Rong, Z.; Huang, G.; Huang, Y.; Chen, P.; Zhang, S.; Liu, L.; Chang, Z. hSef inhibits PC-12 cell differentiation by interfering with Ras-mitogen-activated protein kinase MAPK signaling. *J. Biol. Chem.* **2003**, *278*, 50273–50282. [CrossRef] [PubMed]

252. Rong, Z.; Ren, Y.; Cheng, L.; Li, Z.; Li, Y.; Sun, Y.; Li, H.; Xiong, S.; Chang, Z. Sef-S, an alternative splice isoform of sef gene, inhibits NIH3T3 cell proliferation via a mitogen-activated protein kinases p42 and p44 (ERK1/2)-independent mechanism. *Cell Signal.* **2007**, *19*, 93–102. [CrossRef] [PubMed]

253. Tsang, M.; Friesel, R.; Kudoh, T.; Dawid, I.B. Identification of Sef, a novel modulator of FGF signalling. *Nat. Cell Biol.* **2002**, *4*, 165–169. [CrossRef] [PubMed]

254. Korsensky, L.; Ron, D. Regulation of FGF signaling: Recent insights from studying positive and negative modulators. *Semin. Cell Dev. Biol.* **2016**, *53*, 101–114. [CrossRef] [PubMed]

255. Ziv, I.; Fuchs, Y.; Preger, E.; Shabtay, A.; Harduf, H.; Zilpa, T.; Dym, N.; Ron, D. The human sef-a isoform utilizes different mechanisms to regulate receptor tyrosine kinase signaling pathways and subsequent cell fate. *J. Biol. Chem.* **2006**, *281*, 39225–39235. [CrossRef] [PubMed]

256. Murphy, T.; Darby, S.; Mathers, M.E.; Gnanapragasam, V.J. Evidence for distinct alterations in the FGF axis in prostate cancer progression to an aggressive clinical phenotype. *J. Pathol.* **2010**, *220*, 452–460. [CrossRef] [PubMed]

257. Hori, S.; Wadhwa, K.; Pisupati, V.; Zecchini, V.; Ramos-Montoya, A.; Warren, A.Y.; Neal, D.E.; Gnanapragasam, V.J. Loss of hSef promotes metastasis through upregulation of EMT in prostate cancer. *Int. J. Cancer* **2017**, *140*, 1881–1887. [CrossRef] [PubMed]

258. Katoh, M. Fibroblast growth factor receptors as treatment targets in clinical oncology. *Nat. Rev. Clin. Oncol.* **2019**, *16*, 105–122. [CrossRef] [PubMed]

5

Sprouty3 and Sprouty4, Two Members of a Family Known to Inhibit FGF-Mediated Signaling, Exert Opposing Roles on Proliferation and Migration of Glioblastoma-Derived Cells

Burcu Emine Celik-Selvi, Astrid Stütz, Christoph-Erik Mayer, Jihen Salhi, Gerald Siegwart and Hedwig Sutterlüty *

Institute of Cancer Research, Department of Medicine I, Comprehensive Cancer Center,
Medica University of Vienna, A-1090 Vienna, Austria
* Correspondence: hedwig.sutterluety@meduniwien.ac.at

Abstract: Dysregulation of receptor tyrosine kinase-induced pathways is a critical step driving the oncogenic potential of brain cancer. In this study, we investigated the role of two members of the Sprouty (Spry) family in brain cancer-derived cell lines. Using immunoblot analyses we found essential differences in the pattern of endogenous Spry3 and Spry4 expression. While Spry4 expression was mitogen-dependent and repressed in a number of cells from higher malignant brain cancers, Spry3 levels neither fluctuated in response to serum withdrawal nor were repressed in glioblastoma (GBM)-derived cell lines. In accordance to the well-known inhibitory role of Spry proteins in fibroblast growth factor (FGF)-mediated signaling, both Spry proteins were able to interfere with FGF-induced activation of the MAPK pathway although to a different extent. In response to serum solely, Spry4 exerts its role as a negative regulator of MAPK activation. Ectopic expression of Spry4 inhibited proliferation and migration of GBM-originated cells, positioning it as a tumor suppressor in brain cancer. In contrast, elevated Spry3 levels accelerated both proliferation and migration of these cell lines, while repression of Spry3 levels using shRNA caused a significant diminished growth and migration velocity rate of a GBM-derived cell line. This argues for a tumor-promoting function of Spry3 in GBMs. Based on these data we conclude that Spry3 and Spry4 fulfill different if not opposing roles within the cancerogenesis of brain malignancies.

Keywords: Sprouty proteins; brain cancer; FGF-mediated signaling; tumor suppressor; tumor promoter

1. Introduction

The term brain cancer summarizes multiple subtypes of tumors originating from different tissues of the central nervous system [1]. The most prevalent type of brain tumors are gliomas which arise from glial or precursor cells. They include, among others, lower graded astrocytoma (AC) and oligodendroglioma (ODG), as well as the WHO Grade IV classified glioblastoma multiforme (GBM) and its variant gliosarcoma (GS). GBM are the most common brain tumors and patients have a poor prognosis with a five-year survival rate of only 5.6% [2]. A group of neuronal tumors arising in the central but also in the autonomic nervous system are the rare neuroblastoma (NB) which are the second most common tumors in children [3]. Like in all human cancer cells, malignant transformation in gliomas is driven by typical chromosomal changes. The Cancer Genome Atlas project identified alterations in the network regulated by receptor-tyrosine kinases (RTK) as a frequent molecular cause of these cancers. Important molecules responsible for transducing the signals like the epidermal growth factor receptor (EGFR), the phosphatidylinositol 3-kinases (PI3K), NRAS and BRAF are frequently altered to a more efficient state, while inhibitors of their activities like neurofibromin (NF1) and the Phosphatase and TENsin homolog (PTEN) are often deleted or less effective [4].

Sprouty (Spry) proteins which represent modulators of RTK-driven signaling pathways were first identified as inhibitors of fibroblast growth factor (FGF)-induced signaling in *Drosophila* [5]. In humans, four homologues were described [6]. In contrast to the other Spry family members which are ubiquitously expressed in all tissues [6], the Spry3 encoding gene localizes to the pseudoautosomal region 2 and its expression is rarely documented. Only in brain and glia, Spry3 expression is doubtless detected [7]. Spry proteins fulfill important functions in many RTK-mediated signal transduction cascades. Primarily, they are known to interfere specifically with MAPK-ERK activation [8–10], but in other systems they were shown to influence the PI3K pathway as well [11]. Additionally, Spry proteins are able to interfere with phospholipase C-induced pathways [12]. In contrast to their manifold inhibitory function on RTK-mediated pathways, Spry proteins are able to interact with the E3-ubiquitin ligase c-Cbl and thereby constrict the degradation of some RTKs as shown for the EGFR [13]. Considering their functions in fine tuning of the cellular response to RTK-inducing signals, members of the Spry family are good candidates for an important role in the tumorigenesis of different cells. Accordingly, Spry2 and/or Spry4 are shown to act as tumor-suppressors in cancer originated from, e.g., lung [14–16], liver [17], breast [18,19], prostate [20] and bone [21]. In other types of tumors, members of the Spry protein family fulfill a tumor-promoting task as it was demonstrated for Spry2 in colon carcinoma [22,23] and for Spry1 in rhabdomyosarcoma [24]. In brain tumors, repression of Spry2 has been shown to interfere with proliferation of GBM-derived cell lines and tumor formation [25,26]. Compatible with the tumor-promoting function of Spry2 in brain, the Spry proteins are important for other neuronal processes. Spry2 as well as Spry4 downregulation is associated with promoted axon outgrowth [27,28], and Spry1, Spry2 and Spry4 inhibit FGF-induced processes in the cerebellum [29]. Data generated in *Xenopus* document that Spry3 is important in regulating axon branching of motoneurons [30], and the finding that Spry3 is associated with autism susceptibility indicates a further role in the human brain [7].

In the presented study, we investigated the expression of Spry3 and Spry4 in brain cancer-derived cells and analyzed how a modulation of their expression influences the behavior of glioblastoma-derived cell lines.

2. Material and Methods

2.1. Cell Lines

The astrocytoma-derived cells (SW1088) and both neuroblastoma-derived cell lines (SK-N-DZ and SK-N-FI), as well as the glioblastoma-derived cell lines DBTRG-05MG, T98G and U373 and the oligodendroglioma-derived cell line Hs683 were purchased from the American Type Culture Collection (ATCC). NMC-G1, a cell line established from an astrocytoma, and AM-38, a glioblastoma originated cell line, were obtained from the JCRB cell bank. Cell lines LN40 and LN140 were kindly provided by Dr. Tribolet (Lausanne). Cell lines BTL1529, BTL2177 and BTL53 were established from glioblastoma diagnosed patients and BTL1376 and BTL2175 from gliosarcoma patients at the Neuromed Campus in Linz (NML) as described [31]. The cell line VBT72 was established from a glioblastoma at the Institute for Cancer Research [31]. These cell lines were kindly provided by Walter Berger (Medical University of Vienna). All cells were cultured in the recommended medium containing 10% fetal calf serum (FCS) and supplemented with penicillin (100 U/mL) and streptomycin (100 µg/mL) at 37 °C in 7.5% CO_2.

2.2. Adenoviral Infection of Cells

The coding sequence of human Spry3 was amplified by PCR using Pfx Polymerase (Invitrogen) with upstream primer 5-AGCTCTGGATCCATGGATGCTGCGGTGACAGAT-3 (Spry3-s) and downstream primer 5-TAGCGAATTCCTCGAGTCATACAGACTTT-3 (Spry2-as) to add appropriate cloning sites. The amplified DNA fragments were subsequently cloned via BamHI/EcoRI into a pADlox plasmid to generate pADlox-Spry3. To construct an adenovirus expressing shRNA directed against Spry3, the CMV promoter of pADlox was exchanged by the human U6 promoter of the pSilencer Vector. Two oligonucleotides harboring an shRNA directed against Spry3 were annealed: sh-Spry3 sense 5'-TCG AGC GCA GCT GTT CAA TAG GCA GAA TTT GTT GAA GCT TGA ACA AAT TCT GCC TAT TGA ACA GCT GCG CTC TTT TTT-3' and shSpry3 as 5'-AAT TAA AAA AGA GCG CAG CTG TTC AAT AGG CAG AAT TTG TTC AAG CTT CAA CAA ATT CTG CCT ATT GAA CAG CTG CGC-3'. The double stranded DNA with overlapping XhoI and EcoRI sites was then inserted in the digested pAdloxU6 vector to obtain pADlox-shSpry3. To obtain a virus directed against Spry4, two oligonucleotides (5'-TCGAGCTCAGCTCGCTACCTCCGCGGCGATGTTGAAGCTTGAACATCGCCGCGGAGGTAGC GAGCTGAGCTGTTTTTT-3' and 5'-AATTAAAAAACAGCTCAGCTCGCTACCTCCGCGGCGATGTT CAAGCTTCAACATCGCCGCGGAGGTAGCGAGCTGAGC-3') were annealed and subcloned the same way to construct pADlox-shSpry4. The correct cloning was confirmed by sequencing analysis. Recombinant viruses were produced as described [32]. Adenoviruses expressing Spry4 or control proteins (luciferase, lacZ or CFP) were already generated [21,33].

The optimal concentrations of the viruses for each cell line was determined by infecting the cells with different dilution of adenoviruses expressing Cyan Fluorescence Protein (CFP). The viral concentration of the adenoviruses expressing different proteins were calculated according to their OD_{260}. For infection, viruses were diluted in serum-free medium.

2.3. Cell Signaling Assay

For analyzing ERK phosphorylation, 10^5 cells were seeded into Ø6 cm tissue culture plates in DMEM medium containing 10% FCS. Twenty-four hours later, the cells were washed with and incubated in serum-free medium. Next day cells were infected with adenoviruses and incubated for another 2 days before 20% FCS or 10ng/mL FGF2 were added.

2.4. Scratch Assay

For the scratch assay, 6×10^5 cells were infected with adenoviruses expressing the control proteins, Spry3 or Spry4, respectively. A total of 24 h post infection, cells were transferred into a 6-well plate. The next day, three straight scratches per well were introduced into the monolayer using a sterile yellow pipette tip. To remove debris, cells were washed twice with 1 x PBS. Finally, 3 mL of DMEM supplemented with 10% FCS were added. The closing of the scratch was pictured by the VISITRON Live Cell Imaging System (Visitron, Puchheim, Germany) at 10x magnification using VisiView Software. The running time was set to 40 h and for monitoring a time interval of 30 min was chosen. Using ImageJ software, gap width of three scratches were calculated every two hours. Migration velocity was assessed by applying linear regression using GraphPad Prism software. Migration velocity of three independent experiments were compared.

2.5. Growth Curve

Growth curves were performed and analyzed as described [21]. Each growth curve was counted in triplicate and after assessing the continuity of the growth by depicting it in a semi-logarithmical graph, the doubling time was calculated by applying an exponential growth equation. The calculated doubling times of at least three independent experiments were compared to each other and differences between two groups were calculated using an unpaired t-test.

2.6. Immunoblot

Immunoblotting was carried out as described [34]. The antisera against Spry4 and Spry3 were produced and affinity-purified as described [15]. The Spry3 antibodies were raised against the N-terminal 200 amino acids of the human homolog. As a loading control, antibodies against GAPDH (sc-365062) and ERK 1/2 (sc-514302) were purchased from Santa Cruz. Antibodies against phosphorylated extracellular signal-regulated kinase (pERK) (#9101) were received from Cell Signaling Technology. The horseradisch peroxidase-coupled secondary antibodies were purchased from GE Healthcare.

3. Results

3.1. In Brain Cancer-Derived Cell Lines Spry3 Protein is Commonly Expressed Independent of Mitogen Availability

First, we investigated if Spry3 expression is influenced by the grade of malignancy or the histological background of brain cancer-derived cells. In order to analyze Spry3 protein levels, antibodies had to be produced, affinity-purified and their sensitivity as well as their specificity had to be assessed. To analyze their sensitivity, U373 cells were infected with decreasing amounts of adenovirus expressing Spry3 protein. As depicted in Figure 1A, the antibodies detected a single band at 33 kDa and in cells infected with decreasing titers of Spry3-encoding adenoviruses, the intensity of the detected band corresponded to the amount of introduced viruses while the cellular protein content was comparable. To control the specificity, all four Spry proteins were ectopically expressed by using the respective adenoviruses. Two days after infection, sufficient amounts of all Spry proteins were expressed, but the Spry3 antibody only detected Spry3 (Figure 1B). In the subsequent experiment, we determined the endogenous levels of Spry3 in different brain cancer-derived cell lines. To analyze if, like it was shown for Spry2 and Spry4 [35], Spry3 protein expression is dependent on mitogens in the cellular environment, serum was withdrawn from part of the cells (-), and their Spry3 levels were compared to those of cells cultivated in the presence of serum (+). In only 1/17 cell lines Spry3 protein was undetectable. Most of the cell lines express detectable amounts of Spry3 proteins which appear in a distinguishable pattern of bands. Usually the slower migrating bands were more abundant in the presence of serum indicating that a serum-dependent modification is applied to Spry3 (Figure 1C). Concerning the influence of the histopathologic origin, we observed that in the more advanced GBM-derived cell lines the expression of Spry3 was on average higher than in cells originated from the lower graded ODG and AC (Figure 1D,E). The highest expression of Spry3 was detected in the two NB-derived bone morrow metastases. These observations would favor rather an oncogenic than a tumor-suppressing function of Spry3 in brain cancers. Interestingly, the serum had not the expected influence on Spry3 expression, as half of the cell lines failed to adapt their Spry3 expression in response to mitogen availability. In five of the cell lines, its expression even slightly increased (less than 2-fold) if serum was withdrawn. A more pronounced change of Spry3 in response to serum in form of an increase or decrease was only observed in one cell line each (Figure 1D,F). Therefore, it is unlikely that mitogen-induced signals play an important role in regulating the expression of Spry3.

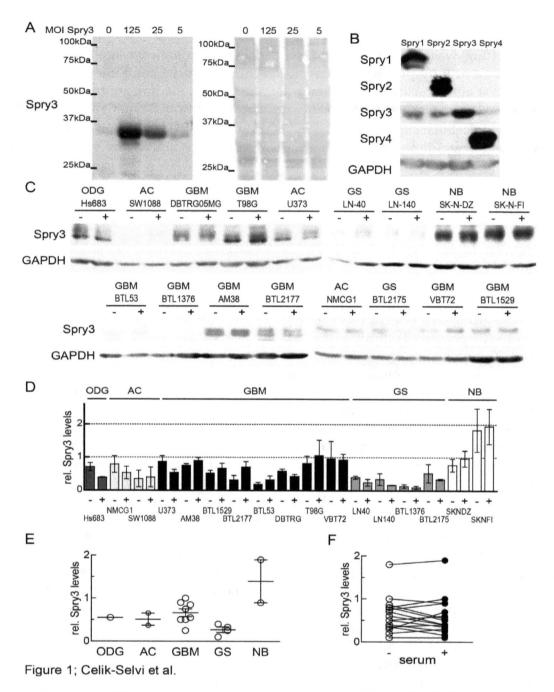

Figure 1; Celik-Selvi et al.

Figure 1. Expression of Spry3 protein in brain cancer-derived cell lines. (**A**) U373 cells were infected with decreasing amounts of adenoviruses expressing Spry3 protein and an immunoblot using Spry3 antibodies was performed. Equal loading was verified by Ponceau S staining of the immunoblot. (**B**) Adenoviruses expressing Spry1, Spry2, Spry3 or Spry4 were introduced into U373 cells. A total of 48 h post-infection cells were harvested and proteins were isolated. An immunoblot sequentially probed with all of the indicated antibodies is depicted. (**C**) Logarithmically growing cell lines derived from oligodendroglioma (ODG), astrocytoma (AC), glioblastoma (GBM), gliosarcoma (GS) and neuroblastoma (NB) were cultured for 24 h without (-) and with (+) serum. Using Western blot, endogenous Spry3 and GAPDH proteins were determined. (**D**) Amounts of Spry3 proteins were measured as ratio to an external control (MG63) by Image Quant software and normalized to GAPDH. Quantification results of 2–3 Western blots depicted as mean ± SEM are shown in a column graph. Cell lines were sorted according to their histopathological origin. (**E**) A scatterplot presenting the Spry3 expression across the histopathological subgroups of brain cancer is shown. (**F**) Calculated Spry3 levels from cells grown in serum-deprived (open circle) and –supplemented (closed circle) mediums are compared.

3.2. Spry4 Protein Expression is Repressed in Cell Lines Derived from More Malignant Brain Tumors, but Usually Still Serum-Dependent

In order to investigate that growth factor-induced signaling in the analyzed cell lines is able to sufficiently influence the negative feedback loop responsible for controlling Spry protein expression, Spry4 protein levels were determined in comparison. In some cell lines, Spry4 expression was very prominent, but in five of them, we were not able to detect Spry4 proteins. Like in the case of Spry3, Spry4 frequently appeared in more than one migrating form (Figure 2A). Compared to the levels detected in cells derived from lower graded patients' tissues, usually Spry4 expression in GBM and GS is strongly repressed, although in few of these cell lines Spry4 protein was definitively abundant (Figure 2B,C). Only in five of the cells lines the expression of Spry4 in serum-free conditions was insignificantly changed when compared to the parallel in serum cultivated cell counterparts. Seven of the brain-derived cell lines displayed a more than twofold decrease of Spry4 protein as a consequence of serum starvation. Moreover, in three of them the detected difference between the serum and non-serum condition exceeded a fivefold dimension (Figure 2B,D). When Spry3 and Spry4 expression in the different brain-derived cell lines were compared (Figure 2E), we found that there was no correlation indicating that Spry3 and Spry4 expression are regulated by independent mechanisms.

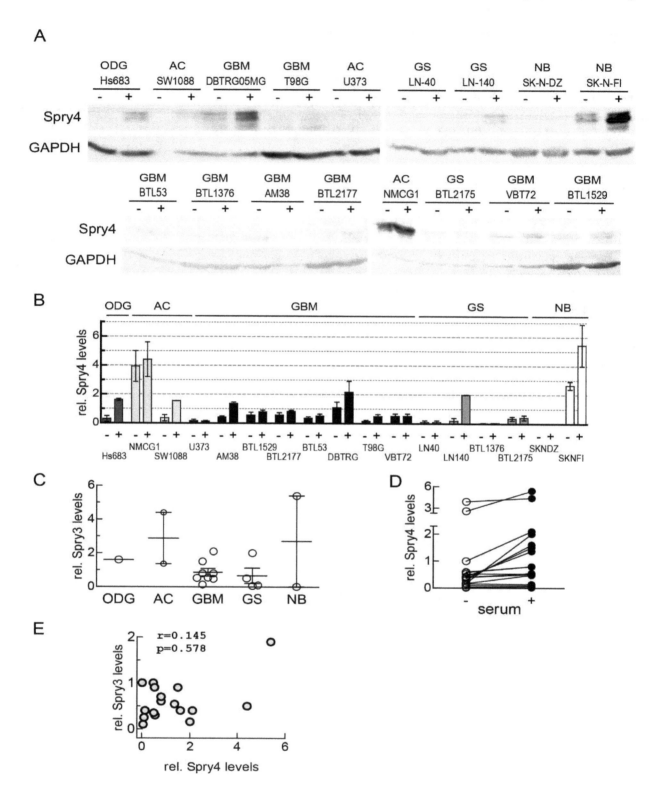

Figure 2. Expression analysis of endogenous Spry4 protein in brain cancer cells. (**A**) Spry4 protein levels in 17 brain cancer-derived cell lines which were cultured devoid of (−) and with (+) serum. GAPDH served as loading control. (**B**) Quantification of Spry4 was performed using Image Quant 5.0. An external control was arbitrarily set as 1 and loading differences were adjusted to GAPDH expression. A column graph summarizes the results of 2–3 independent experiments. (**C**) Spry4 expression in serum-supplemented growth condition was compared. A scattered dot-plot grouping the cells according to the histological origin is depicted. (**D**) A comparison of Spry4 levels detected in starved (open circle) and stimulated (closed circle) cells is presented. (**E**) Correlation of Spry3 and Spry4 expression was calculated using GraphPad Prism.

3.3. MAPK Activation in Response to FGF and Serum is Effectively Inhibited by Spry4 While Spry3 Failed to Fulfill This Function in the Presence of Serum

To analyze if Spry3 and Spry4 are able to interfere with FGF-induced signaling, U373 cells were serum-deprived for 3 days before FGF2 was added for 5, 10 and 20 min. Within the starvation period a portion of cells were infected with viruses expressing Spry3, Spry4 or a control protein. In control treated cells, adding of FGF induced the MAPK pathway after 10 min as measured by determining the fraction of pERK (Figure 3A). In cells expressing excessive amounts of Spry3, like in the control cells, activation of the MAPK pathway was also observed after 10 min, but the extent of phosphorylation was less pronounced (Figure 3A,C). In case of ectopic Spry4 expression we detected that the proportion of activated ERK in serum-starved conditions was clearly less distinct. The addition of FGF caused an augmentation of the pERK levels, which was less intense than in the other two groups (Figure 3A,C). These data evince the inhibitory role of Spry proteins on FGF-mediated signaling, but demonstrated that Spry4 was more potent concerning interference with MAPK activation than Spry3.

In order to asses if the two Spry forms differ concerning their potential to inhibit MAPK activation in response to serum, a respective cell signaling assay was applied. In response to serum, ERK was immediately phosphorylated to a much higher extent (at least 10 times the value observed in starved cells) than in FGF-treated cells (two- to threefold induction). When Spry3 was expressed, the induction was slightly delayed but the amplitude was not significantly diminished. In contrast, Spry4 inhibited ERK phosphorylation significantly. As already observed in case of FGF induction, the basal pERK levels of cells cultivated in the absence of mitogens was clearly diminished, but also the maximal levels of pERK phosphorylation were reduced in comparison to the cells expressing a control or Spry3 protein. These data demonstrate that Spry4 can potently interfere with induction of the MAPK and indicate that Spry4 was more potent concerning interference with MAPK activation than Spry3.

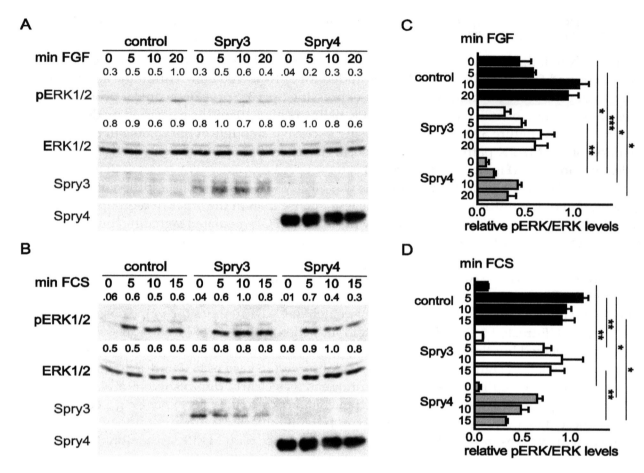

Figure 3. Influence of Spry3 and Spry4 proteins on ERK activation by FGF2 and serum. Glioblastoma (GBM)-derived cells (U373) were serum-starved for 24 h and then infected with adenoviruses expressing either a control protein (luciferase), Spry3 or Spry4. Two days later, cells were incubated with FGF2 (**A**) or serum (**B**) for the indicated times. Representative immunoblots of an experiment using antibodies recognizing pERK1/2 and total ERK1/2 are shown. Expression of Spry3 and Spry4 were verified by the respective antibodies. Using ImageQuant 5.0, the pERK1/2 bands detected were quantified and normalized to the corresponding values obtained for the ERK expression. The highest values were arbitrarily set as 1. The results of the quantification for the presented blots are depicted. (**C**) A summary of calculated mean values ± SEM of the pERK/ERK values from three experiments using FGF2 to stimulate the cells is depicted. Significance between the three groups was calculated by using a one-way ANOVA test in GraphPad prism. (**D**) The bands of pERK and ERK in response to serum were densitometrically quantified using ImageQuant 5.0, and the highest values of each experiment were set as 1. The graph summarizes three experiments. Significance was determined by a one-way ANOVA test in using GraphPad prism software. * $p < 0.05$; ** $p < 0.01$; *** $p < 0.001$.

3.4. In Brain Cancer-Derived Cell Lines, Spry3 and Spry4 Expressions Have an Opposing Effect on Cell Proliferation

To investigate if Spry3 and Spry4 interfere with cell proliferation in brain cancer-derived cells, we selected DBTRG-05MG and U373 cell lines to apply ectopic overexpression of the respective Spry proteins. Both of these cell lines were easy to infect by adenoviruses as tested by using CFP expressing adenoviruses (data not shown). Furthermore, in DBTRG-05MG Spry3 appears mainly in its slower migrating form and Spry4 levels are pronounced while in U373 Spry3 mainly appears in its faster migrating form and a shift is only detected after serum addition. Spry4 was not detected in this cell line (Figures 1C and 2A). To measure cell proliferation, growth curve analyses were performed. In DBTRG-05MG, Spry3 expressing cells double significantly faster (0.9 ± 0.01 doublings per day) than control treated cells (0.8 ± 0.02) while Spry4 expression decelerate the proliferation process to only

0.69 doublings per day (Figure 4A,B). Corroborating in U373, Spry3 accelerate cell proliferation from 0.56 ± 0.01 to 0.63 ± 0.01 doublings per day and Spry4 expression inhibits cell expansion to 0.51 ± 0.01 (Figure 4C,D). In both cell lines, Spry3 and Spry4 proteins are clearly overexpressed if the respective adenoviruses are applied (Figure 4E).

These data demonstrate that cell proliferation is promoted by Spry3 and suppressed by Spry4 expression arguing for an opposing effect of these Spry members.

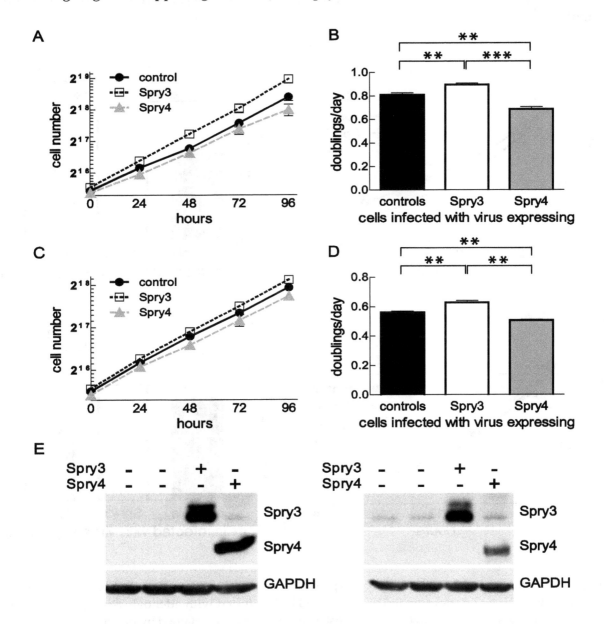

Figure 4. Influence of ectopic Spry3 and Spry4 expression on cell proliferation. Proliferation of cells overexpressing the indicated proteins was assessed by growth curve analysis. (**A**) The number of DBTRG-05MG cells were counted every 24 h for 5 days and are depicted as growth curves using a semi-logarithmical scale. A representative growth curve of three replicates is depicted. (**B**) Using GraphPad Prism, doubling times of at least three independent growth curve analyses performed with DBTRG-05MG cells were calculated and presented as mean doublings per day ± SEM. (**C**) A representative growth curve of U373 cell line is shown. (**D**) Using exponential growth equations, doubling times of U373MG cells were calculated and shown as doublings per day. Significance was assessed using an unpaired t-test in GraphPad Prism and mean ± SEM are shown. * $p < 0.05$; ** $p < 0.01$; *** $p < 0.001$ (**E**) Overexpression of Spry3 and Spry4 in the GBM cell lines DBTRG-05MG (left) and U373 (right) were verified by immunoblotting.

3.5. Spry3 and Spry4 Exert a Contrary Effect on the Migratory Capabilities of Brain Cells

Aberrant cell migration is another RTK-mediated process contributing to the malignancy of cancer cells. Therefore, we next investigated if the expression of Spry3 and Spry4 proteins modulate the closure of the gap in a scratch assay. In DBTRG-05MG, ectopic expression of the Spry3 protein has a prominent influence on cell migration by augmenting their velocity from 26.1 ± 1.4 to 36.1 ± 0.5 µm/h. In contrast, Spry4 expression slows down these cells to 21.3 ± 0.99 µm/h (Figure 5A,B). Both effects were significant.

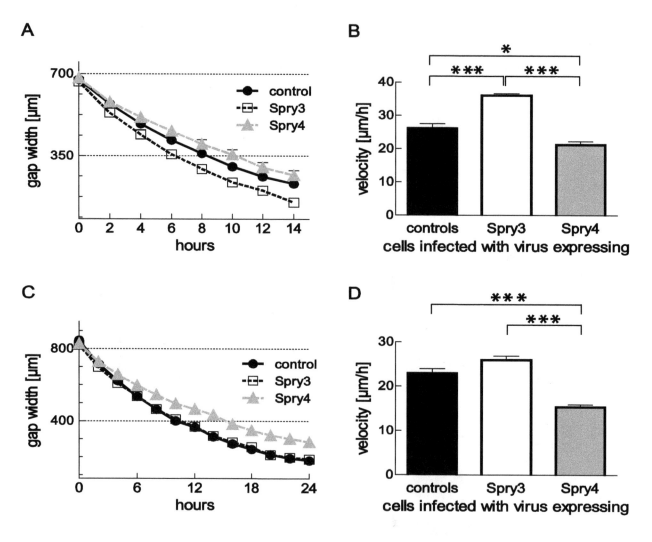

Figure 5. Influence of Spry3 and Spry4 expression on cell migration of GBM-derived cell lines. (**A**) Scratch assay was performed in DBTRG-05MG cells infected with adenoviruses expressing the indicated proteins. Representative curves of distance coverage were obtained by measuring decreasing gap widths of three replicative scratches at every two-hour time points using ImageJ. (**B**) Using linear regression, migration velocities were calculated. Means of at least three experimentations ± SEM are summarized as column bars. (**C**) Representative measurements of replicative gap closure in a close layer of U-373 MG cell expressing the indicated proteins are shown. (**D**) Velocities of at least three experiments were calculated using linear regression in GraphPad Prism and summarized in a graph depicting means ± SEM. An unpaired t-test was used to acquire significance. $p < 0.05$; ** $p < 0.01$; *** $p < 0.001$.

Compared to DBTRG-05MG, U373 cells are slower migrating and the effect of the Spry proteins was less developed. Spry3 has no significant effect on the velocity of gap closure although a slightly faster calculated average velocity points towards a positive effect of its expression on cell migration. In accordance with the data obtained in DBTRG-05MG, Spry4 expression in U373 delays the closure of the gap significantly. The control-treated cells move with a speed of 22.9 ± 1.0 μm/h towards the opposite front, while in the presence of Spry4 as an average speed only 15.4 ± 0.5 μm/h were calculated (Figure 5C,D).

These observations indicate that in brain cells, Spry3 and Spry4 exert different effects not only on cell proliferation but also on cell migration.

3.6. Repressed Expression of Spry3 Inhibits Cell Proliferation and Migration

To further verify our observations, we wanted to investigate if lowered Spry expressions would influence cell proliferation in the opposite way than their overexpression. Therefore, an adenovirus expressing a shRNA directed against Spry3 was introduced into DBTRG-05MG and Spry3 levels were compared to the ones in control-treated and Spry3 overexpressing cells. As depicted in Figure 6A, expression of shSpry3 failed to modulate Spry3 levels, while the overexpression was successfully applied. In contrast, in U373 cells, expression of shRNA targeting Spry3 mRNA resulted in clearly lowered levels of Spry3 protein (Figure 6B). Since Spry4 is not expressed in detectable amounts in U373, it was just useful to express shSpry4 in DBTRG-05MG, where similar to the application of shSpry3, the endogenous expression of the protein was unaffected by expressing a shRNA directed against Spry4 (data not shown). Next, we performed a growth curve analysis using shSpry3 in U373 cells. Reduced Spry3 expression caused an inhibition of cell proliferation while in parallel overexpression accelerated the doubling of these cells (Figure 6C). Compared to control cells, doubling of shSpry3-treated cells was reduced from 0.58 to 0.50 doublings per day, substantiating an oncogenic effect of Spry3 in brain cancer (Figure 6D). To evaluate if a repression of Spry3 in addition to its interference with proliferation is also influencing cell migration, Spry 3 levels of U373 cells were modulated by treatment with the respective adenoviruses and a scratch assay was performed. The time to close the gap was significantly delayed when Spry3 levels were lowered (Figure 6E). On average, cells expressing a shRNA targeting Spry3 cover a distance of 21 μm in an hour, while control treated cells move about 1.2-fold faster, while Spry3 overexpression had no significant influence on the velocity of gap closure (Figure 6F). These data demonstrate that like proliferation, cell migration of GBM-derived cells is hindered if less Spry3 proteins are present confirming the tumor-promoting function of Spry3.

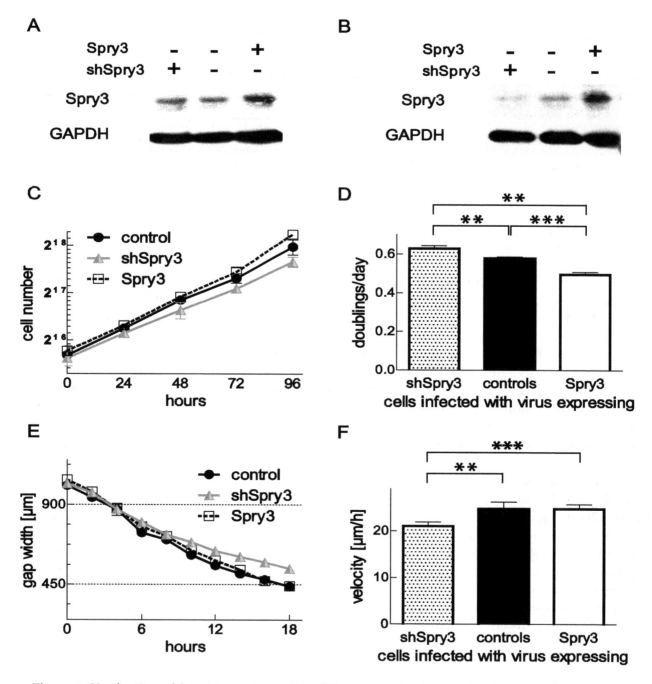

Figure 6. Verification of Spry3 impact on cell proliferation and migration by downregulation of the endogenous protein levels. DBTRG-05MG (**A**) and U373 (**B**) cells infected with adenoviruses expressing Spry3, shSpry3 or a control protein were analyzed concerning their Spry3 protein levels. (**C**) Three days after infection with the indicated viruses a growth curve analysis was performed in U373. (**D**) The doubling time of three experiments was calculated by performing an exponential growth equation and the mean doublings per day ± SEM are depicted. (**E**) U373 cell expressing the indicated proteins were cultured to form a close layer before a scratch assay was performed. Measurements of three replicative gaps were performed every two hours and a representative experiment is shown. (**F**) Velocities of three experiments were calculated using linear regression in GraphPad Prism and a summary is depicted. Using an unpaired t-test, significance was determined. ** $p < 0.01$; *** $p < 0.001$.

4. Discussion

Deregulated signal transduction is one of the most frequent alterations contributing to malignancy of brain cancer. In this study we provide data showing that Spry3 and Spry4 expression may be altered in brain cancer and affect cell proliferation and migration in opposing ways. Both Spry proteins are expressed in most of the brain cancer-derived cell lines, and two bands with a slightly different migration velocity can be detected. In case of Spry3, the faster migrating band is more frequently detected in the cells cultivated in the absence of serum. Although posttranslational modification of Spry3 is not reported, it is likely that analog to the other family members [36] the protein is phosphorylated at serine and/or tyrosine residues and that one of these potential modification is causing a shift in the gel.

A comparison of the Spry3 and Spry4 protein levels in the different cell lines revealed that the expression of these two Spry family members does not correlate. While Spry3 expression was on average elevated in cell lines originated from higher malignant tumors, Spry4 tended to be repressed in GBM and GS compared to cells derived from lower graded cancers. In accordance with our observation, an earlier report describes that Spry4 is often missing or deleted in gliomas [37]. Indeed, in several of the GBMs analyzed we were unable to detect this Spry isoform. Similarly, in lung [14,38] and breast cancer [39], a repression of Spry4 is coinciding with a postulated tumor-suppressive function. With regard to Spry3, in normal brain tissue its expression is well documented, but due to its low abundance in other tissues, expression data in cancers are rarely available [6,7]. To our knowledge, only one report by Sirivatanauksorn et al. investigated the RNA level of Spry3 as well as Spry4 in hepatocellular cancer, and comparable to our observations in the brain, Spry4 mRNA levels were downregulated in liver cancer-associated tissue, while Spry3 expression was unaltered [40]. With respect to Spry2, data generated on RNA level clearly points towards an upregulation of this Spry member in GBM when compared to non-tumor tissue [25], while a study exploring protein levels in immunohistochemistry suggests downregulation of Spry2 in higher malignant brain cancers when compared to lower graded tumors [41]. Another obvious different variable concerning regulation of Spry3 and Spry4 is their dependency on mitogen availability. Like Spry1 in lung cancer cells [34], Spry3 protein levels fail to fluctuate in response to serum-withdrawal in brain cancer-derived cells. In contrast, Spry4 expression is usually manifold augmented when serum-containing factors are supplied. This is in accordance with observations in lung [35,42], prostate and osteosarcoma [42]. Additionally, it is reported that in neuronal cells derived from the dorsal root ganglion, Spry4 can be induced as a consequence of FGF2 as well as NFG supplementation [28]. Differences in Spry3 and Spry4 expression control are furthermore reported in bovine ovarian granulosa cells where Spry4 was increased in response to FGF1 and FGF4, while in parallel Spry3 levels were lowered [43].

Concerning their impact on the cellular behavior, we observed that ectopic expression of Spry3 is augmenting the growth and migration rate of different GBM-derived cell lines. Corroborating, repression of its protein levels as achieved by introducing a specific shRNA resulted in diminished cell proliferation. This would suggest that this Spry protein member exerts a tumor-promoting role in brain cancers. Accordingly, two different reports suggest that Spry2 is advantageous for the malignancy of GBM [25,26]. Knock-down of its expression decelerates cell proliferation of GBM cell lines [25,26] while astrocytes were unaffected by modulated Spry2 levels [25]. Additionally Spry2 was identified as prognostic marker for GBM patients survival [25]. Although in most tissues Spry proteins fulfill the function of tumor-suppressors, individual Spry proteins are promoting tumorigenic potential here and there [44]. Spry2, for example, is above its function in GBM, shown to promote colon cancer malignancy by increasing proliferation, migration, tumor growth [23] and invasion [22] of colon cancer cells. In case of Spry1, an oncogenic function of the protein was demonstrated in the embryonal subtype of rhabdomyosarcoma [24].

In contrast to Spry3 and Spry2, Spry4 expression is inhibiting cell migration and proliferation of GBM-derived cell lines and is able to inhibit ERK phosphorylation in FGF2- and serum-induced as well as in unstimulated GBM-derived cells. An opposing role of Spry4 to other Spry proteins is already signified in colon carcinomas. Zhou et al. [45] demonstrated that Spry4 expression interferes with in vitro and in vivo cell proliferation of colon cancer cells. In contrast, Spry2 and Spry1 are fulfilling oncogenic functions in these tumors [22,23,46]. Additionally, in osteosarcoma [21] and in ovarian cancer [47], a tumor-suppressing role for Spry2 but not Spry4 was explicitly highlighted. Nonetheless, cell migration is specifically inhibited by Spry4 expression in prostate [48], pancreatic [49] and endothelial cells [50]. A concordant inhibition of proliferation and migration in case of Spry4 expression is reported for breast [18] and lung cancer cells [14]. Additionally, Spry4 can fulfill a tumor-suppressing role by interfering with angiogenic signals and thereby inhibits neovascularization and tumor growth [51].

5. Conclusion

In summary, our study describes that Spry3 and Spry4 exert different roles in brain cancer. Spry3 potentiates the tumorigenic potential of glioblastoma cells and Spry4 functions as tumor-suppressing protein in this entity.

Author Contributions: Conceptualization, B.E.C.-S., C.-E.M. and H.S.; methodology, B.E.C.-S. A.S., C.-E.M., J.S., and G.S.; validation, B.E.C.S., A.S., C.E.M., and J.S.; formal analysis, B.E.C.S. and H.S.; investigation, B.E.C.S., A.S. and J.S.; resources, C.E.M., J.S., and G.S.; data curation, H.S.; writing—original draft preparation, B.E.C.S. and H.S.; writing—review and editing, B.E.C.S., A.S. and H.S.; visualization, B.E.C.S. and H.S.; supervision, H.S.; project administration, H.S.; and funding acquisition, H.S.

Acknowledgments: We thank Walter Berger and Lisa Gabler for providing the cell lines. Daniel Valcanover, Gabriel Kaufmann, Dorra Boumaiza, Gerald Timelthaler and Angelina Doriguzzi for their technical assistance and Alina Bayer for her help with producing a Spry3 antigen.

References

1.   Furnari, F.B.; Fenton, T.; Bachoo, R.M.; Mukasa, A.; Stommel, J.M.; Stegh, A.; Hahn, W.C.; Ligon, K.L.; Louis, D.N.; Brennan, C.; et al. Malignant astrocytic glioma: Genetics, biology, and paths to treatment. *Genes Dev.* **2007**, *21*, 2683–2710. [CrossRef] [PubMed]
2.   Kruchko, C.; Ostrom, Q.T.; Boscia, A.; Truitt, G.; Gittleman, H.; Barnholtz-Sloan, J.S. CBTRUS Statistical Report: Primary Brain and Other Central Nervous System Tumors Diagnosed in the United States in 2011–2015. *Neuro-Oncol.* **2018**, *20*, iv1–iv86. [CrossRef]
3.   Schilling, F.H.; Spix, C.; Berthold, F.; Erttmann, R.; Fehse, N.; Hero, B.; Klein, G.; Sander, J.; Schwarz, K.; Treuner, J.; et al. Neuroblastoma screening at one year of age. *N. Engl. J. Med.* **2002**, *346*, 1047–1053. [CrossRef] [PubMed]
4.   Cancer Genome Atlas Research, N. Comprehensive genomic characterization defines human glioblastoma genes and core pathways. *Nature* **2008**, *455*, 1061–1068. [CrossRef] [PubMed]
5.   Hacohen, N.; Kramer, S.; Sutherland, D.; Hiromi, Y.; Krasnow, M.A. Sprouty encodes a novel antagonist of FGF signaling that patterns apical branching of the Drosophila airways. *Cell* **1998**, *92*, 253–263. [CrossRef]
6.   Minowada, G.; Jarvis, L.A.; Chi, C.L.; Neubuser, A.; Sun, X.; Hacohen, N.; Krasnow, M.A.; Martin, G.R. Vertebrate Sprouty genes are induced by FGF signaling and can cause chondrodysplasia when overexpressed. *Development* **1999**, *126*, 4465–4475. [PubMed]
7.   Ning, Z.; McLellan, A.S.; Ball, M.; Wynne, F.; O'Neill, C.; Mills, W.; Quinn, J.P.; Kleinjan, D.A.; Anney, R.J.; Carmody, R.J.; et al. Regulation of SPRY3 by X chromosome and PAR2-linked promoters in an autism susceptibility region. *Hum. Mol. Genet.* **2015**, *24*, 7450. [CrossRef]
8.   Sasaki, A.; Taketomi, T.; Kato, R.; Saeki, K.; Nonami, A.; Sasaki, M.; Kuriyama, M.; Saito, N.; Shibuya, M.; Yoshimura, A. Mammalian Sprouty4 suppresses Ras-independent ERK activation by binding to Raf1. *Nat. Cell Biol.* **2003**, *5*, 427–432. [CrossRef]
9.   Hanafusa, H.; Torii, S.; Yasunaga, T.; Nishida, E. Sprouty1 and Sprouty2 provide a control mechanism for the Ras/MAPK signalling pathway. *Nat. Cell Biol.* **2002**, *4*, 850–858. [CrossRef]
10. Gross, I.; Morrison, D.J.; Hyink, D.P.; Georgas, K.; English, M.A.; Mericskay, M.; Hosono, S.; Sassoon, D.; Wilson, P.D.; Little, M.; et al. The receptor tyrosine kinase regulator Sprouty1 is a target of the tumor suppressor WT1 and important for kidney development. *J. Biol. Chem.* **2003**, *278*, 41420–41430. [CrossRef]

11. Edwin, F.; Singh, R.; Endersby, R.; Baker, S.J.; Patel, T.B. The tumor suppressor PTEN is necessary for human Sprouty 2-mediated inhibition of cell proliferation. *J. Biol. Chem.* **2006**, *281*, 4816–4822. [CrossRef] [PubMed]

12. Akbulut, S.; Reddi, A.L.; Aggarwal, P.; Ambardekar, C.; Canciani, B.; Kim, M.K.; Hix, L.; Vilimas, T.; Mason, J.; Basson, M.A.; et al. Sprouty proteins inhibit receptor-mediated activation of phosphatidylinositol-specific phospholipase C. *Mol. Biol. Cell* **2010**, *21*, 3487–3496. [CrossRef] [PubMed]

13. Wong, E.S.; Lim, J.; Low, B.C.; Chen, Q.; Guy, G.R. Evidence for direct interaction between Sprouty and Cbl. *J. Biol. Chem.* **2001**, *276*, 5866–5875. [CrossRef] [PubMed]

14. Tennis, M.A.; Van Scoyk, M.M.; Freeman, S.V.; Vandervest, K.M.; Nemenoff, R.A.; Winn, R.A. Sprouty-4 inhibits transformed cell growth, migration and invasion, and epithelial-mesenchymal transition, and is regulated by Wnt7A through PPARgamma in non-small cell lung cancer. *Mol. Cancer Res.* **2010**, *8*, 833–843. [CrossRef] [PubMed]

15. Sutterluty, H.; Mayer, C.E.; Setinek, U.; Attems, J.; Ovtcharov, S.; Mikula, M.; Mikulits, W.; Micksche, M.; Berger, W. Down-regulation of Sprouty2 in non-small cell lung cancer contributes to tumor malignancy via extracellular signal-regulated kinase pathway-dependent and -independent mechanisms. *Mol. Cancer Res.* **2007**, *5*, 509–520. [CrossRef]

16. Shaw, A.T.; Meissner, A.; Dowdle, J.A.; Crowley, D.; Magendantz, M.; Ouyang, C.; Parisi, T.; Rajagopal, J.; Blank, L.J.; Bronson, R.T.; et al. Sprouty-2 regulates oncogenic K-ras in lung development and tumorigenesis. *Genes Dev.* **2007**, *21*, 694–707. [CrossRef]

17. Fong, C.W.; Chua, M.S.; McKie, A.B.; Ling, S.H.; Mason, V.; Li, R.; Yusoff, P.; Lo, T.L.; Leung, H.Y.; So, S.K.; et al. Sprouty 2, an inhibitor of mitogen-activated protein kinase signaling, is down-regulated in hepatocellular carcinoma. *Cancer Res.* **2006**, *66*, 2048–2058. [CrossRef]

18. Vanas, V.; Muhlbacher, E.; Kral, R.; Sutterluty-Fall, H. Sprouty4 interferes with cell proliferation and migration of breast cancer-derived cell lines. *Tumour Biol.* **2014**, *35*, 4447–4456. [CrossRef]

19. Lo, T.L.; Yusoff, P.; Fong, C.W.; Guo, K.; McCaw, B.J.; Phillips, W.A.; Yang, H.; Wong, E.S.; Leong, H.F.; Zeng, Q.; et al. The ras/mitogen-activated protein kinase pathway inhibitor and likely tumor suppressor proteins, sprouty 1 and sprouty 2 are deregulated in breast cancer. *Cancer Res.* **2004**, *64*, 6127–6136. [CrossRef]

20. Kwabi-Addo, B.; Wang, J.; Erdem, H.; Vaid, A.; Castro, P.; Ayala, G.; Ittmann, M. The expression of Sprouty1, an inhibitor of fibroblast growth factor signal transduction, is decreased in human prostate cancer. *Cancer Res.* **2004**, *64*, 4728–4735. [CrossRef]

21. Rathmanner, N.; Haigl, B.; Vanas, V.; Doriguzzi, A.; Gsur, A.; Sutterluty-Fall, H. Sprouty2 but not Sprouty4 is a potent inhibitor of cell proliferation and migration of osteosarcoma cells. *FEBS Lett.* **2013**, *587*, 2597–2605. [CrossRef]

22. Barbachano, A.; Ordonez-Moran, P.; Garcia, J.M.; Sanchez, A.; Pereira, F.; Larriba, M.J.; Martinez, N.; Hernandez, J.; Landolfi, S.; Bonilla, F.; et al. SPROUTY-2 and E-cadherin regulate reciprocally and dictate colon cancer cell tumourigenicity. *Oncogene* **2010**, *29*, 4800–4813. [CrossRef]

23. Holgren, C.; Dougherty, U.; Edwin, F.; Cerasi, D.; Taylor, I.; Fichera, A.; Joseph, L.; Bissonnette, M.; Khare, S. Sprouty-2 controls c-Met expression and metastatic potential of colon cancer cells: Sprouty/c-Met upregulation in human colonic adenocarcinomas. *Oncogene* **2010**, *29*, 5241–5253. [CrossRef]

24. Schaaf, G.; Hamdi, M.; Zwijnenburg, D.; Lakeman, A.; Geerts, D.; Versteeg, R.; Kool, M. Silencing of SPRY1 triggers complete regression of rhabdomyosarcoma tumors carrying a mutated RAS gene. *Cancer Res.* **2010**, *70*, 762–771. [CrossRef]

25. Park, J.W.; Wollmann, G.; Urbiola, C.; Fogli, B.; Florio, T.; Geley, S.; Klimaschewski, L. Sprouty2 enhances the tumorigenic potential of glioblastoma cells. *Neuro-Oncol.* **2018**, *20*, 1044–1054. [CrossRef]

26. Walsh, A.M.; Kapoor, G.S.; Buonato, J.M.; Mathew, L.K.; Bi, Y.; Davuluri, R.V.; Martinez-Lage, M.; Simon, M.C.; O'Rourke, D.M.; Lazzara, M.J. Sprouty2 Drives Drug Resistance and Proliferation in Glioblastoma. *Mol. Cancer Res.* **2015**, *13*, 1227–1237. [CrossRef]

27. Hausott, B.; Vallant, N.; Schlick, B.; Auer, M.; Nimmervoll, B.; Obermair, G.J.; Schwarzer, C.; Dai, F.; Brand-Saberi, B.; Klimaschewski, L. Sprouty2 and -4 regulate axon outgrowth by hippocampal neurons. *Hippocampus* **2012**, *22*, 434–441. [CrossRef]

28. Hausott, B.; Vallant, N.; Auer, M.; Yang, L.; Dai, F.; Brand-Saberi, B.; Klimaschewski, L. Sprouty2 down-regulation promotes axon growth by adult sensory neurons. *Mol. Cell. Neurosci.* **2009**, *42*, 328–340. [CrossRef]

29.	Yu, T.; Yaguchi, Y.; Echevarria, D.; Martinez, S.; Basson, M.A. Sprouty genes prevent excessive FGF signalling in multiple cell types throughout development of the cerebellum. *Development* **2011**, *138*, 2957–2968. [CrossRef]

30.	Panagiotaki, N.; Dajas-Bailador, F.; Amaya, E.; Papalopulu, N.; Dorey, K. Characterisation of a new regulator of BDNF signalling, Sprouty3, involved in axonal morphogenesis in vivo. *Development* **2010**, *137*, 4005–4015. [CrossRef]

31.	Benke, S.; Agerer, B.; Haas, L.; Stoger, M.; Lercher, A.; Gabler, L.; Kiss, I.; Scinicariello, S.; Berger, W.; Bergthaler, A.; et al. Human tripartite motif protein 52 is required for cell context-dependent proliferation. *Oncotarget* **2018**, *9*, 13565–13581. [CrossRef]

32.	Sutterluty, H.; Chatelain, E.; Marti, A.; Wirbelauer, C.; Senften, M.; Muller, U.; Krek, W. p45SKP2 promotes p27Kip1 degradation and induces S phase in quiescent cells. *Nat. Cell Biol.* **1999**, *1*, 207–214. [CrossRef]

33.	Vanas, V.; Haigl, B.; Stockhammer, V.; Sutterluty-Fall, H. MicroRNA-21 Increases Proliferation and Cisplatin Sensitivity of Osteosarcoma-Derived Cells. *PLoS ONE* **2016**, *11*, e0161023. [CrossRef]

34.	Kral, R.M.; Mayer, C.E.; Vanas, V.; Gsur, A.; Sutterluty-Fall, H. In non-small cell lung cancer mitogenic signaling leaves Sprouty1 protein levels unaffected. *Cell. Biochem. Funct.* **2014**, *32*, 96–100. [CrossRef]

35.	Mayer, C.E.; Haigl, B.; Jantscher, F.; Siegwart, G.; Grusch, M.; Berger, W.; Sutterluty, H. Bimodal expression of Sprouty2 during the cell cycle is mediated by phase-specific Ras/MAPK and c-Cbl activities. *Cell. Mol. Life Sci.* **2010**, *67*, 3299–3311. [CrossRef]

36.	Impagnatiello, M.A.; Weitzer, S.; Gannon, G.; Compagni, A.; Cotten, M.; Christofori, G. Mammalian sprouty-1 and -2 are membrane-anchored phosphoprotein inhibitors of growth factor signaling in endothelial cells. *J. Cell Biol.* **2001**, *152*, 1087–1098. [CrossRef]

37.	Zhang, W.; Lv, Y.; Xue, Y.; Wu, C.; Yao, K.; Zhang, C.; Jin, Q.; Huang, R.; Li, J.; Sun, Y.; et al. Co-expression modules of NF1, PTEN and sprouty enable distinction of adult diffuse gliomas according to pathway activities of receptor tyrosine kinases. *Oncotarget* **2016**, *7*, 59098–59114. [CrossRef]

38.	Doriguzzi, A.; Salhi, J.; Sutterluty-Fall, H. Sprouty4 mRNA variants and protein expressions in breast and lung-derived cells. *Oncol. Lett.* **2016**, *12*, 4161–4166. [CrossRef]

39.	Faratian, D.; Sims, A.H.; Mullen, P.; Kay, C.; Um, I.; Langdon, S.P.; Harrison, D.J. Sprouty 2 is an independent prognostic factor in breast cancer and may be useful in stratifying patients for trastuzumab therapy. *PLoS ONE* **2011**, *6*, e23772. [CrossRef]

40.	Sirivatanauksorn, Y.; Sirivatanauksorn, V.; Srisawat, C.; Khongmanee, A.; Tongkham, C. Differential expression of sprouty genes in hepatocellular carcinoma. *J. Surg. Oncol.* **2012**, *105*, 273–276. [CrossRef]

41.	Kwak, H.J.; Kim, Y.J.; Chun, K.R.; Woo, Y.M.; Park, S.J.; Jeong, J.A.; Jo, S.H.; Kim, T.H.; Min, H.S.; Chae, J.S.; et al. Downregulation of Spry2 by miR-21 triggers malignancy in human gliomas. *Oncogene* **2011**, *30*, 2433–2442. [CrossRef]

42.	Doriguzzi, A.; Haigl, B.; Gsur, A.; Sutterluty-Fall, H. The increased Sprouty4 expression in response to serum is transcriptionally controlled by Specific protein 1. *Int. J. Biochem. Cell Biol.* **2015**, *64*, 220–228. [CrossRef]

43.	Jiang, Z.; Price, C.A. Differential actions of fibroblast growth factors on intracellular pathways and target gene expression in bovine ovarian granulosa cells. *Reproduction* **2012**, *144*, 625–632. [CrossRef]

44.	Masoumi-Moghaddam, S.; Amini, A.; Morris, D.L. The developing story of Sprouty and cancer. *Cancer Metastasis Rev.* **2014**, *33*, 695–720. [CrossRef]

45.	Zhou, X.; Xie, S.; Yuan, C.; Jiang, L.; Huang, X.; Li, L.; Chen, Y.; Luo, L.; Zhang, J.; Wang, D.; et al. Lower Expression of SPRY4 Predicts a Poor Prognosis and Regulates Cell Proliferation in Colorectal Cancer. *Cell. Physiol. Biochem.* **2016**, *40*, 1433–1442. [CrossRef]

46.	Zhang, Q.; Wei, T.; Shim, K.; Wright, K.; Xu, K.; Palka-Hamblin, H.L.; Jurkevich, A.; Khare, S. Atypical role of sprouty in colorectal cancer: Sprouty repression inhibits epithelial-mesenchymal transition. *Oncogene* **2016**, *35*, 3151–3162. [CrossRef]

47.	Masoumi-Moghaddam, S.; Amini, A.; Wei, A.Q.; Robertson, G.; Morris, D.L. Sprouty 2 protein, but not Sprouty 4, is an independent prognostic biomarker for human epithelial ovarian cancer. *Int. J. Cancer* **2015**, *137*, 560–570. [CrossRef]

48.	Wang, J.; Thompson, B.; Ren, C.; Ittmann, M.; Kwabi-Addo, B. Sprouty4, a suppressor of tumor cell motility, is down regulated by DNA methylation in human prostate cancer. *Prostate* **2006**, *66*, 613–624. [CrossRef]

49.	Jaggi, F.; Cabrita, M.A.; Perl, A.K.; Christofori, G. Modulation of endocrine pancreas development but not beta-cell carcinogenesis by Sprouty4. *Mol. Cancer Res.* **2008**, *6*, 468–482. [CrossRef]

50. Gong, Y.; Yang, X.; He, Q.; Gower, L.; Prudovsky, I.; Vary, C.P.; Brooks, P.C.; Friesel, R.E. Sprouty4 regulates endothelial cell migration via modulating integrin beta3 stability through c-Src. *Angiogenesis* **2013**, *16*, 861–875. [CrossRef]

51. Taniguchi, K.; Ishizaki, T.; Ayada, T.; Sugiyama, Y.; Wakabayashi, Y.; Sekiya, T.; Nakagawa, R.; Yoshimura, A. Sprouty4 deficiency potentiates Ras-independent angiogenic signals and tumor growth. *Cancer Sci.* **2009**, *100*, 1648–1654. [CrossRef] [PubMed]

6

Fibroblast Growth Factor Receptors (FGFRs): Structures and Small Molecule Inhibitors

Shuyan Dai [1], Zhan Zhou [1], Zhuchu Chen [1], Guangyu Xu [2,*] and Yongheng Chen [1,*]

[1] NHC Key Laboratory of Cancer Proteomics & Laboratory of Structural Biology, Xiangya Hospital, Central South University, Changsha 410008, Hunan, China; syandai@hotmail.com (S.D.); zhouzhan285@163.com (Z.Z.); chenzhuchu@126.com (Z.C.)

[2] Key Laboratory of Chemical Biology and Traditional Chinese Medicine Research (Ministry of Education), College of Chemistry and Chemical Engineering, Hunan Normal University, Changsha 410081, Hunan, China

* Correspondence: gyxu@hunnu.edu.cn (G.X.); yonghenc@163.com (Y.C.)

Abstract: Fibroblast growth factor receptors (FGFRs) are a family of receptor tyrosine kinases expressed on the cell membrane that play crucial roles in both developmental and adult cells. Dysregulation of FGFRs has been implicated in a wide variety of cancers, such as urothelial carcinoma, hepatocellular carcinoma, ovarian cancer and lung adenocarcinoma. Due to their functional importance, FGFRs have been considered as promising drug targets for the therapy of various cancers. Multiple small molecule inhibitors targeting this family of kinases have been developed, and some of them are in clinical trials. Furthermore, the pan-FGFR inhibitor erdafitinib (JNJ-42756493) has recently been approved by the U.S. Food and Drug Administration (FDA) for the treatment of metastatic or unresectable urothelial carcinoma (mUC). This review summarizes the structure of FGFR, especially its kinase domain, and the development of small molecule FGFR inhibitors.

Keywords: fibroblast growth factor receptors; structure; kinase inhibitor; targeted therapy

1. Introduction

The human fibroblast growth factor receptor (FGFR) family consists of four members: FGFR1 to FGFR4. Despite being encoded by separate genes, the four members share high homology, with their sequence identity varying from 56% to 71% [1]. Similar to other receptor tyrosine kinases (RTKs), FGFRs are expressed on the cell membrane and can be stimulated and activated by extracellular signals. The native ligand of FGFRs is fibroblast growth factors (FGFs) [2–4]. The binding of FGFs drives the dimerization of FGFRs; subsequently, a transautophosphorylation event of the intracellular kinase domain is induced, followed by the activation of downstream transduction pathways [5,6]. Through triggering downstream signaling pathways, FGFRs participate in various vital physiological processes, such as proliferation, differentiation, cell migration and survival [7–9].

Aberrant expression of FGFRs has been shown in various kinds of solid tumors, and moreover, the aberrancy is considered an oncogenic signaling pathway [10–12]. It is believed that small molecules that competitively bind to the adenosine triphosphate (ATP) pocket of aberrant FGFRs while exhibiting little or no toxicity provide limitless prospects for the treatment of relevant tumors. The structure of FGFRs, especially the kinase domain, and the design of small molecular inhibitors have attracted intensive study in the past two decades. Multiple small molecule inhibitors have been developed, and some of them are currently being used in clinical trials, such as FGF401, which targets FGFR4 for the treatment of hepatocarcinoma (HCC) [13]; AZD4547, which targets FGFR1-3 for the treatment of a variety of tumors [14]. Moreover, erdafitinib (JNJ-42756493) [15] has been approved recently by U.S.

Food and Drug Administration (FDA) for the treatment of mUC. More than 20 FGFR kinase/inhibitor complex structures have been determined to-date, and these structures have yielded extensive insights into the understanding of inactivation of FGFRs for related disease therapy.

2. Organization of FGFR

FGFRs share a canonical RTK architecture. From the N- to the C-terminus, all four FGFR members contain a large extracellular ligand-binding domain that comprises three immunoglobulin (Ig)-like subunits (D1, D2 and D3) followed by a single transmembrane helix and an intracellular tyrosine kinase domain [1,16] (Figure 1A). The linker region between D1 and D2 contains a highly conserved motif that is rich in aspartate acids, called the acid box [17]. The detailed function of those structural units will be further introduced below.

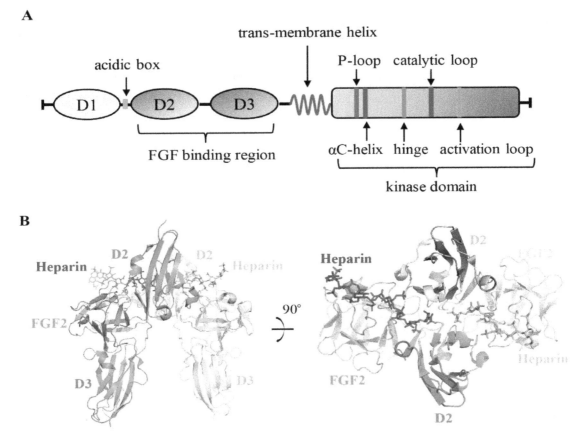

Figure 1. Schematic diagram of FGFRs and the structure of the FGFR extracellular domain. (**A**) Organization of FGFRs. Important functional elements are highlighted. (**B**) Crystal structure of the FGF2:FGFR1:heparin ternary complex (PDB ID 1FQ9). The two copies of FGFR1 molecules are colored in green and light blue respectively. Heparin molecules are shown in red stick representation; FGF2 (colored in orange) and FGFR1 are shown in cartoon representation.

FGFs are the native ligand for this family of kinases. Through its extracellular domain, FGFR recognizes and is stimulated by specific FGFs. The FGF binding pocket is formed by the D2 and D3 subregions [18]. There have been contradicting views regarding the stoichiometry of the FGF/FGFR complex. Schlessinger, J. et al. solved the ternary complex structure of FGF2/FGFR1/heparin [19]. With the help of heparin, FGFR1 was dimerized after the binding of FGF2 to form the complex at a symmetric 2:2:2 stoichiometry ratio; both the FGF2 and heparin molecules simultaneously contacted the two FGFR1 monomers (Figure 1B). In the FGF1/FGFR2/heparin crystal structure solved by Pellegrini et al., the complex was assembled by asymmetric 2:2:1 stoichiometry [20]. By utilizing nuclear magnetic resonance, Saxena et al. studied the interactions of FGF1(FGF2)/FGFR4/HM (HM: heparin mimetics) complex, and their results supported the formation of the symmetric mode of FGF/FGFR dimerization in solution [21]. Interestingly, although all FGFs have a heparin sulfate binding site on their surface [22,23], endocrine FGFs such as FGF21 and FGF23 show a lower binding affinity to heparin sulfate [23] and require Klotho coreceptors instead to act as cofactors for FGFR activation [24–26].

In addition to acting as the ligand sensor, the extracellular domain also undertakes an autoinhibitory role, which relies on regulation by D1 and the acid box [27,28]. Several studies have proposed that the acid box could competitively bind to the heparin binding site of D2 to suppress heparin binding, while D1 forms intramolecular contacts with D2-D3, thus blocking the binding of FGFs [28–30]. Nevertheless, the mechanisms of autoinhibition need to be further clarified.

3. Structure of FGFR Kinase Domain

The intracellular tyrosine kinase domain is the most well studied region of the FGFR protein. This domain exhibits the canonical bilobed architecture of protein kinases [31–35]. The fold of the N-terminal small lobe (N-lobe, ~100 amino acid residues) consists of a five-stranded antiparallel β-sheet (β1–β5) and the αC-helix, an important regulatory element. The C-terminal large lobe (C-lobe, ~200 amino acid residues) predominately comprises seven a helices (αD, αE, αEF, αF-αI) (Figure 2A). The active site, which is responsible for ATP and substrate protein binding, is located in a clef between the two lobes (Figure 2B).

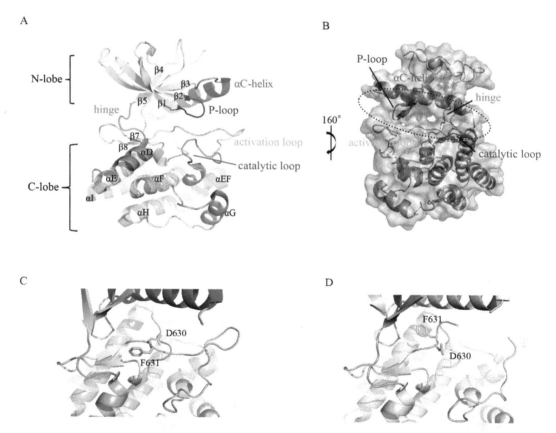

Figure 2. Structure of the FGFR kinase domain. (**A**) Overall crystal structure of FGFR4 in cartoon representation. The five β-sheets of the N-lobe are labeled in cyan, and the helixes of the C-lobe are colored in salmon. The αC helix (red), P-loop (blue), catalytic loop (magenta), activation loop (bright orange) and hinge (green) are highlighted. (**B**) Surface presentation of FGFR4. The ATP binding pocket located between the N- and C-lobe is indicated by the dashed circle. (**C**) DFG-out conformation of the FGFR4 activation loop. The side chains of D630 and F631 are shown in stick representation. (**D**) DFG-in status of the FGFR4 activation loop. (**A–C**) were prepared from PDB ID 4UXQ; (**D**) was prepared from PDB ID 5JKG.

The C-lobe folds tightly with the αF-helix to form a hydrophobic core, around which the other secondary segments are packed. In addition to the primary helixes, the C-lobe contains a short helix located between the activation loop (A-loop) and the αF-helix named the αEF-helix, which is conserved among all FGFR members as well as other protein kinases [36]. Two short β-strands (β7 and β8) (Figure 2A) between the catalytic loop and activation loop (introduced below) interact with each other and are believed to participate in the regulation of FGFR activation [37]. In contrast to the C-lobe, the N-lobe folds in a more flexible manner, which benefits the binding and release of ATP/ADP and substrates.

There are several functionally important loops in both lobes. The loop between β8 and the αEF-helix is an activation loop (A-loop), which is essential for kinase activation [38–40]. The conformation of the highly conserved Asp-Phe-Gly motif (DFG-motif) in the A-loop is an indicator of kinase activity status [39,41]. Generally, the DFG-motif exists in two states: the active DFG-in and inactive DFG-out conformations [42,43] (Figure 2C,D). In the DFG-in state, the aspartate residue of the DFG-motif plays an essential role in ATP binding through the coordination of all three phosphate groups of ATP, either directly or via magnesium ions, while these interactions are sterically impossible when the motif is flipped into the DFG-out conformation. Phosphorylation is catalyzed by the conserved aspartate of the His-Arg-Asp (HRD) motif in the catalytic (αE–β7) loop [44]. The glycine rich P-loop (also called the nucleotide binding loop), located between the β1- and β2-strands, folds over to enclose ATP for phosphotransfer [45]. The so-called molecular brake located at the hinge region that connects the N- and C-lobes plays a critical role in the regulation of autoinhibition and activation [46].

The catalytic activity of the kinase domain is precisely controlled. There are two general conformations for all protein kinases, including those of the FGFR family. Activation typically involves changes in the orientation of the αC-helix in the small lobe and the activation loop in the C-lobe. During the catalytic cycle, the active kinase toggles between open and closed conformations. In the open form, the kinase binds MgATP and the protein substrate, while during catalysis, the kinase adopts the closed form. Once catalysis is completed, the MgADP and phosphorylated substrate are released, and the enzyme recovers to the open conformation, preparing for the next catalytic cycle [16,47].

4. Characteristics of FGFR/Inhibitor Interaction

As noted above, aberrantly expressed FGFRs have been implicated in various tumors. Therefore, extensive work has been performed on the development of FGFR inhibitors. The inhibitors that are in clinical trials or approved by the FDA for clinical use are summarized in Table 1, and the chemical structures of those inhibitors are shown in Figure 3.

Table 1. FGFR inhibitors that are in clinical trials or approved by the FDA.

Inhibitor Name	Binding Features	IC50 (nM)	PDB ID	Clinical Trial Phase/Number	Reference
JNJ-42756493 (Erdafitinib)	Pan-FGFR Reversible Type I	FGFR1: 1.2 FGFR2: 2.5 FGFR3: 3.0 FGFR4: 5.7	n/a	FDA approved	[15]
AZD4547	Pan-FGFR Reversible Type I	FGFR1: 0.2 FGFR2: 2.5 FGFR3: 1.8 FGFR4: 165	4V05	Phase I/II NCT02824133	[14,34]
Ly2874455	Pan-FGFR Reversible Type I	FGFR1: 2.8 FGFR2: 2.6 FGFR3: 6.4 FGFR4: 6	5JKG	Phase I NCT01212107	[33,48]

Table 1. *Cont.*

Inhibitor Name	Binding Features	IC50 (nM)	PDB ID	Clinical Trial Phase/Number	Reference
CH5183284	Pan-FGFR Reversible Type I	FGFR1: 9.3 FGFR2: 7.6 FGFR3: 22 FGFR4: 290	5N7V	Phase II/III NCT03344536	[49]
NVP-BGJ398	Pan-FGFR Reversible Type I	FGFR1: 0.9 FGFR2: 1.4 FGFR3: 1 FGFR4: 60	3TT0	Phase II NCT02706691	[50]
INCB054828	Pan-FGFR Reversible Type I	FGFR1: 0.4 FGFR2: 0.5 FGFR3: 1.2 FGFR4: 30	n/a	Phase II NCT03011372	[51]
Rogaratinib	Pan-FGFR Reversible Type I	FGFR1: 12–15 FGFR2: <1 FGFR3: 19 FGFR4: 33	n/a	Phase II/III NCT03410693	[52]
PRN1371	Pan-FGFR Irreversible Type I	FGFR1: 0.6 FGFR2: 1.3 FGFR3: 4.1 FGFR4: 19.3	n/a	Phase I NCT02608125	[53]
TAS-120	Pan-FGFR Irreversible Type I	FGFR1: 3.9 FGFR2: 1.3 FGFR3: 1.6 FGFR4: 8.3	6M2Q	Phase I/II NCT02052778	[54]
BLU-554	FGFR4 selective Irreversible, Type I	FGFR1: 624 FGFR2: 1202 FGFR3: 2203 FGFR4: 5	n/a	Phase I NCT02508467	[55]
H3B-6527	FGFR4 selective Irreversible, Type I	FGFR1: 320 FGFR2: 1290 FGFR3: 1060 FGFR4: <1.2	5VND	Phase I NCT02834780	[56]
FGF401	FGFR4 selective Reversible Covalent, Type I	FGFR1-3: >10,000 FGFR4: 1.1	6JPJ	Phase I/II NCT02325739	[13,57]

n/a stands for not available.

FGFR inhibitors can generally be divided into two groups according to their binding behaviors, namely, type I and type II inhibitors [58,59]. Type I inhibitors bind FGFRs in the DFG-in enzymatic active conformation in an ATP-competitive manner, while the binding of type II requires the DFG-motif to be flipped to the DFG-out state [60,61]. The X-ray crystallographic structures of AZD4547 (PDB ID 4V05) [34] and PD173074 (PDB ID 2FGI) [62] bound to FGFR1 demonstrate that these two inhibitors are type I inhibitors. Taking the FGFR1/AZD4547 structure as an example, the AZD4547 occupying the

ATP pocket of FGFR1 forms a hydrogen bond with the backbone nitrogen atom of the DFG aspartate (Asp641) and forms three hydrogen bonds with the hinge residues [34] (Figure 4A). In contrast, both the DFG-motifs of FGFR4 and FGFR1 are flipped out into an inactive conformation in the FGFR4/ponatinib (PDB ID 4UXQ) and FGFR1/ponatinib (PDB ID 4V04) structures [34]. In addition to the basal interactions, a hydrogen bond formed between ponatinib and the side chain of the strictly conserved glutamate from the αC-helix (Glu520 in FGFR4 and Glu531 in FGFR1) was also observed, which is characteristic of a type II inhibitor [61] (Figure 4B). Thus, these structures reveal ponatinib to be a type II inhibitor for FGFRs. As a consequence, the flip of the phenylalanine side chain breaks the regulatory spine and creates an additional induced-fit hydrophobic pocket that allows deeper binding of the inhibitor and provides better selectivity [34,43] as well as slower dissociation kinetics [34,63].

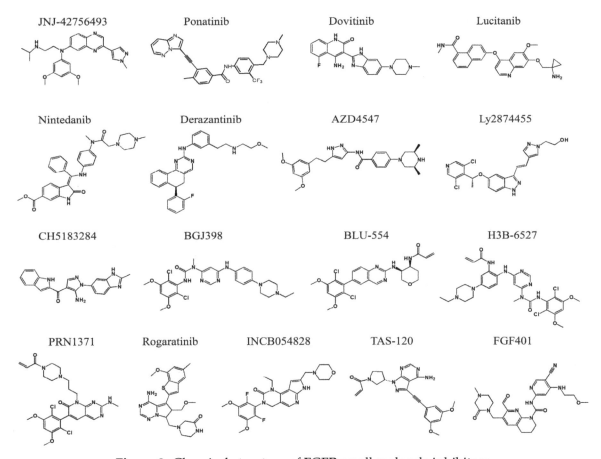

Figure 3. Chemical structure of FGFR small molecule inhibitors.

The interaction between a small molecule inhibitor and protein kinase can be covalent (irreversible) or noncovalent (reversible) [64,65]. Typically, covalent inhibitors have a functional group known as the warhead, which can improve binding affinity and selectivity through covalent interaction with a certain residue of the target kinase [65,66]. Moreover, a well-designed warhead could provide better performance against drug resistance than reversible inhibitors [67,68]. The reported covalent reactive residues in protein kinases include cysteine [69], aspartic acid [70], lysine [71] and others [72]. For FGFRs, the conserved cysteine of the P-loop (C488 in FGFR1, C491 in FGFR2, C482 in FGFR3 and C477 in FGFR4) and the unique C552 in FGFR4 from the hinge region are the covalent binding sites. The FGFR4/FIIN-2 complex structure (PDB accession number 4QQC) is the first solved irreversible structure, where FIIN-2 formed a covalent bond through its reactive acrylamide group with the hydrosulfonyl side chain of FGFR4 C488 [73] (Figure 4C).

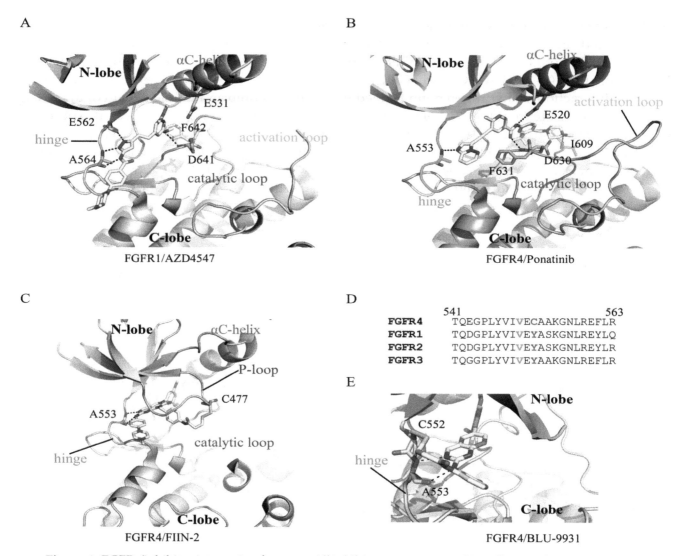

Figure 4. FGFRs/inhibitor interaction features. All inhibitors are presented in yellow stick representation. (**A**) Structure of AZD4547 bound to FGFR1. The side chains of A564, E562 and D641, which directly form hydrogen bonds with the inhibitor and F642 of the DFG-motif, are shown. The hydrogen bonds are indicated by dashed lines. AZD4547 binds FGFR1 into DFG-in status, the side chain of F642 points out from the ATP pocket. (**B**) The structure of FGFR4 in a complex with ponatinib. The DFG-motif of FGFR4 flipped to an out conformation with F631 benzene ring flipped into the ATP-pocket and D630 point out from the pocket. (**C**) Covalent interaction of FIIN-2 and FGFR4. The side chains of A553 and C477, which interact with ponatinib, are shown in stick representation. Covalent bond formed between C477 and the acrylamide group of FIIN-2. (**D**) Sequence alignment of the FGFR hinge region. The C552 in FGR4 is replaced by a tyrosine in the other 3 members. The gatekeeper residues which locate at the kinases hinge region and play an essential role in determining pocket accessibility for inhibitors are highlighted in green. (**E**) Structure of BLU-9931 in complex with FGFR4. Unlike the pan-FGFR covalent inhibitors, BLU-9931 targets the unique C552 of FGFR4 to form covalent interactions.

5. Current Status of Small Molecule FGFR Inhibitor Development

Designing specific small molecule inhibitors targeting protein kinases is challenging because the ATP binding pockets of the human kinome are similar [74,75]. Inhibitor research for FGFRs has gone through several stages. Initially, nonselective multiple-kinase inhibitors were developed to treat FGFR aberrations. Those nonselective inhibitors, including ponatinib [76], dovitinib [77] lucitanib [78] and nintedanib [79] (see Table 1 for details), were originally designed for other kinases and then proved to have potent inhibition activity toward FGFRs. For instance, the type II inhibitor ponatinib was originally developed to overcome the BCR-ABL T315I gatekeeper mutant and showed single-digit nanomolar binding strength to FGFR1-4 in later researches [76]. Although nonselective inhibitors might be clinically beneficial and achieve therapeutic success to some extent, the development of those inhibitors has been restricted due to the undesirable off-target toxicities [80,81].

To overcome the off-target effects of nonselective inhibitors, efforts have been made to develop FGFR-selective inhibitors (pan-FGFR inhibitors). In the earlier stage, multiple noncovalent pan-FGFR inhibitors were developed, including the well-known AZD4547 [14] and LY2874455 [48] (see Table 1 for details). Those pan-FGFR inhibitors are typically type I inhibitors. For example, AZD4547 is capable of potently inhibiting FGFR1-3 but shows negligible binding affinity to FGFR4. AZD4547 is currently in phase II clinical trials. However, preclinical data show that AZD4547 is not able to overcome the gatekeeper mutation V555M in FGFR3 [82]. Unlike AZD4547, the inhibitor LY2874455 shows inhibition efficacy against all 4 FGFRs, and the crystal structure as well as in vitro and vivo experiments confirmed that this inhibitor maintained equal inhibitory ability against the gatekeeper mutant V550M/V550L of FGFR4 [32].

Given that covalent inhibitors confer better binding kinetics and selectivity than noncovalent ones, developing irreversible inhibitors of FGFRs has attracted intensive pharmaceutical and academic attention in recent years. Since the first covalent inhibitor FIIN-1 [83], this field has achieved much progress. A number of FGFR covalent inhibitors have been developed, and some of those agents are already in clinical trials (see Table 1 for details). Moreover, several irreversible inhibitor/FGFR structures have been revealed by crystal structures, including FGFR4/BLU9931 (PDB ID: 4XCU) [84], FGFR4/FIIN-3 (PDB ID: 4R6V) [73], FGFR4/FIIN-2 (PDB ID: 4QQ5) [85], FGFR4/CGA159527 (PDB ID: 5NUD) [86], FGFR1(Y563C)/H3B-6527 (PDB ID: 5VND) [86], and FGFR1/TAS-120 (PDB ID: 6MZW) [54].

The kinase domains of FGFRs are highly homologous, with sequence identity varying from 74% to 77% [34]. Unexpectedly, most of the reported pan-FGFR reversible inhibitors tend to bind FGFR1-3 but exhibit greatly reduced potency toward FGFR4 [87]. The underlying mechanism is not quite clear. Tucker et al. proposed that the innate flexibility of the FGFR4 kinase domain might be responsible for the decrease in binding ability [34]. This feature, together with the unique C552 of FGFR4, which replaces a tyrosine in FGFR1-3, confers the opportunity to develop FGFR4-selective inhibitors [88,89] (Figure 4D). Indeed, H3B-6527 [56], BLU-9931 [84], BLU-554 [84] and FGF401 [90] were developed as FGFR4-selective covalent inhibitors that target C552 for irreversible binding (Figure 4E). Among these four molecules, FGF401 is the most interesting because the covalent bond it forms is reversible, which reduces the off-target effect and prolongs the residence time [91,92]. The crystal structure of FGF401/FGFR4 was recently reported by our laboratory (PDB ID 6JPJ) [57], and its potential utility is currently under intensive research.

In addition to the kinase domain, the ectodomains of FGFRs have also attracted intensive interests for drug discovery. Unlike the highly conserved kinase domain, the ectodomains of FGFRs are less conserved; targeting this domain may offer better isoform selectivity. The dominant strategy to target FGFR ectodomains is using monoclonal antibody/antibody-drug conjugate [93]. Several anti-FGFR monoclonal antibodies have been developed, and some of them are in clinical trials [94–98]. In addition, efforts have been made in the development of small molecule inhibitor targeting FGFR ectodomains. An inhibitor, SSR128129E, which allosterically binds to the ectodomain of FGFR in a non-FGF competitive manner, has been reported to inhibit FGF-induced signaling [99,100].

6. FGFR Gatekeeper Mutation and Drug Resistance

The long-term efficacy of kinase inhibitors in cancer treatment is often disturbed by acquired resistance. One common mechanism of resistance is generated by mutating the so-called gatekeeper residue of the kinase domain [101]. The gatekeeper mutation has been reported in various kinases, such as Bcr-Abl (T315I) [102], EGFR (T790M) [68], PDGFR (T674I) [103], FGFR1 (V561M) [104] and FGFR2 (V565I) [105]. The gatekeeper residue lies at the beginning of the hinge region and controls the accessibility of the hydrophobic pocket. Most protein kinases harbor a threonine that plays a major role in the interaction with the inhibitor by forming a critical hydrogen bond via its side chain hydroxyl oxygen. The mutation of this residue to a bulky hydrophobic amino acid, either Met or Ile, breaks the hydrogen interaction and introduces steric hindrance for inhibitor binding [76,106].

The drug resistance of FGFR gatekeeper mutations has been extensively verified, both in vitro and in vivo. For instance, the V564M mutation of FGFR2 confers the ability to resist dovitinib and BGJ398 [107]. Furthermore, an array of FGFR gatekeeper mutations have been identified in clinical samples. For example, the FGFR4 V550M mutation was detected in 13% of neuroendocrine breast carcinomas [108]. In FGFRs, the gatekeeper residue is a valine (Figure 4D); as a consequence, its side chain cannot form hydrogen interactions with inhibitors (see above context), and the resistance thus arises mainly through the introduction of steric hindrance. Several FGFR inhibitors have been shown to have the ability to overcome FGFR gatekeeper mutations. For example, Ly2874455 has almost equal binding affinity to wild-type FGFR4, FGFR4 (V550M), FGFR4 (V550L) FGFR1 (V561M), FGFR2 (V564F) and FGFR3 (V555M) [32]; FGF401 has similar affinity to wild-type FGFR4, FGFR4 (V550M) and FGFR4 (V550L) [57]; FIIN-2 shows a binding affinity loss of ~10-fold for FGFR4 (V550L) compared with the wild-type kinase [85].

7. Conclusions

Increasing evidence indicates that aberrant Fibroblast growth factor receptors (FGFR)signaling plays a crucial role in tumorigenesis and progression. Now, small molecule inhibitors targeting FGFRs offer a novel and effective strategy for the therapy of cancers caused by FGFR aberrations. Efforts and progress have been made in the field of FGFR inhibitor development. Some small molecules show promising antitumor activity and are evaluated in clinical trials. Recently, the pan-FGFR inhibitor erdafitinib has been approved by the Food and Drug Administration for the treatment of mUC, making it the first approved FGFR-targeted drug. However, there are still challenges in the field of FGFR inhibitor development, such as the need for more potent and selective FGFR inhibitors, and inhibitors with the ability to overcome gatekeeper mutations. Most FGFR inhibitors currently under evaluation are typical type I inhibitors that occupy only the ATP binding pocket. Development of type II FGFR inhibitors, which could be inserted deeper into the pocket, could confer better potency and selectivity. In addition, the development of covalent irreversible or covalent reversible FGFR inhibitors might be another strategy to improve safety and efficacy for cancer treatment.

Author Contributions: All authors were involved in the design of the review, and in writing or revising the manuscript. All authors approved the submitted version.

References

1.   Itoh, N.; Ornitz, D.M. Evolution of the Fgf and Fgfr gene families. *Trends Genet.* **2004**, *20*, 563–569. [CrossRef] [PubMed]
2.   Schlessinger, J. Cell signaling by receptor tyrosine kinases. *Cell* **2000**, *103*, 211–225. [CrossRef]
3.   Weiner, H.L.; Zagzag, D. Growth factor receptor tyrosine kinases: Cell adhesion kinase family suggests a novel signaling mechanism in cancer. *Cancer Investig.* **2000**, *18*, 544–554. [CrossRef]
4.   Lemmon, M.A.; Schlessinger, J. Cell signaling by receptor tyrosine kinases. *Cell* **2010**, *141*, 1117–1134. [CrossRef] [PubMed]
5.   Schlessinger, J. Cell signaling by receptor tyrosine kinases: From basic concepts to clinical applications. *Eur. J. Cancer Suppl.* **2006**, *4*, 3–26. [CrossRef]
6.   Schlessinger, J. Receptor Tyrosine Kinases: Legacy of the First Two Decades. *Cold Spring Harb. Perspect. Biol.* **2014**. [CrossRef] [PubMed]
7.   Ornitz, D.M.; Itoh, N. The Fibroblast Growth Factor signaling pathway. *Wiley Interdiscip. Rev. Dev. Biol.* **2015**, *4*, 215–266. [CrossRef] [PubMed]

8.   Andrew, B.; Moosa, M. The FGF family: Biology, pathophysiology and therapy. *Nat. Rev. Drug Discov.* **2009**, *8*, 235–253.

9.   Karel, D.; Enrique, A. FGF signalling: Diverse roles during early vertebrate embryogenesis. *Development* **2010**, *137*, 3731–3742.

10. Gowardhan, B.; Douglas, D.A.; Mathers, M.E.; McKie, A.B.; McCracken, S.R.C.; Robson, C.N.; Leung, H.Y. Evaluation of the fibroblast growth factor system as a potential target for therapy in human prostate cancer. *Br. J. Cancer* **2005**, *92*, 320–327. [CrossRef]

11. Brooks, A.N.; Kilgour, E.; Smith, P.D. Molecular pathways: Fibroblast growth factor signaling: A new therapeutic opportunity in cancer. *Clin. Cancer Res.* **2012**, *18*, 1855–1862. [CrossRef] [PubMed]

12. Turner, N.; Grose, R. Fibroblast growth factor signalling: From development to cancer. *Nat. Rev. Cancer* **2010**, *10*, 116–129. [CrossRef] [PubMed]

13. Porta, D.G.; Weiss, A.; Fairhurst, R.A.; Wartmann, M.; Stamm, C.; Reimann, F.; Buhles, A.; Kinyamu-Akunda, J.; Sterker, D.; Murakami, M. Abstract 2098: NVP-FGF401, a first-in-class highly selective and potent FGFR4 inhibitor for the treatment of HCC. *Cancer Res.* **2017**. [CrossRef]

14. Gavine, P.R.; Mooney, L.; Kilgour, E.; Thomas, A.P.; Al-Kadhimi, K.; Beck, S.; Rooney, C.; Coleman, T.; Baker, D.; Mellor, M.J.; et al. AZD4547: An orally bioavailable, potent, and selective inhibitor of the fibroblast growth factor receptor tyrosine kinase family. *Cancer Res.* **2012**, *72*, 2045–2056. [CrossRef] [PubMed]

15. Perera, T.P.S.; Jovcheva, E.; Mevellec, L.; Vialard, J.; De Lange, D.; Verhulst, T.; Paulussen, C.; Van De Ven, K.; King, P.; Freyne, E.; et al. Discovery and Pharmacological Characterization of JNJ-42756493 (Erdafitinib), a Functionally Selective Small-Molecule FGFR Family Inhibitor. *Mol. Cancer Ther.* **2017**, *16*, 1010–1020. [CrossRef]

16. Farrell, B.; Breeze, A.L. Structure, activation and dysregulation of fibroblast growth factor receptor kinases: Perspectives for clinical targeting. *Biochem. Soc. Trans.* **2018**, *46*, 1753–1770. [CrossRef] [PubMed]

17. Sanchez-Heras, E.; Howell, F.V.; Williams, G.; Doherty, P. The fibroblast growth factor receptor acid box is essential for interactions with N-cadherin and all of the major isoforms of neural cell adhesion molecule. *J. Biol. Chem.* **2006**, *281*, 35208–35216. [CrossRef]

18. Wang, F.; Kan, M.; Xu, J.; Yan, G.; McKeehan, W.L. Ligand-specific structural domains in the fibroblast growth factor receptor. *J. Biol. Chem.* **1995**, *270*, 10222–10230. [CrossRef]

19. Schlessinger, J.; Plotnikov, A.N.; Ibrahimi, O.A.; Eliseenkova, A.V.; Yeh, B.K.; Yayon, A.; Linhardt, R.J.; Mohammadi, M. Crystal structure of a ternary FGF-FGFR-heparin complex reveals a dual role for heparin in FGFR binding and dimerization. *Mol. Cell* **2000**, *6*, 743–750. [CrossRef]

20. Pellegrini, L.; Burke, D.F.; von Delft, F.; Mulloy, B.; Blundell, T.L. Crystal structure of fibroblast growth factor receptor ectodomain bound to ligand and heparin. *Nature* **2000**, *407*, 1029–1034. [CrossRef]

21. Saxena, K.; Schieborr, U.; Anderka, O.; Duchardt-Ferner, E.; Elshorst, B.; Gande, S.L.; Janzon, J.; Kudlinzki, D.; Sreeramulu, S.; Dreyer, M.K.; et al. Influence of heparin mimetics on assembly of the FGF.FGFR4 signaling complex. *J. Biol. Chem.* **2010**, *285*, 26628–26640. [CrossRef] [PubMed]

22. Eriksson, A.E.; Cousens, L.S.; Weaver, L.H.; Matthews, B.W. Three-dimensional structure of human basic fibroblast growth factor. *Proc. Natl. Acad. Sci. USA* **1991**, *88*, 3441–3445. [CrossRef] [PubMed]

23. Goetz, R.; Mohammadi, M. Exploring mechanisms of FGF signalling through the lens of structural biology. *Nat. Rev. Mol. Cell Biol.* **2013**, *14*, 166–180. [CrossRef] [PubMed]

24. Razzaque, M.S. The FGF23-Klotho axis: Endocrine regulation of phosphate homeostasis. *Nat. Rev. Endocrinol.* **2009**, *5*, 611–619. [CrossRef] [PubMed]

25. Chen, G.; Liu, Y.; Goetz, R.; Fu, L.; Jayaraman, S.; Hu, M.C.; Moe, O.W.; Liang, G.; Li, X.; Mohammadi, M. alpha-Klotho is a non-enzymatic molecular scaffold for FGF23 hormone signalling. *Nature* **2018**, *553*, 461–466. [CrossRef] [PubMed]

26. Lee, S.; Choi, J.; Mohanty, J.; Sousa, L.P.; Tome, F.; Pardon, E.; Steyaert, J.; Lemmon, M.A.; Lax, I.; Schlessinger, J. Structures of beta-klotho reveal a 'zip code'-like mechanism for endocrine FGF signalling. *Nature* **2018**, *553*, 501–505. [CrossRef] [PubMed]

27. Mohammadi, M.; Olsen, S.K.; Ibrahimi, O.A. Structural basis for fibroblast growth factor receptor activation. *Cytokine Growth Factor Rev.* **2005**, *16*, 107–137. [CrossRef]

28. Kalinina, J.; Dutta, K.; Ilghari, D.; Beenken, A.; Goetz, R.; Eliseenkova, A.V.; Cowburn, D.; Mohammadi, M. The alternatively spliced acid box region plays a key role in FGF receptor autoinhibition. *Structure* **2012**, *20*, 77–88. [CrossRef]

29.   Olsen, S.K.; Ibrahimi, O.A.; Raucci, A.; Zhang, F.; Eliseenkova, A.V.; Yayon, A.; Basilico, C.; Linhardt, R.J.; Schlessinger, J.; Mohammadi, M. Insights into the molecular basis for fibroblast growth factor receptor autoinhibition and ligand-binding promiscuity. *Proc. Natl. Acad. Sci. USA* **2004**, *101*, 935–940. [CrossRef]

30.   Wang, F.; Kan, M.; Yan, G.; Xu, J.; McKeehan, W.L. Alternately spliced NH2-terminal immunoglobulin-like Loop I in the ectodomain of the fibroblast growth factor (FGF) receptor 1 lowers affinity for both heparin and FGF-1. *J. Biol. Chem.* **1995**, *270*, 10231–10235. [CrossRef]

31.   Mohammadi, M.; Schlessinger, J.; Hubbard, S.R. Structure of the FGF receptor tyrosine kinase domain reveals a novel autoinhibitory mechanism. *Cell* **1996**, *86*, 577–587. [CrossRef]

32.   Wu, D.; Guo, M.; Min, X.; Dai, S.; Li, M.; Tan, S.; Li, G.; Chen, X.; Ma, Y.; Li, J.; et al. LY2874455 potently inhibits FGFR gatekeeper mutants and overcomes mutation-based resistance. *Chem. Commun. (Camb.)* **2018**, *54*, 12089–12092. [CrossRef] [PubMed]

33.   Wu, D.; Guo, M.; Philips, M.A.; Qu, L.; Jiang, L.; Li, J.; Chen, X.; Chen, Z.; Chen, L.; Chen, Y. Crystal Structure of the FGFR4/LY2874455 Complex Reveals Insights into the Pan-FGFR Selectivity of LY2874455. *PLoS ONE* **2016**, *11*, e0162491. [CrossRef] [PubMed]

34.   Tucker, J.A.; Klein, T.; Breed, J.; Breeze, A.L.; Overman, R.; Phillips, C.; Norman, R.A. Structural insights into FGFR kinase isoform selectivity: Diverse binding modes of AZD4547 and ponatinib in complex with FGFR1 and FGFR4. *Structure* **2014**, *22*, 1764–1774. [CrossRef] [PubMed]

35.   Ni, F.; Kung, A.; Duan, Y.; Shah, V.; Amador, C.D.; Guo, M.; Fan, X.; Chen, L.; Chen, Y.; McKenna, C.E.; et al. Remarkably Stereospecific Utilization of ATP alpha,beta-Halomethylene Analogues by Protein Kinases. *J. Am. Chem. Soc.* **2017**, *139*, 7701–7704. [CrossRef] [PubMed]

36.   Knighton, D.R.; Zheng, J.H.; Ten Eyck, L.F.; Ashford, V.A.; Xuong, N.H.; Taylor, S.S.; Sowadski, J.M. Crystal structure of the catalytic subunit of cyclic adenosine monophosphate-dependent protein kinase. *Science* **1991**, *253*, 407–414. [CrossRef] [PubMed]

37.   Yang, Y.; Ye, Q.; Jia, Z.; Cote, G.P. Characterization of the Catalytic and Nucleotide Binding Properties of the alpha-Kinase Domain of Dictyostelium Myosin-II Heavy Chain Kinase A. *J. Biol. Chem.* **2015**, *290*, 23935–23946. [CrossRef]

38.   Furdui, C.M.; Lew, E.D.; Schlessinger, J.; Anderson, K.S. Autophosphorylation of FGFR1 kinase is mediated by a sequential and precisely ordered reaction. *Mol. Cell* **2006**, *21*, 711–717. [CrossRef]

39.   Kornev, A.P.; Taylor, S.S.; Ten Eyck, L.F. A helix scaffold for the assembly of active protein kinases. *Proc. Natl. Acad. Sci. USA* **2008**, *105*, 14377–14382. [CrossRef]

40.   Duan, Y.; Chen, L.; Chen, Y.; Fan, X.G. c-Src binds to the cancer drug Ruxolitinib with an active conformation. *PLoS ONE* **2014**, *9*, e106225. [CrossRef]

41.   Hu, J.; Ahuja, L.G.; Meharena, H.S.; Kannan, N.; Kornev, A.P.; Taylor, S.S.; Shaw, A.S. Kinase regulation by hydrophobic spine assembly in cancer. *Mol. Cell. Biol.* **2015**, *35*, 264–276. [CrossRef] [PubMed]

42.   Hari, S.B.; Merritt, E.A.; Maly, D.J. Sequence determinants of a specific inactive protein kinase conformation. *Chem. Biol.* **2013**, *20*, 806–815. [CrossRef] [PubMed]

43.   Vijayan, R.S.; He, P.; Modi, V.; Duong-Ly, K.C.; Ma, H.; Peterson, J.R.; Dunbrack, R.L., Jr.; Levy, R.M. Conformational analysis of the DFG-out kinase motif and biochemical profiling of structurally validated type II inhibitors. *J. Med. Chem.* **2015**, *58*, 466–479. [CrossRef] [PubMed]

44.   Klein, T.; Vajpai, N.; Phillips, J.J.; Davies, G.; Holdgate, G.A.; Phillips, C.; Tucker, J.A.; Norman, R.A.; Scott, A.D.; Higazi, D.R.; et al. Structural and dynamic insights into the energetics of activation loop rearrangement in FGFR1 kinase. *Nat. Commun.* **2015**. [CrossRef]

45.   Guimaraes, C.R.; Rai, B.K.; Munchhof, M.J.; Liu, S.; Wang, J.; Bhattacharya, S.K.; Buckbinder, L. Understanding the impact of the P-loop conformation on kinase selectivity. *J. Chem. Inf. Model.* **2011**, *51*, 1199–1204. [CrossRef] [PubMed]

46.   Chen, H.; Ma, J.; Li, W.; Eliseenkova, A.V.; Xu, C.; Neubert, T.A.; Miller, W.T.; Mohammadi, M. A molecular brake in the kinase hinge region regulates the activity of receptor tyrosine kinases. *Mol. Cell* **2007**, *27*, 717–730. [CrossRef] [PubMed]

47.   Roskoski, R., Jr. Src protein-tyrosine kinase structure, mechanism, and small molecule inhibitors. *Pharm. Res.* **2015**, *94*, 9–25. [CrossRef] [PubMed]

48.   Zhao, G.; Li, W.Y.; Chen, D.; Henry, J.R.; Li, H.Y.; Chen, Z.; Zia-Ebrahimi, M.; Bloem, L.; Zhai, Y.; Huss, K.; et al. A novel, selective inhibitor of fibroblast growth factor receptors that shows a potent broad spectrum of antitumor activity in several tumor xenograft models. *Mol. Cancer. Ther.* **2011**, *10*, 2200–2210. [CrossRef]

49. Nakanishi, Y.; Akiyama, N.; Tsukaguchi, T.; Fujii, T.; Sakata, K.; Sase, H.; Isobe, T.; Morikami, K.; Shindoh, H.; Mio, T.; et al. The fibroblast growth factor receptor genetic status as a potential predictor of the sensitivity to CH5183284/Debio 1347, a novel selective FGFR inhibitor. *Mol. Cancer. Ther.* **2014**, *13*, 2547–2558. [CrossRef]

50. Guagnano, V.; Furet, P.; Spanka, C.; Bordas, V.; Le Douget, M.; Stamm, C.; Brueggen, J.; Jensen, M.R.; Schnell, C.; Schmid, H.; et al. Discovery of 3-(2,6-dichloro-3,5-dimethoxy-phenyl)-1-{6-[4 -(4-ethyl-piperazin-1-yl)-phenylamin o]-pyrimidin-4-yl}-1-methyl-urea (NVP-BGJ398), a potent and selective inhibitor of the fibroblast growth factor receptor family of receptor tyrosine kinase. *J. Med. Chem.* **2011**, *54*, 7066–7083. [CrossRef]

51. Karkera, J.D.; Cardona, G.M.; Bell, K.; Gaffney, D.; Portale, J.C.; Santiago-Walker, A.; Moy, C.H.; King, P.; Sharp, M.; Bahleda, R.; et al. Oncogenic Characterization and Pharmacologic Sensitivity of Activating Fibroblast Growth Factor Receptor (FGFR) Genetic Alterations to the Selective FGFR Inhibitor Erdafitinib. *Mol. Cancer. Ther.* **2017**, *16*, 1717–1726. [CrossRef] [PubMed]

52. Collin, M.P.; Lobell, M.; Hubsch, W.; Brohm, D.; Schirok, H.; Jautelat, R.; Lustig, K.; Bomer, U.; Vohringer, V.; Heroult, M.; et al. Discovery of Rogaratinib (BAY 1163877): A pan-FGFR Inhibitor. *Chem. Med. Chem.* **2018**, *13*, 437–445. [CrossRef] [PubMed]

53. Brameld, K.A.; Owens, T.D.; Verner, E.; Venetsanakos, E.; Bradshaw, J.M.; Phan, V.T.; Tam, D.; Leung, K.; Shu, J.; LaStant, J.; et al. Discovery of the Irreversible Covalent FGFR Inhibitor 8-(3-(4-Acryloylpiperazin-1-yl) propyl)-6-(2,6-dichloro-3,5-dimethoxyphenyl)-2-(me thylamino)pyrido[2,3-d]pyrimidin-7(8H)-one (PRN1371) for the Treatment of Solid Tumors. *J. Med. Chem.* **2017**, *60*, 6516–6527. [CrossRef] [PubMed]

54. Kalyukina, M.; Yosaatmadja, Y.; Middleditch, M.J.; Patterson, A.V.; Smaill, J.B.; Squire, C.J. TAS-120 Cancer Target Binding: Defining Reactivity and Revealing the First Fibroblast Growth Factor Receptor 1 (FGFR1) Irreversible Structure. *ChemMedChem* **2019**, *14*, 494–500. [CrossRef] [PubMed]

55. Kim, R.; Sharma, S.; Meyer, T.; Sarker, D.; Macarulla, T.; Sung, M.; Choo, S.P.; Shi, H.; Schmidt-Kittler, O.; Clifford, C.; et al. First-in-human study of BLU-554, a potent, highly-selective FGFR4 inhibitor designed for hepatocellular carcinoma (HCC) with FGFR4 pathway activation. *Eur. J. Cancer* **2016**, *69*, S41. [CrossRef]

56. Joshi, J.J.; Coffey, H.; Corcoran, E.; Tsai, J.; Huang, C.L.; Ichikawa, K.; Prajapati, S.; Hao, M.H.; Bailey, S.; Wu, J.; et al. H3B-6527 Is a Potent and Selective Inhibitor of FGFR4 in FGF19-Driven Hepatocellular Carcinoma. *Cancer Res.* **2017**, *77*, 6999–7013. [CrossRef]

57. Zhou, Z.; Chen, X.; Fu, Y.; Zhang, Y.; Dai, S.; Li, J.; Chen, L.; Xu, G.; Chen, Z.; Chen, Y. Characterization of FGF401 as a reversible covalent inhibitor of fibroblast growth factor receptor 4. *Chem. Commun. (Camb.)* **2019**. [CrossRef]

58. Roskoski, R., Jr. ERK1/2 MAP kinases: Structure, function, and regulation. *Pharm. Res.* **2012**, *66*, 105–143.

59. Dar, A.C.; Shokat, K.M. The evolution of protein kinase inhibitors from antagonists to agonists of cellular signaling. *Annu. Rev. Biochem.* **2011**, *80*, 769–795. [CrossRef]

60. Norman, R.A.; Schott, A.K.; Andrews, D.M.; Breed, J.; Foote, K.M.; Garner, A.P.; Ogg, D.; Orme, J.P.; Pink, J.H.; Roberts, K.; et al. Protein-ligand crystal structures can guide the design of selective inhibitors of the FGFR tyrosine kinase. *J. Med. Chem.* **2012**, *55*, 5003–5012. [CrossRef]

61. Liu, Y.; Gray, N.S. Rational design of inhibitors that bind to inactive kinase conformations. *Nat. Chem. Biol.* **2006**, *2*, 358–364. [CrossRef] [PubMed]

62. Mohammadi, M.; Froum, S.; Hamby, J.M.; Schroeder, M.C.; Panek, R.L.; Lu, G.H.; Eliseenkova, A.V.; Green, D.; Schlessinger, J.; Hubbard, S.R. Crystal structure of an angiogenesis inhibitor bound to the FGF receptor tyrosine kinase domain. *EMBO J.* **1998**, *17*, 5896–5904. [CrossRef] [PubMed]

63. Davis, M.I.; Hunt, J.P.; Herrgard, S.; Ciceri, P.; Wodicka, L.M.; Pallares, G.; Hocker, M.; Treiber, D.K.; Zarrinkar, P.P. Comprehensive analysis of kinase inhibitor selectivity. *Nat. Biotechnol.* **2011**, *29*, 1046–1051. [CrossRef]

64. Baillie, T.A. Targeted Covalent Inhibitors for Drug Design. *Angew. Chem. Int. Ed.* **2016**, *55*, 13408–13421. [CrossRef] [PubMed]

65. Awoonor-Williams, E.; Walsh, A.G.; Rowley, C.N. Modeling covalent-modifier drugs. *Biochim. Biophys. Acta* **2017**, *1865*, 1664–1675. [CrossRef]

66. Liu, Q.; Sabnis, Y.; Zhao, Z.; Zhang, T.; Buhrlage, S.J.; Jones, L.H.; Gray, N.S. Developing irreversible inhibitors of the protein kinase cysteinome. *Chem. Biol.* **2013**, *20*, 146–159. [CrossRef] [PubMed]

67. Serafimova, I.M.; Pufall, M.A.; Krishnan, S.; Duda, K.; Cohen, M.S.; Maglathlin, R.L.; McFarland, J.M.; Miller, R.M.; Frodin, M.; Taunton, J. Reversible targeting of noncatalytic cysteines with chemically tuned electrophiles. *Nat. Chem. Biol.* **2012**, *8*, 471–476. [CrossRef]

68. Zhou, W.; Ercan, D.; Chen, L.; Yun, C.H.; Li, D.; Capelletti, M.; Cortot, A.B.; Chirieac, L.; Iacob, R.E.; Padera, R.; et al. Novel mutant-selective EGFR kinase inhibitors against EGFR T790M. *Nature* **2009**, *462*, 1070–1074. [CrossRef]

69. Zhang, J.; Yang, P.L.; Gray, N.S. Targeting cancer with small molecule kinase inhibitors. *Nat. Rev. Cancer* **2009**, *9*, 28–39. [CrossRef]

70. Powis, G.; Bonjouklian, R.; Berggren, M.M.; Gallegos, A.; Abraham, R.; Ashendel, C.; Zalkow, L.; Matter, W.F.; Dodge, J.; Grindey, G.; et al. Wortmannin, a potent and selective inhibitor of phosphatidylinositol-3-kinase. *Cancer Res.* **1994**, *54*, 2419–2423.

71. Fox, T.; Fitzgibbon, M.J.; Fleming, M.A.; Hsiao, H.M.; Brummel, C.L.; Su, M.S. Kinetic mechanism and ATP-binding site reactivity of p38gamma MAP kinase. *FEBS Lett.* **1999**, *461*, 323–328. [CrossRef]

72. Shannon, D.A.; Weerapana, E. Covalent protein modification: The current landscape of residue-specific electrophiles. *Curr. Opin. Chem. Biol.* **2015**, *24*, 18–26. [CrossRef] [PubMed]

73. Tan, L.; Wang, J.; Tanizaki, J.; Huang, Z.; Aref, A.R.; Rusan, M.; Zhu, S.J.; Zhang, Y.; Ercan, D.; Liao, R.G.; et al. Development of covalent inhibitors that can overcome resistance to first-generation FGFR kinase inhibitors. *Proc. Natl. Acad. Sci. USA* **2014**, *111*, E4869–E4877. [CrossRef] [PubMed]

74. Paul, S.M.; Mytelka, D.S.; Dunwiddie, C.T.; Persinger, C.C.; Munos, B.H.; Lindborg, S.R.; Schacht, A.L. How to improve R&D productivity: The pharmaceutical industry's grand challenge. *Nat. Rev. Drug Discov.* **2010**, *9*, 203–214. [CrossRef] [PubMed]

75. Fedorov, O.; Muller, S.; Knapp, S. The (un)targeted cancer kinome. *Nat. Chem. Biol.* **2010**, *6*, 166–169. [CrossRef] [PubMed]

76. O'Hare, T.; Shakespeare, W.C.; Zhu, X.; Eide, C.A.; Rivera, V.M.; Wang, F.; Adrian, L.T.; Zhou, T.; Huang, W.S.; Xu, Q.; et al. AP24534, a pan-BCR-ABL inhibitor for chronic myeloid leukemia, potently inhibits the T315I mutant and overcomes mutation-based resistance. *Cancer Cell* **2009**, *16*, 401–412. [CrossRef] [PubMed]

77. Trudel, S.; Li, Z.H.; Wei, E.; Wiesmann, M.; Chang, H.; Chen, C.; Reece, D.; Heise, C.; Stewart, A.K. CHIR-258, a novel, multitargeted tyrosine kinase inhibitor for the potential treatment of t(4;14) multiple myeloma. *Blood* **2005**, *105*, 2941–2948. [CrossRef]

78. Bello, E.; Colella, G.; Scarlato, V.; Oliva, P.; Berndt, A.; Valbusa, G.; Serra, S.C.; D'Incalci, M.; Cavalletti, E.; Giavazzi, R.; et al. E-3810 is a potent dual inhibitor of VEGFR and FGFR that exerts antitumor activity in multiple preclinical models. *Cancer Res.* **2011**, *71*, 1396–1405. [CrossRef]

79. Hilberg, F.; Roth, G.J.; Krssak, M.; Kautschitsch, S.; Sommergruber, W.; Tontsch-Grunt, U.; Garin-Chesa, P.; Bader, G.; Zoephel, A.; Quant, J.; et al. BIBF 1120: Triple angiokinase inhibitor with sustained receptor blockade and good antitumor efficacy. *Cancer Res.* **2008**, *68*, 4774–4782. [CrossRef]

80. Nogova, L.; Sequist, L.V.; Perez Garcia, J.M.; Andre, F.; Delord, J.P.; Hidalgo, M.; Schellens, J.H.; Cassier, P.A.; Camidge, D.R.; Schuler, M.; et al. Evaluation of BGJ398, a Fibroblast Growth Factor Receptor 1–3 Kinase Inhibitor, in Patients with Advanced Solid Tumors Harboring Genetic Alterations in Fibroblast Growth Factor Receptors: Results of a Global Phase I, Dose-Escalation and Dose-Expansion Study. *J. Clin. Oncol.* **2017**, *35*, 157–165. [CrossRef]

81. Degirolamo, C.; Sabba, C.; Moschetta, A. Therapeutic potential of the endocrine fibroblast growth factors FGF19, FGF21 and FGF23. *Nat. Rev. Drug Discov.* **2016**, *15*, 51–69. [CrossRef]

82. Chell, V.; Balmanno, K.; Little, A.S.; Wilson, M.; Andrews, S.; Blockley, L.; Hampson, M.; Gavine, P.R.; Cook, S.J. Tumour cell responses to new fibroblast growth factor receptor tyrosine kinase inhibitors and identification of a gatekeeper mutation in FGFR3 as a mechanism of acquired resistance. *Oncogene* **2013**, *32*, 3059–3070. [CrossRef] [PubMed]

83. Zhou, W.; Hur, W.; McDermott, U.; Dutt, A.; Xian, W.; Ficarro, S.B.; Zhang, J.; Sharma, S.V.; Brugge, J.; Meyerson, M.; et al. A structure-guided approach to creating covalent FGFR inhibitors. *Chem. Biol.* **2010**, *17*, 285–295. [CrossRef] [PubMed]

84. Hagel, M.; Miduturu, C.; Sheets, M.; Rubin, N.; Weng, W.; Stransky, N.; Bifulco, N.; Kim, J.L.; Hodous, B.; Brooijmans, N.; et al. First Selective Small Molecule Inhibitor of FGFR4 for the Treatment of Hepatocellular Carcinomas with an Activated FGFR4 Signaling Pathway. *Cancer Discov.* **2015**, *5*, 424–437. [CrossRef] [PubMed]

85. Huang, Z.; Tan, L.; Wang, H.; Liu, Y.; Blais, S.; Deng, J.; Neubert, T.A.; Gray, N.S.; Li, X.; Mohammadi, M. DFG-out mode of inhibition by an irreversible type-1 inhibitor capable of overcoming gate-keeper mutations in FGF receptors. *ACS Chem. Biol.* **2015**, *10*, 299–309. [CrossRef] [PubMed]

86. Fairhurst, R.A.; Knoepfel, T.; Leblanc, C.; Buschmann, N.; Gaul, C.; Blank, J.; Galuba, I.; Trappe, J.; Zou, C.; Voshol, J.; et al. Approaches to selective fibroblast growth factor receptor 4 inhibition through targeting the ATP-pocket middle-hinge region. *MedChemComm* **2017**, *8*, 1604–1613. [CrossRef]

87. Ho, H.K.; Yeo, A.H.; Kang, T.S.; Chua, B.T. Current strategies for inhibiting FGFR activities in clinical applications: Opportunities, challenges and toxicological considerations. *Drug Discov. Today* **2014**, *19*, 51–62. [CrossRef]

88. Katoh, M. FGFR inhibitors: Effects on cancer cells, tumor microenvironment and whole-body homeostasis (Review). *Int. J. Mol. Med.* **2016**, *38*, 3–15. [CrossRef]

89. Lu, X.; Chen, H.; Patterson, A.V.; Smaill, J.B.; Ding, K. Fibroblast Growth Factor Receptor 4 (FGFR4) Selective Inhibitors as Hepatocellular Carcinoma Therapy: Advances and Prospects. *J. Med. Chem.* **2019**, *62*, 2905–2915. [CrossRef]

90. Hierro, C.; Rodon, J.; Tabernero, J. Fibroblast Growth Factor (FGF) Receptor/FGF Inhibitors: Novel Targets and Strategies for Optimization of Response of Solid Tumors. *Semin. Oncol.* **2015**, *42*, 801–819. [CrossRef]

91. Bradshaw, J.M.; McFarland, J.M.; Paavilainen, V.O.; Bisconte, A.; Tam, D.; Phan, V.T.; Romanov, S.; Finkle, D.; Shu, J.; Patel, V.; et al. Prolonged and tunable residence time using reversible covalent kinase inhibitors. *Nat. Chem. Biol.* **2015**, *11*, 525–531. [CrossRef] [PubMed]

92. Knoepfel, T.; Furet, P.; Mah, R.; Buschmann, N.; Leblanc, C.; Ripoche, S.; Graus-Porta, D.; Wartmann, M.; Galuba, I.; Fairhurst, R.A. 2-Formylpyridyl Ureas as Highly Selective Reversible-Covalent Inhibitors of Fibroblast Growth Factor Receptor 4. *ACS Med. Chem. Lett.* **2018**, *9*, 215–220. [CrossRef] [PubMed]

93. Shabani, M.; Hojjat-Farsangi, M. Targeting Receptor Tyrosine Kinases Using Monoclonal Antibodies: The Most Specific Tools for Targeted-Based Cancer Therapy. *Curr. Drug Targets* **2016**, *17*, 1687–1703. [CrossRef] [PubMed]

94. Pierce, K.L.; Deshpande, A.M.; Stohr, B.A.; Gemo, A.T.; Patil, N.S.; Brennan, T.J.; Bellovin, D.I.; Palencia, S.; Giese, T.; Huang, C.; et al. FPA144, a humanized monoclonal antibody for both FGFR2-amplified and nonamplified, FGFR2b-overexpressing gastric cancer patients. *J. Clin. Oncol.* **2014**. [CrossRef]

95. Sommer, A.; Kopitz, C.; Schatz, C.A.; Nising, C.F.; Mahlert, C.; Lerchen, H.G.; Stelte-Ludwig, B.; Hammer, S.; Greven, S.; Schuhmacher, J.; et al. Preclinical Efficacy of the Auristatin-Based Antibody-Drug Conjugate BAY 1187982 for the Treatment of FGFR2-Positive Solid Tumors. *Cancer Res.* **2016**, *76*, 6331–6339. [CrossRef] [PubMed]

96. Schatz, C.A.; Kopitz, C.; Wittemer-Rump, S.; Sommer, A.; Lindbom, L.; Osada, M.; Yamanouchi, H.; Huynh, H.; Krahn, T.; Asadullah, K. Abstract 4766: Pharmacodynamic and stratification biomarker for the anti-FGFR2 antibody (BAY1179470) and the FGFR2-ADC. *Cancer Res.* **2014**. [CrossRef]

97. Trudel, S.; Bergsagel, P.L.; Singhal, S.; Niesvizky, R.; Comenzo, R.L.; Bensinger, W.I.; Lebovic, D.; Choi, Y.; Lu, D.; French, D.; et al. A Phase I Study of the Safety and Pharmacokinetics of Escalating Doses of MFGR1877S, a Fibroblast Growth Factor Receptor 3 (FGFR3) Antibody, in Patients with Relapsed or Refractory t(4;14)-Positive Multiple Myeloma. *Blood* **2012**, *120*, 4029.

98. Blackwell, C.; Sherk, C.; Fricko, M.; Ganji, G.; Barnette, M.; Hoang, B.; Tunstead, J.; Skedzielewski, T.; Alsaid, H.; Jucker, B.M.; et al. Inhibition of FGF/FGFR autocrine signaling in mesothelioma with the FGF ligand trap, FP-1039/GSK3052230. *Oncotarget* **2016**, *7*, 39861–39871. [CrossRef]

99. Bono, F.; De Smet, F.; Herbert, C.; De Bock, K.; Georgiadou, M.; Fons, P.; Tjwa, M.; Alcouffe, C.; Ny, A.; Bianciotto, M.; et al. Inhibition of tumor angiogenesis and growth by a small-molecule multi-FGF receptor blocker with allosteric properties. *Cancer Cell* **2013**, *23*, 477–488. [CrossRef]

100. Herbert, C.; Schieborr, U.; Saxena, K.; Juraszek, J.; De Smet, F.; Alcouffe, C.; Bianciotto, M.; Saladino, G.; Sibrac, D.; Kudlinzki, D.; et al. Molecular mechanism of SSR128129E, an extracellularly acting, small-molecule, allosteric inhibitor of FGF receptor signaling. *Cancer Cell* **2013**, *23*, 489–501. [CrossRef]

101. Babina, I.S.; Turner, N.C. Advances and challenges in targeting FGFR signalling in cancer. *Nat. Rev. Cancer* **2017**, *17*, 318–332. [CrossRef] [PubMed]

102. Cheetham, G.M.; Charlton, P.A.; Golec, J.M.; Pollard, J.R. Structural basis for potent inhibition of the Aurora kinases and a T315I multi-drug resistant mutant form of Abl kinase by VX-680. *Cancer Lett.* **2007**, *251*, 323–329. [CrossRef] [PubMed]

103. Weisberg, E.; Choi, H.G.; Ray, A.; Barrett, R.; Zhang, J.; Sim, T.; Zhou, W.; Seeliger, M.; Cameron, M.; Azam, M.; et al. Discovery of a small-molecule type II inhibitor of wild-type and gatekeeper mutants of BCR-ABL,

PDGFRalpha, Kit, and Src kinases: Novel type II inhibitor of gatekeeper mutants. *Blood* **2010**, *115*, 4206–4216. [CrossRef] [PubMed]

104. Ryan, M.R.; Sohl, C.D.; Luo, B.; Anderson, K.S. The FGFR1 V561M Gatekeeper Mutation Drives AZD4547 Resistance through STAT3 Activation and EMT. *Mol. Cancer Res.* **2019**, *17*, 532–543. [CrossRef] [PubMed]

105. Byron, S.A.; Chen, H.; Wortmann, A.; Loch, D.; Gartside, M.G.; Dehkhoda, F.; Blais, S.P.; Neubert, T.A.; Mohammadi, M.; Pollock, P.M. The N550K/H mutations in FGFR2 confer differential resistance to PD173074, dovitinib, and ponatinib ATP-competitive inhibitors. *Neoplasia* **2013**, *15*, 975–988. [CrossRef] [PubMed]

106. Yoza, K.; Himeno, R.; Amano, S.; Kobashigawa, Y.; Amemiya, S.; Fukuda, N.; Kumeta, H.; Morioka, H.; Inagaki, F. Biophysical characterization of drug-resistant mutants of fibroblast growth factor receptor 1. *Genes Cells* **2016**, *21*, 1049–1058. [CrossRef]

107. Goyal, L.; Saha, S.K.; Liu, L.Y.; Siravegna, G.; Leshchiner, I.; Ahronian, L.G.; Lennerz, J.K.; Vu, P.; Deshpande, V.; Kambadakone, A.; et al. Polyclonal Secondary FGFR2 Mutations Drive Acquired Resistance to FGFR Inhibition in Patients with FGFR2 Fusion-Positive Cholangiocarcinoma. *Cancer Discov.* **2017**, *7*, 252–263. [CrossRef]

108. Ang, D.; Ballard, M.; Beadling, C.; Warrick, A.; Schilling, A.; O'Gara, R.; Pukay, M.; Neff, T.L.; West, R.B.; Corless, C.L.; et al. Novel mutations in neuroendocrine carcinoma of the breast: Possible therapeutic targets. *Appl. Immunohistochem. Mol. Morphol.* **2015**, *23*, 97–103. [CrossRef]

7

Targeting the Oncogenic FGF-FGFR Axis in Gastric Carcinogenesis

Jinglin Zhang [1,2,3], Patrick M. K. Tang [1], Yuhang Zhou [1,2,3], Alfred S. L. Cheng [4], Jun Yu [2,5], Wei Kang [1,2,3,*] and Ka Fai To [1,2,3,*]

[1] Department of Anatomical and Cellular Pathology, State Key Laboratory of Translational Oncology, Prince of Wales Hospital, The Chinese University of Hong Kong, Hong Kong, China; jinglinzhang@cuhk.edu.hk (J.Z.); patrick.tang@cuhk.edu.hk (P.M.K.T.); zyhjoe@gmail.com (Y.Z.)

[2] Institute of Digestive Disease, State Key Laboratory of Digestive Disease, The Chinese University of Hong Kong, Hong Kong, China; junyu@cuhk.edu.hk

[3] Li Ka Shing Institute of Health Science, Sir Y.K. Pao Cancer Center, The Chinese University of Hong Kong, Hong Kong, China

[4] School of Biomedical Sciences, The Chinese University of Hong Kong, Hong Kong, China; alfredcheng@cuhk.edu.hk

[5] Department of Medicine and Therapeutics, The Chinese University of Hong Kong, Hong Kong, China

[*] Correspondence: weikang@cuhk.edu.hk (W.K.); kfto@cuhk.edu.hk (K.F.T.)

Abstract: Gastric cancer (GC) is one of the most wide-spread malignancies in the world. The oncogenic role of signaling of fibroblast growing factors (FGFs) and their receptors (FGFRs) in gastric tumorigenesis has been gradually elucidated by recent studies. The expression pattern and clinical correlations of FGF and FGFR family members have been comprehensively delineated. Among them, FGF18 and FGFR2 demonstrate the most prominent driving role in gastric tumorigenesis with gene amplification or somatic mutations and serve as prognostic biomarkers. FGF-FGFR promotes tumor progression by crosstalking with multiple oncogenic pathways and this provides a rational therapeutic strategy by co-targeting the crosstalks to achieve synergistic effects. In this review, we comprehensively summarize the pathogenic mechanisms of FGF-FGFR signaling in gastric adenocarcinoma together with the current targeted strategies in aberrant FGF-FGFR activated GC cases.

Keywords: FGF; FGFR; gastric cancer; monoclonal antibody; small molecule

1. Introduction

Gastric cancer (GC), the third leading cause of cancer death globally, is considered a heterogeneous disease. Although the prevalence has declined over the past decades, more than half of newly diagnosed cases are found to possess local advancement or metastasis [1,2]. Late diagnosis and lack of effective therapeutics still make GC a challenge globally. For decades, researchers have been dedicated to uncover the mysteries behind GC, not only the medication strategies to alleviate or cure the disease, but the key factors for detecting the challenging disease at its early stage. It has been proven that environmental, etiological, and genetic factors largely contribute to GC development, for example, high salt diets, *H. pylori* infections [3], and *CDH1* mutations [4,5]. Systematically, in-depth and comprehensive mechanistic studies revealed the crosstalk of oncogenic signaling pathways during GC progression as well as pre-cancerous gastric lesion development [6–9]. Of note, inactivation of the Hippo pathway has been substantially demonstrated in the pathogenesis of GC, via the accumulation of nuclear YAP1 in an uncontrollable manner [10–12]. Moreover, recent studies have further uncovered the emerging roles of fibroblast growing factors (FGFs) and their receptors (FGFRs) in the carcinogenesis of some GC subtypes, owing to their molecular characteristics [13]. It has been well documented that the FGF and FGFR families are important regulators for biological development [14,15]. Aberration of FGF-FGFR signaling substantially results in skeletal disorders as well as cancer development, including GC [16]. Since genetic aberrations of FGFR2 have been recently defined, it serves as a diagnostic marker and clinical drug target for GC [17–19]. However, development of FGFR2-targeted therapy has been largely decelerated due to recently reported disadvantages. Thus, further investigation of the FGF-FGFR must be continued in order to identify drug targets for GC therapy. This review aims to summarize the updated discoveries and discuss the further prospects of FGF-FGFR signaling in GC pathogenesis and therapy development.

2. Emerging Role of FGF-FGFR in Solid Tumors

2.1. FGF Family Induces Tumor Growth

FGFRs belong to the receptor tyrosine kinases (RTKs) superfamily. Most of the RTKs are membrane receptors with high affinity to multiple growth factors, cytokines, and hormones, and they contain intracellular domains with tyrosine kinase activity. Canonically, FGFRs are monomers in their inactivation state. Dimerization of the intracellular part occurs after binding with their ligand FGFs. Functional binding of FGF and FGFR leads to cross-phosphorylation and activation of the receptor. Activated FGFRs then transduce biochemical signals into cytosolic activities [20]. Indeed, the FGF family comprises 22 secreted factors that are generally divided into seven subgroups in terms of their phylogenetic relation, homology, and biochemical function [21]. As reported, five FGF subfamilies are released in paracrine and autocrine manners, including FGF1 (FGF1, FGF2), FGF4 (FGF4, FGF5, FGF6), FGF7 (FGF3, FGF7, FGF10, FGF22), FGF8 (FGF8, FGF17, FGF18), and FGF9 (FGF9, FGF16, FGF20). In contrast, the FGF15 (FGF15, FGF19, FGF21, FGF23) subfamily is secreted through endocrine glands as a hormone for metabolic modulation with α- and β-Klotho family proteins. Nevertheless, there are intracellular FGFs (FGF11, FGF12, FGF13, FGF14) that lack secretory N-terminal peptides, which execute their functions independent of FGFRs [22].

FGFs not only show regulatory roles in cell fate and survival, but also exerts biological functions in tissue regeneration and repair [23,24]. In the last few decades, clinical reports have highlighted the importance of FGFs in tumorigenesis, including excessive cell growth and angiogenesis. For example, basic fibroblast growth factor (bFGF) promotes angiogenesis for hepatoma progression [25], and a follow-up study suggested serum bFGF as a biological indicator for invasive and recurrent hepatocellular carcinoma (HCC) [26]. The clinical significance of bFGF was first recognized in patients who received surgical removal of colorectal cancer (CRC) at serological and pathological levels, where expression of bFGF indicated the independency in lymphatic invasion [27]. In addition, *FGF* amplification rated 10% in human malignancies, as overproduction of FGFs enables the communication between epithelial cells and stromal cells in the tumor microenvironment for tumorigenesis [28,29].

2.2. FGFR Family Drives Oncogenesis

2.2.1. Functional Structures of FGFR

Interestingly, FGF ligands interact with only four FGFRs (FGFR1–4), which are highly conserved in mammals, although FGFs harbor many family members. In general, FGFRs can be classified into three major domains based on their location relative to the cell membrane: (1) a ternary extracellular immunoglobulin (Ig) (domain I, II, III) that is in charge of binding with ligands; (2) a signal-pass transmembrane helix that acts as a connection; and (3) an intracellular tyrosine kinase (TK) that conveys the signals [30,31]. Generally speaking, the extracellular part of the FGFR provides binding sites for ligand binding, while the intracellular part is responsible for potentiating the relevant signaling pathways. Between the extracellular domains I and II, there is an acidic box region for the FGFR to interact with some molecules other than FGFs, while domains II and III possess the heparin binding site and FGF binding site [32]. The Ig domain III in FGFR1-3 has alternative splicing sites. The domain IIIa remains invariant while the other half varies according to the encoded exon IIIb or IIIc, which are based on tissue-dependent expression [33,34]. This means that the FGFRs only differ between certain parts of the Ig that governs the affinity and specificity of their ligands. There is a single-pass transmembrane domain connecting the Ig domains and the intracellular FGFR domains. The intracellular part of the FGFR includes a juxtamembrane domain for phosphotyrosine binding of adaptors, and two tyrosine kinase (TK) domains. As soon as the TK domains are phosphorylated, the downstream cascades are activated to further expand the signal [35]. Special FGFRs devoid of TK activity, namely FGFR5 or FGFRL1, have been identified and proposed as decoys, interfering with downstream signaling pathways [36]. Due to the diversity of receptor structure and transcript sequence, there are a number of FGFR variants that have been identified. For example, the FGFR2 IIIb isoform has high binding affinity to FGF3, FGF7, and FGF10, while its IIIc form is much preferable to FGF2, FGF4, and FGF20 [20,29]. Further investigation may lead to the discovery of a potential FGFR variant for GC management.

2.2.2. Mechanisms of FGFR in Driving Cancer

Recently, the oncogenic roles of FGFRs have been extensively demonstrated, and somatic alterations and differential expression patterns of FGFR have been seen in different human cancers. Helsten et al. recently depicted a landscape of FGFR aberrations from a large-cohort high-throughput sequencing of cancer patients. In total, FGFR aberrations were detected in 7.1% of the malignancies, including gene amplification (66%), mutations (26%), and rearrangements (8%), suggesting the occurrence of FGFR aberration in most cancer types [37]. Mechanistically, FGFR disorder drives oncogenesis mainly via the following mechanisms: (1) *FGFR* gene amplification: It makes up the majority of the genetic alterations and results in abundant membrane FGFRs, which further augment the activation of its downstream signaling. Gene amplification is common in FGFR1, followed by FGFR2, but rare in FGFR3 and FGFR4. (2) Activating mutations: Most of the mutations exist in the extracellular receptor domains and cause constitutive activation of FGFRs automatically, without the participation of ligands. Activating mutations are frequently found in FGFR2 and FGFR3. (3) FGFRs fusion protein via chromosomal translocation: In this mechanism, the final exon at the C-terminus of the FGFR is replaced by another gene, which results in increasing dimerization and constitutive kinase activity, while ligands are also not required in this manner. (4) Hyperactivation of FGFRs under FGF overproduction from cancer and stromal cells: Additionally, the alternative splicing reconstitutes FGFRs from IIIb to IIIc isoforms, the binding specificity and affinity between FGF and FGFR is altered accordingly. (5) Apart from the genetic alterations of FGF and FGFR, more and more evidence supports that the differential expression of their downstream partners also evidently contributes to the oncogenic progression in multiple cancers [35].

2.3. Partner Proteins Mediate FGF-FGFR Signal Transduction

Signal transduction of FGF-FGFR cannot proceed without the participation of partner proteins. Cell adhesion molecules (CAMs), other types of RTKs, and G-Protein-Coupled Receptors (GPCRs) have been found to interact with FGFR family members and regulate a broad range of cell behaviors [38]. Intrinsically, FGFs can be anchored to the extracellular matrix by heparan sulfate proteoglycans (HPSGs) and thus avoid degradation by proteases. FGFs then bind to certain cell-surface FGFRs to form a ternary complex FGF-FGFR-HPSG [39,40]. Otherwise, a deficiency of HPSG results in the enhanced FGF ligand diffusion and failure of the FGF-FGFR signal transduction, which imposes a restriction on cell polarity and motility [41]. As the complex is formed, intracellular tyrosine kinases of FGFR dimerize and cross-phosphorylate on their tyrosine residues at the activation chain. The main intracellular substrates of FGFR are known as phospholipase C (PLCγ) (FRS1), FGFR substrate 2 (FRS2α), and FGFR substrate 3 (FRS2β) [42,43]. These proteins function as adaptors and are directly phosphorylated by the activating FGFRs [44,45]. FRS2 is a lipid-anchored protein and is located on the juxtamembrane domain to recruit signaling components toward the receptor in response to stimulation by ligands [46]. The functional domain of the FRS2 recruits growth factor receptor-bound 2 (GRB2) by four main phosphorylation sites (Tyr196, Tyr306, Tyr349, Tyr392) [47]. GRB2 then enrolls either the guanine nucleotide exchange factor son of sevenless (SOS) or the GRB2-associated binding protein 1 (GAB1) [42]. These proteins form a scaffold for initiating downstream signaling and compose a significant part for signal transduction of the FGF-FGFR signaling. It is noted that some negative regulators exist on the cell surface to counteract the effect of FGFR. One such family is called similar expression to FGF (SEF), members of this family interact with the intracellular domain of FGFRs and inhibit downstream responses. In tumors, the expression of SEF is significantly decreased [48,49].

2.4. Signaling Pathways Respond to FGF-FGFR Activation

Upon the recruitment and activation of the FGF-FGFR complex, extracellular signals are turned into intracellular events. Cytosolic signaling pathways aroused by the FGF-FGFR complex are recognized as downstream of FGF-FGFR. It has been well-defined that the Ras-dependent mitogen-activated protein kinase (RAS-MAPK), Ras-independent phosphoinositide 3-kinase (PI3K-Akt), PLCγ-Ca^{2+}-PKC, and Janus kinase-signal transducers and activators of transcription (JAK-STAT) act as canonical downstream signaling pathways of FGF-FGFR [50–53]. On one hand, phosphorylation of FRS2 and GRB2 further initiates the RAS-MAPK and PI3K-AKT signaling pathways by recruiting SOS and GAB1 to the protein complex, respectively. RAS phosphorylates a series of MAPKs such as extracellular signal-regulated kinase 1 (ERK1) and ERK2, which potentiate E26 transformation-specific (ETS) transcription factors to interact and regulate their target genes related to cell proliferation, survival, and transformation [50,54]. As a feedback, inhibitory factors can also be induced by FGF signals. Sprouty (SPRY) interrupts the activation of GRB2, and MAPK phosphatase 3 (MKP3) dephosphorylates ERK1/2 [15]. The PI3K-AKT signaling pathway works differently. After FGF stimulation, GRB2 phosphorylates PI3K-AKT and then inhibits nuclear localization of a pro-apoptotic effector, promoting expression of genes associating with cell survival [55]. In contrast, inhibiting FGFR impairs the function of this pathway and leads to retardation of tumor growth and metastasis [51]. On the other hand, phosphorylation of PLCγ by the FGFR kinase domain hydrolyzes phosphatidylinositol 4,5-bisphosphate to produce inositol triphosphate (IP3) and diacylglycerol (DAG), which support intracellular calcium release and activate protein kinase C (PKC), respectively [56]. Moreover, it has been suggested that amplification of FGFR is required for the signal transducers and activators of transcription-3 (STAT3) activation in cancers. The interaction of FGFR and STAT3 depends on the involvement of JAK [57]. It should be noted that FGF-FGFR signaling cascades also cooperate with other signaling pathways, including Notch [58], Wnt [59], Hedgehog [60], and BMP signaling [61]. Fine-tuning of the cascades ensures homeostasis among normal cells, but their dysfunction may induce multiple diseases and even cancers.

3. Deregulation of the FGF-FGFR Signaling in Gastric Carcinogenesis

3.1. Significance of FGFR2 in Gastric Tissues

FGF-FGFR signaling exerts multiple biological functions and effects. FGFR2 isoforms IIIb and IIIc are predominantly expressed in the epithelial and mesenchymal tissues [62–64]. Along with the understanding of FGFR2, their FGF ligands have been gradually identified. Structurally, FGFR2-IIIb bonds to FGF1, FGF3, FGF7, FGF10, and FGF22 in epithelial tissues; while FGFR2-IIIc responds to a number of FGFs (i.e., FGF1, FGF2, FGF4, FGF5, FGF6, FGF8, FGF9, FGF16, FGF17, FGF18, FGF19, and FGF20) in mesenchymal cells [65,66]. Interestingly, different FGFs will result in various downstream effects via FGFR activation. In gastric tissue, FGFR2 is involved in early epithelial growth before differentiation, and FGF10 and FGFR2-IIIb promote proliferation and patterning of the forestomach. In contrast, silence of both FGF10 and FGFR2 severely induces abnormal lining of gastric epithelium [67].

3.2. Aberrant FGF-FGFRs Advance Gastric Tumorigenesis

FGFR2 not only has physiological roles in normal gastric tissue, but also contributes to the development and progression of GC. Indeed, *FGFR2* amplification was detected in GC cells three decades ago [68,69]. The understanding of FGFR2 is extensive, especially in terms of its abnormal genetic alterations that are rare in other FGFR members. *FGFR* amplification is the main genetic alteration in GC, accounting for up to 9% in western populations and 1.2–4.9% in Asian cohorts [13,70]. Nevertheless, mutation and fusion genes are rare in GC patients. From tissue-based studies, incidence of *FGFR2* amplification is equivalent to that of *ERBB2* and *KRAS*, ranging from 2% to 9% according to different methodologies and geographies. Clinical data also manifest that the frequency of *FGFR2* amplification basically contributes to diffuse-type GC [71–74]. In addition, amplification of *FGFR2* in GC is mutually exclusive with *HER2* and *KRAS* amplification by FISH assay [18,75], suggesting they are independent prognostic biomarkers. Gene amplification is a common cause for mRNA overexpression. In fact, a recent in situ analysis showed that FGFR2 mRNA is highly correlated with *FGFR2* amplification in primary cases clinically, where a high expression level of FGFR2 is associated with poor survival rate of GC patients [76]. Recently, FGFR2 overexpression has been detected in a great portion of GC cases by immunohistochemistry staining, the high level FGFR2-IIIb isoform predicts poor overall survival in patients [19]. A retrospective study revealed that FGFR2 expression was negatively associated with relapse-free survival in a Japanese diffuse-type GC cohort. In that study, although association between FGFR2 expression and survival outcomes in patients with stage II/III GC after surgery and S-1 chemotherapy was insignificant, patients with recurrence after five years of treatment made up a relatively large proportion of the high FGFR2 levels, implying the FGFR2 overexpression may be relevant to GC development [77]. FGFR2 may also contribute to drug resistance of GC. A GC model with *FGFR2* amplification was sensitive to a FGFR inhibitor AZD4547. However, another study questions the efficacy and safety of AZD4547 in GC patients since their progression-free survival rate did not significantly improve with AZD4547 monotherapy compared with paclitaxel, which may due to the intratumor heterogeneity of the *FGFR2* copy-number aberration [78]. Based on these studies, aberrant FGFR2 is largely involved in gastric tumorigenesis and is a candidate to be a diagnostic marker and has the potential to be a therapeutic target for GC treatment. However, challenges will exist until the complexity of the FGFR2 signaling network is resolved.

Autocrine and paracrine FGFs constitute an important functional role in the FGFR2 signaling cascade. In the last two decades, FGF ligands have been reported in multiple cancers, but only a few FGFs were investigated in GC. For example, gastric fibroblast-derived FGF7 increases scirrhous GC cell proliferation in a paracrine manner. Although intrinsic levels of FGF7 are low in GC cells, its corresponding receptor FGFR2 is highly expressed. Subsequently, FGF7 was reported to interact with FGFR2 to promote cell migration and invasion in GC [79,80]. On the other hand, a study found that FGF9 triggers proliferation and inhibits apoptosis of GC cells in an autocrine manner in a Chinese GC cohort [81]. At the genetic level, amplification of *FGF* genes may lead to their overproduction in GC, specifically, *FGF10* amplification has been reported in 3% of GC and in 5.7% of gastric adenocarcinomas [82,83]. FGF10 is correlated to GC cell invasion and has been suggested as a prognostic biomarker and potential drug target in gastric adenocarcinoma [84]. In our recent study, we explored the FGF mRNA profiling in 10 GC cell lines by microarray analysis, where FGF18 showed the highest expression among all the FGF members. This study also identified clinical correlation of FGF18 and highlighted FGF18 as a potent diagnostic indicator in GC. Upon FGF18 stimulation, cell growth is facilitated by activation of SMAD2/3 and suppression of ATM signaling [85]. Nevertheless, the molecular network of FGF-FGFRs responsible for GC progression remains to be revealed.

3.3. FGFR2 Crosstalk in GC

It is believed that the FGFR2 aberration fundamentally contributes to GC development, but how FGFR2 coordinates with other regulatory signaling remains unclear. Investigations of FGFR2 and other oncogenic signaling have been conducted to decipher the comprehensive network.

The amplification of FGFR2 has been implied to facilitate cell growth in GC through crosstalking with other RTKs. It is reported that activated epidermal growth factor receptor (EGFR), human epidermal growth factor receptor 3 (HER3), and MET correlate with drug hyposensitivity of GC cells with *FGFR2* amplification. Interestingly, a combination of an FGFR2 inhibitor and EGFR neutralizing antibody partially enhanced drug sensitivity of GC in vitro and in vivo, suggesting these RTKs may cause drug resistance in cancer cells under FGFR2 inhibition. Eventually, a novel mechanism was identified whereby RTKs can coexpress with FGFR2 and synergistically promote the growth of GC [86]. In contrast, another study reported that HER2, MET, and FGFR2 are mutually exclusive oncogenic drivers, where a large number of HER2-negative patients were highly sensitive to the MET- and FGFR2-targeted therapies [87]. However, these contradictory conclusions may be attributable to the differences of the GC cohorts and the experimental models applied in the studies. One possible reason is that the former study focused on *FGFR2* amplification cases where patients were hyposensitive to AZD4547, while the later one concerned both gene amplification and overexpression. Nevertheless, these results examined the potential relationship between FGFR2 and other RTKs, though the underlying molecular mechanisms are not fully understood. Combined therapy for targeting both FGFR2 and RTKs may be a new strategy for clinically treating GC.

In addition, several signaling pathways have been highlighted as downstream of FGFR2 that may also be involved in GC development (Figure 1). Lau et al. revealed a survival mechanism for developing acquired-resistance under FGFR inhibition. They established drug resistance on both primary and patient derived xenograft (PDX) models of various GCs with different *FGFR2* amplification levels by applying FGFR2 inhibitors. Interestingly, they observed that MAPK and AKT signaling pathways were dispensable for drug resistance, but the constitutive inhibition of GSK3β, which depends on activation of PKC, was required for cell survival [88]. Therefore, the FGFR2-PKC-GSK3β axis is considered as the main mechanism causing resistance in GC during anti-FGFR2 therapy. Additionally, PI3K-Akt-mTOR signaling contributes to the oncogenic activity of FGFR signaling in GC. Huang et al. recently suggested that FGFR2 signaling promotes GC by regulating the expression of Thrombospondin 1 (THBS1) and THBS4 via the PI3K-Akt-mTOR pathway. They indicated that FGF7-FGFR2 signaling upregulates THBS1, while THBS4 is decreased by the FGFR2-Akt cascade [80,89]. These studies established that PI3K-Akt signaling partially contributes to the tumor-promoting function of FGFR2 in GC, although the contribution of the THBS family to GC is still not fully understood. Therefore, further studies are required to reveal the detailed mechanisms. Moreover, epithelial mesenchymal transition (EMT) is a well-known mechanism that facilitates tumor cell transformation and distant metastasis during oncogenic progression. FGF-FGFR signaling has been shown to potentiate EMT [20]. The basic components of EMT, WNT signaling, and Twist-related protein 1 (Twist1) have been found to upregulate FGFR2 in GC cell lines. In turn, FGFR2 further amplifies Twist1 mediated EMT and cell invasion, implying dual inhibition of these pathways is needed for GC therapy [90]. Of note, under the FGFR2 signaling cascade, nuclear accumulation of β-catenin and EMT transcription factors, such as SNAIL, have also been proposed [91].

Current findings have uncovered the complicated interactions between FGFR2 signaling and other RTKs and oncogenic signaling pathways in GC. These signaling networks trigger primary and secondary resistance of GC cells under treatment and eventually lead to the advanced stage of disease. Fortunately, a better understanding of the FGFR signaling network will gradually help in the development of novel therapeutic options for GC.

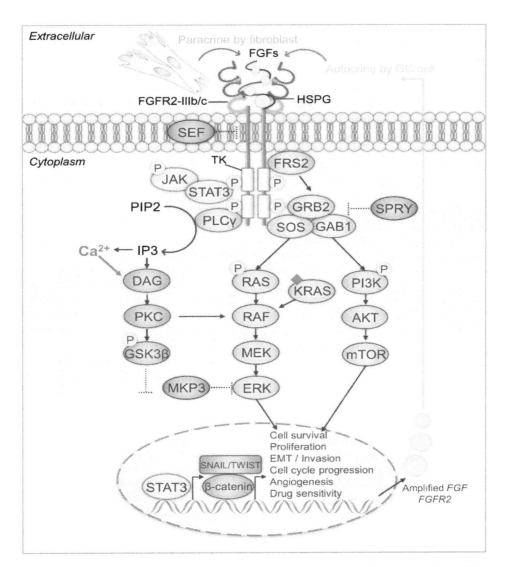

Figure 1. The FGF-FGFR cascade interplays with the downstream signaling network in GC progression. Firstly, FGFR is aberrantly activated in GC cells. FGFs can be released mainly in two ways, in a paracrine manner by gastric fibroblasts, and in an autocrine manner from cancer cells. Gene amplification of *FGFs* and *FGFRs* leads to overproduction of FGFs and FGFRs. FGFs are stabilized and bind to FGFR via HSPG. Alternative splicing of FGFR induces two isoforms that highly are expressed in GC. The isoforms show different affinity to FGFs and contribute to diverse cellular processes. The intracellular region of the FGFR has tyrosine kinase (TK) activity. FGF stimulation leads to dimerization, phosphorylation, and activation of FGFR. The inhibitory effect of SEF is attenuated in GC cells. Secondly, after FGFR activation, adaptor proteins are recruited and also activated by phosphorylation. FRS2 further recruits GRB2, GAB1, and SOS to form a complex. The complex activates RAS-MAPK and PI3K-Akt-mTOR signaling pathways and transduces FGF stimulation into transcriptional regulation to forward tumorigenesis. The inhibitory effects of SPRY and MKP3 are abrogated in GC cells. PLCγ hydrolyzes PIP2 to IP3, increases Ca^{2+} levels, triggers DAG-PKC signaling, and phosphorylates GSK3β. Then, GSK3β is decreased and β-catenin is released to the nuclei. β-catenin and other EMT transcription factors, SNAIL and TWIST, initiate expression of oncogenes that are required for GC progression. Besides, JAK-STAT3 is activated by FGFR and contributes to transcriptional regulation of GC progression. (Arrows represent the activation or release routes; dash dots indicate the weakening of inhibitory effects).

4. Targeting Aberrant FGF-FGFR Activation in GC by Specific Antibodies or Small Molecules

As the FGF-FGFR singling plays an oncogenic role in tumorigenesis by crosstalking with or regulating multiple crucial other pathways, targeting of FGF-FGFR by specifically designed therapeutic agents has shed light on the precision of medicine [92]. These agents include specific anti-FGFR monoclonal antibodies, FGF traps [93], non-selective RTK inhibitors, and selective RTK inhibitors.

4.1. Specific Antibodies and FGF Traps

In the aberrant FGF-FGFR-activation GC cases, anti-FGF (FGF traps) or anti-FGFR monoclonal antibodies might exert anti-cancer effects for the treatment (Table 1). Compared with tyrosine kinase inhibitors, the specific antibodies targeting FGFs or FGFRs have more specificity and less toxicity because they can avoid the off-target effects.

Table 1. A list of anti-FGFR monoclonal antibodies and FGF traps potentially employed in GC.

Monoclonal Antibodies	Targets	References
GAL-FR21 and GAL-FR22	FGFR2	[94]
FPA144 (Bemarituzumab)	FGFR2 amplification or overexpression	[98]
BAY 1179470	FGFR2 amplification or overexpression	[95]
FGF traps		
GSK3052230	FGFs	[93,96]
NSC12	FGFs	[97]

The specific monoclonal antibodies generated and effectively employed for targeting FGFRs in GC research are quite limited [22]. They include GAL-FR21 and GAL-FR22 antibodies. GAL-FR21 binds only the FGFR2IIIb isoform, whereas GAL-FR22 and GAL-FR23 can directly bind to both the FGFR2IIIb and FGFR2IIIc isoforms, with binding regions respectively in the D3, D2-D3, and D1 domains of FGFR2. GAL-FR21 and GAL-FR22 block the binding of FGF2, FGF7, and FGF10 to FGFR2IIIb. GAL-FR21 inhibits FGF2- and FGF7-induced phosphorylation of FGFR2, and both antibodies dramatically down-modulate the activation of FGFR2 in SNU16 cells (with FGFR2 amplification). These monoclonal antibodies also effectively inhibit the tumor growth of established SNU16 and OCUM-2M xenografts in mice [94]. Another FGFR2b-specific antibody, FPA144, can not only treat GC patients with FGFR2 amplification, but also patients with FGFR2b overexpression who lack FGFR2 gene amplification. FPA144 is still being evaluated in a phase III clinical trial of GC. Another novel antibody-drug conjugate (ADC), namely BAY 1179470, provides preclinical efficacy. It consists of a fully human FGFR2 monoclonal antibody, which binds to the FGFR2 isoforms FGFR2-IIIb and FGFR2-IIIc, conjugated through a noncleavable linker to a novel derivative of the microtubule-disrupting cytotoxic drug auristatin (FGFR2-ADC). Functional studies demonstrated that FGFR2-ADC administration leads to a significant tumor growth inhibition or tumor regression of cell line-based or patient-derived xenograft models of human gastric or breast cancer. Similar to FPA144, FGFR2 amplification or mRNA overexpression predicted high response to BAY 1179470 treatment [95].

As some FGF members, such as FGF18, are abundant in gastric carcinogenesis, using FGF ligand traps is another strategy to neutralize FGF and quench malignancies [85]. An FGF "ligand trap" is comprised of a fusion protein of an immunoglobulin Fc fragment and a soluble FGFR extracellular domain that competitively binds with FGF1, 2, 3, 7, and 10 to suppress ligand-dependent FGFR signaling [93]. For example, the FGF traps FP-1039 (GSK3052230) and sFGFR3 are soluble proteins that contain the extracellular regions of FGFR1 and FGFR3, respectively [96], thus they can successfully neutralize the oncogenic role of FGFs. Another good example is NSC12, acting as an extracellular FGF trap. It can be employed in anti-angiogenic and anti-vascular endothelial growth factor therapy as an FGF antagonist [97].

4.2. Small Molecules: Non-Selective and Selective FGFR Inhibitors

Apart from the antibodies or traps, small molecules can also generally and effectively inhibit tyrosine kinase receptor-related signaling (non-selective FGFR inhibitors). SOMCL-085 is a novel

FGFR-dominant multi-target kinase inhibitor. This compound can simultaneously inhibit the angiogenesis kinases such as vascular endothelial growth factor receptor (VEGFR) and platelet-derived growth factor receptor (PDGFR). SOMCL-085 potently inhibits FGFR1, FGFR2, and FGFR3 kinase activity, with IC_{50} values of 1.8, 1.9, and 6.9 nmol/L, respectively [99]. In the FGFR1-amplified lung cancer cell line H1581-xenograft mice and FGFR2-amplified GC cell line SNU16-xenograft mice, oral administration of SOMCL-085 for 21 days substantially inhibited tumor growth without loss of body weight. Nintedanib, a triple-angiokinase inhibitor, is a potent and selective inhibitor for tumor angiogenesis through the blocking of the tyrosine kinase activities of VEGFR1-3, PDGFR-alpha and -beta, together with FGFR1-3 [100]. In combination with docetaxel, nintedanib has been approved for the second-line treatment of adenocarcinoma non-small cell lung cancer (NSCLC). In human GC cell lines driven by an FGFR2 amplification, such as KatoIII, nintedanib is also confirmed to be highly effective. Regorafenib has also demonstrated survival benefits in patients with metastatic colorectal and gastrointestinal stromal tumors. More importantly, FGFR2 amplification was the only genetic alteration associated with in vitro sensitivity to regorafenib. Regorafenib induces G1 phase cell cycle arrest in SNU16 and KATOIII GC cells and suppresses their xenograft formation abilities [101]. S49076 is a novel and potent inhibitor of MET, AXL/MER, and FGFR1/2/3. S49076 potentially blocks cellular phosphorylation of MET, AXL, and FGFRs and inhibits downstream signaling pathways in vitro and in vivo. S49076 alone can cause tumor growth arrest in bevacizumab-resistant cancer cells. Based on the favorable and novel pharmacologic profile of S49076, a phase I study is currently being conducted in patients with advanced solid tumors [102]. Ponatinib (AP24534), an oral multitargeted tyrosine kinase inhibitor, has been explored in a pivotal phase II trial in patients with chronic myelogenous leukemia due to its potent ability against BCR-ABL. It has also been shown to inhibit the in vitro kinase activity of all four FGFRs. In a panel of 14 cell lines representing multiple tumor types (endometrial, bladder, gastric, breast, lung, and colon) and containing FGFRs dysregulated by amplification, overexpression, or mutation, ponatinib inhibited FGFR-mediated signaling with IC_{50} values below 40 nmol/L, supporting it as a potent pan-FGFR inhibitor in patients with FGFR-driven cancers [103].

To avoid the off-target effects of non-selective inhibitors, novel selective FGFR inhibitors were generated and employed for specifically blocking the FGF-FGFR cascade in GC. Among all the selective FGFR inhibitors, AZD4547 is the most famous [17]. It is a selective FGFR1, 2, 3 tyrosine kinase inhibitor with potent preclinical activity in FGFR2-amplified gastric adenocarcinoma SNU16 and SGC083 xenograft animal models, together with the patient-derived cells (PDCs) [104]. The randomized phase II SHINE study (NCT01457846) investigated whether AZD4547 improved clinical outcome versus paclitaxel as a second-line treatment in patients with advanced gastric adenocarcinoma displaying FGFR2 polysomy or gene amplification detected by fluorescence in situ hybridization (FISH). However, the final results indicated AZD4547 failed to significantly improve progression-free survival compared with paclitaxel in GC patients with FGFR2 amplification or polysomy [78]. The related molecular mechanism needs to be further addressed. LY2874455, a potent oral selective pan-FGFR inhibitor, was investigated for its efficacy in a phase I clinical trial. LY2874455 was gradually absorbed and generally showed linear pharmacokinetics. The effective half-life span was approximately 12 h. In 15 GC patients, one patient had a partial response, while 12 patients had stable disease. Thus, LY2874455 has a recommended phase II dosing of 16 mg BID in solid-organ cancer patients [105]. However, in FGFR2-amplified GC patients, some will eventually develop an acquired LY2874455 resistance due to a novel FGFR2-ACSL5 fusion protein that is formed [106]. Based on the structure, medicinal chemistry optimization, and unique ADME assays of a covalent drug discovery program, a novel compound, namely PRN1371, was discovered to serve as a highly selective and potent FGFR1-4 inhibitor [107]. In combination with the de novo synthesis program 'SYNOPSIS' to generate high scoring and synthetically accessible compounds, alofanib (RPT835) was found to be an effective inhibitor of the FGF/FGFR2 pathway. RPT835 potently inhibited growth of KATOIII GC cells with a GI_{50} value of 10 nmol/L [108]. ARQ 087 is a novel, ATP competitive, small molecule, multi-kinase inhibitor with potent in vitro and in vivo activity against FGFR-addicted cell lines and tumors. It exhibited IC_{50}

values of 1.8 nM for FGFR2, and 4.5 nM for FGFR1 and 3. ARQ 087 has anti-proliferative activity in cell lines driven by FGFR dysregulation, including amplifications, fusions, and mutations, such as the SNU16 cell line. It is currently being investigated in a phase I/II clinical trial [109]. BGJ398, a pan-FGFR inhibitor, was also investigated in a GC model. In vitro, FGFR inhibition was most effective in KKLS cells (high FGFR1, FGFR2IIIc, no FGFR2IIIb expression) with inhibition of growth and motility. BGJ398 also showed partial activity in MKN45 cells (intermediate FGFR1, high FGFR2IIIb, low FGFR2IIIc expression), while TMK-1 cells (low FGFR1, no FGFR2IIIb and FGFR2IIIc expression) showed a negative response to this drug [110]. Some of the non-selective and selective FGFR inhibitors that have been investigated in gastric adenocarcinoma are listed in Table 2.

Table 2. The list of non-selective and selective FGFR tyrosine kinase inhibitors reported in GC.

Non-Selective FGFR Inhibitors	Main Targets	References
SOMCL-085	FGFR, VEGFR, and PDGFR	[99]
Nintedanib	FGFR, VEGFR, and PDGFR	[100]
Regorafenib	FGFR2, VEGFR1-3, PDGFR, c-Kit, and RET	[101]
S49076	MET, AXL/MER, and FGFR1-3	[102]
Ponatinib	BCR-ABL, VEGFR2-3, and FGFR1-4	[103]
Selective FGFR inhibitors		
AZD4547	FGFR1, FGFR2 and FGFR3	[17,78]
LY2874455	FGFR1, FGFR2, FGFR3 and FGFR4	[105]
PRN1371	FGFR1, FGFR2, FGFR3 and FGFR4	[107]
RPT835	FGFR2	[108]
ARQ 087	FGFR1, FGFR2 and FGFR3	[109]
BGJ398	FGFR1, FGFR2 and FGFR3	[110]

5. Conclusions and Future Directions

Although we have made great progress in understanding the molecular mechanisms and crosstalk of FGF-FGFR in gastric carcinogenesis, and are even trying to employ small molecules or specific antibodies to block the oncogenic-driven role of FGF-FGFR signaling, several important issues need to be addressed urgently in future studies. First of all, GC can be subgrouped into intestinal and diffuse type from the histological classification, and it can also be stratified as four molecular subtypes according to TCGA molecular classification, Epstein-Barr virus (EBV)-positive tumors, microsatellite instable (MSI) tumors, genomically stable (GS) tumors, and tumors with chromosomal instability (CIN) [13]. Each subtype has its distinct molecular features and the etiology together with pathological processes are quite different among the subtypes. Thus, we need to re-evaluate the genetic and epigenetic changes and clinical correlations in a large cohort of FGF-FGFR for each subgroup to confirm the impact of different genetic backgrounds on FGF-FGFR activation. For example, in a small size cohort study, high FGFR4 expression correlated with tumor progression and survival in both diffuse and intestinal GC, whereas high expression of FGFR1 and 2 correlated with tumor progression and survival only in diffuse type GC [111]. Secondly, as FGF-FGFR crosstalks with multiple signaling pathways, such as the RAS-MAPK pathway, PI3K-Akt-mTOR pathway [112], and PKC-GSK3β pathway, we need to stratify our primary samples again according to different crosstalks by the immunohistochemistry method combined with FISH analysis. We will re-evaluate the clinical significance and perform co-administration of multiple anti-cancer drugs to achieve synergistic effects. The successful development of highly specific anti-FGFR personalized strategies will rely on our deeper knowledge of the key alterations that drive oncogenesis in GC [113]. Based on the identification of novel key downstream effectors of the FGF-FGFR cascade in gastric carcinogenesis, we aim to effectively and accurately target FGFR-related signaling in this precision medicine era.

Author Contributions: W.K. and K.F.T. provided direction and instruction in preparing this manuscript. J.Z., Y.Z. and W.K. reviewed the literature and drafted this manuscript. P.M.K.T., A.S.L.C. and J.Y. reviewed the manuscript and made significant revisions on the drafts.

Acknowledgments: We acknowledge the support from the Core Utilities of Cancer Genomics and Pathobiology of Department of Anatomical and Cellular Pathology, The Chinese University of Hong Kong.

References

1. Bray, F.; Ferlay, J.; Soerjomataram, I.; Siegel, R.L.; Torre, L.A.; Jemal, A. Global cancer statistics 2018: GLOBOCAN estimates of incidence and mortality worldwide for 36 cancers in 185 countries. *CA Cancer J. Clin.* **2018**, *68*, 394–424. [CrossRef] [PubMed]

2. Correa, P. Human gastric carcinogenesis: A multistep and multifactorial process—First American Cancer Society Award Lecture on Cancer Epidemiology and Prevention. *Cancer Res.* **1992**, *52*, 6735–6740. [PubMed]

3. Uemura, N.; Okamoto, S.; Yamamoto, S.; Matsumura, N.; Yamaguchi, S.; Yamakido, M.; Taniyama, K.; Sasaki, N.; Schlemper, R.J. Helicobacter pylori infection and the development of gastric cancer. *N. Engl. J. Med.* **2001**, *345*, 784–789. [CrossRef] [PubMed]

4. Pharoah, P.D.; Guilford, P.; Caldas, C. International Gastric Cancer Linkage C: Incidence of gastric cancer and breast cancer in CDH1 (E-cadherin) mutation carriers from hereditary diffuse gastric cancer families. *Gastroenterology* **2001**, *121*, 1348–1353. [CrossRef] [PubMed]

5. Hansford, S.; Kaurah, P.; Li-Chang, H.; Woo, M.; Senz, J.; Pinheiro, H.; Schrader, K.A.; Schaeffer, D.F.; Shumansky, K.; Zogopoulos, G.; et al. Hereditary Diffuse Gastric Cancer Syndrome: CDH1 Mutations and Beyond. *JAMA Oncol.* **2015**, *1*, 23–32. [CrossRef] [PubMed]

6. Yanai, K.; Nakamura, M.; Akiyoshi, T.; Nagai, S.; Wada, J.; Koga, K.; Noshiro, H.; Nagai, E.; Tsuneyoshi, M.; Tanaka, M.; et al. Crosstalk of hedgehog and Wnt pathways in gastric cancer. *Cancer Lett.* **2008**, *263*, 145–156. [CrossRef]

7. Corso, S.; Ghiso, E.; Cepero, V.; Sierra, J.R.; Migliore, C.; Bertotti, A.; Trusolino, L.; Comoglio, P.M.; Giordano, S. Activation of HER family members in gastric carcinoma cells mediates resistance to MET inhibition. *Mol. Cancer* **2010**, *9*, 121. [CrossRef]

8. Fu, Y.F.; Gui, R.; Liu, J. HER-2-induced PI3K signaling pathway was involved in the pathogenesis of gastric cancer. *Cancer Gene Ther.* **2015**, *22*, 145–153. [CrossRef]

9. Riquelme, I.; Saavedra, K.; Espinoza, J.A.; Weber, H.; García, P.; Nervi, B.; Garrido, M.; Corvalán, A.H.; Roa, J.C.; Bizama, C. Molecular classification of gastric cancer: Towards a pathway-driven targeted therapy. *Oncotarget* **2015**, *6*, 24750–24779. [CrossRef]

10. Cui, Z.-L.; Han, F.-F.; Peng, X.-H.; Chen, X.; Luan, C.-Y.; Han, R.-C.; Xu, W.-G.; Guo, X.-J. Yes-Associated Protein 1 Promotes Adenocarcinoma Growth and Metastasis through Activation of the Receptor Tyrosine Kinase Axl. *Int. J. Immunopathol. Pharmacol.* **2012**, *25*, 989–1001. [CrossRef]

11. Kang, W.; Cheng, A.S.; Yu, J.; To, K.F. Emerging role of Hippo pathway in gastric and other gastrointestinal cancers. *World J. Gastroenterol.* **2016**, *22*, 1279–1288. [CrossRef] [PubMed]

12. Zhou, Y.; Huang, T.; Zhang, J.; Wong, C.C.; Zhang, B.; Dong, Y.; Wu, F.; Tong, J.H.M.; Wu, W.K.K.; Cheng, A.S.L.; et al. TEAD1/4 exerts oncogenic role and is negatively regulated by miR-4269 in gastric tumorigenesis. *Oncogene* **2017**, *36*, 6518–6530. [CrossRef] [PubMed]

13. The Cancer Genome Atlas Research Network. Comprehensive molecular characterization of gastric adenocarcinoma. *Nature* **2014**, *513*, 202–209. [CrossRef] [PubMed]

14. Friesel, R.; Neilson, K.M. Ligand-independent Activation of Fibroblast Growth Factor Receptors by Point Mutations in the Extracellular, Transmembrane, and Kinase Domains. *J. Boil. Chem.* **1996**, *271*, 25049–25057.

15. Thisse, B.; Thisse, C. Functions and regulations of fibroblast growth factor signaling during embryonic development. *Dev. Biol.* **2005**, *287*, 390–402. [CrossRef] [PubMed]

16. Dieci, M.V.; Arnedos, M.; Andre, F.; Soria, J.C. Fibroblast Growth Factor Receptor Inhibitors as a Cancer Treatment: From a Biologic Rationale to Medical Perspectives. *Cancer Discov.* **2013**, *3*, 264–279. [CrossRef] [PubMed]

17. Xie, L.; Su, X.; Zhang, L.; Yin, X.; Tang, L.; Zhang, X.; Xu, Y.; Gao, Z.; Liu, K.; Zhou, M.; et al. FGFR2 gene amplification in gastric cancer predicts sensitivity to the selective FGFR inhibitor AZD4547. *Clin. Cancer Res.* **2013**, *19*, 2572–2583. [CrossRef]

18. Su, X.; Zhan, P.; Gavine, P.R.; Morgan, S.; Womack, C.; Ni, X.; Shen, D.; Bang, Y.-J.; Im, S.-A.; Kim, W.H.; et al. FGFR2 amplification has prognostic significance in gastric cancer: Results from a large international multicentre study. *Br. J. Cancer* **2014**, *110*, 967–975. [CrossRef]

19. Ahn, S.; Lee, J.; Hong, M.; Kim, S.T.; Park, S.H.; Choi, M.G.; Lee, J.-H.; Sohn, T.S.; Bae, J.M.; Kim, S.; et al. FGFR2 in gastric cancer: Protein overexpression predicts gene amplification and high H-index predicts poor survival. *Mod. Pathol.* **2016**, *29*, 1095–1103. [CrossRef]

20. Katoh, M.; Nakagama, H. FGF receptors: Cancer biology and therapeutics. *Med. Res. Rev.* **2014**, *34*, 280–300. [CrossRef]

21. Itoh, N.; Ornitz, D.M. Fibroblast growth factors: From molecular evolution to roles in development, metabolism and disease. *J. Biochem.* **2011**, *149*, 121–130. [CrossRef]

22. Ghedini, G.C.; Ronca, R.; Presta, M.; Giacomini, A. Future applications of FGF/FGFR inhibitors in cancer. *Expert Rev. Anticancer. Ther.* **2018**, *18*, 861–872. [CrossRef] [PubMed]

23. Yu, P.; Wilhelm, K.; Dubrac, A.; Tung, J.K.; Alves, T.C.; Fang, J.S.; Xie, Y.; Zhu, J.; Chen, Z.; De Smet, F.; et al. FGF-dependent metabolic control of vascular development. *Nature* **2017**, *545*, 224–228. [CrossRef] [PubMed]

24. Maddaluno, L.; Urwyler, C.; Werner, S. Fibroblast growth factors: Key players in regeneration and tissue repair. *Development* **2017**, *144*, 4047–4060. [CrossRef]

25. Hsu, P.I.; Chow, N.H.; Lai, K.H.; Yang, H.B.; Chan, S.H.; Lin, X.Z.; Cheng, J.S.; Huang, J.S.; Ger, L.P.; Huang, S.M.; et al. Implications of serum basic fibroblast growth factor levels in chronic liver diseases and hepatocellular carcinoma. *Anticancer. Res.* **1997**, *17*, 2803–2809. [PubMed]

26. Poon, R.T.-P.; Ng, I.O.-L.; Lau, C.; Yu, W.-C.; Fan, S.-T.; Wong, J. Correlation of serum basic fibroblast growth factor levels with clinicopathologic features and postoperative recurrence in hepatocellular carcinoma. *Am. J. Surg.* **2001**, *182*, 298–304. [CrossRef]

27. Jibiki, N.; Saito, N.; Kameoka, S.; Kobayashi, M. Clinical Significance of Fibroblast Growth Factor (FGF) Expression in Colorectal Cancer. *Int. Surg.* **2014**, *99*, 493–499. [CrossRef]

28. Helsten, T.; Schwaederle, M.; Kurzrock, R. Fibroblast growth factor receptor signaling in hereditary and neoplastic disease: Biologic and clinical implications. *Cancer Metastasis Rev.* **2015**, *34*, 479–496. [CrossRef]

29. Korc, M.; Friesel, R. The Role of Fibroblast Growth Factors in Tumor Growth. *Curr. Cancer Drug Targets* **2009**, *9*, 639–651. [CrossRef]

30. Lee, P.; Johnson, D.; Cousens, L.; Fried, V.; Williams, L. Purification and complementary DNA cloning of a receptor for basic fibroblast growth factor. *Sci.* **1989**, *245*, 57–60. [CrossRef]

31. Johnson, D.E.; Williams, L.T. Structural and functional diversity in the FGF receptor multigene family. *Adv. Cancer Res.* **1993**, *60*, 1–41.

32. Haugsten, E.M.; Wiedlocha, A.; Olsnes, S.; Wesche, J. Roles of Fibroblast Growth Factor Receptors in Carcinogenesis. *Mol. Cancer Res.* **2010**, *8*, 1439–1452. [CrossRef]

33. Holzmann, K.; Grunt, T.; Heinzle, C.; Sampl, S.; Steinhoff, H.; Reichmann, N.; Kleiter, M.; Hauck, M.; Marian, B. Alternative Splicing of Fibroblast Growth Factor Receptor IgIII Loops in Cancer. *J. Nucleic Acids* **2012**, *2012*, 950508. [CrossRef]

34. Yeh, B.K.; Igarashi, M.; Eliseenkova, A.V.; Plotnikov, A.N.; Sher, I.; Ron, D.; Aaronson, S.A.; Mohammadi, M. Structural basis by which alternative splicing confers specificity in fibroblast growth factor receptors. *Proc. Natl. Acad. Sci. USA* **2003**, *100*, 2266–2271. [CrossRef]

35. Babina, I.S.; Turner, N.C. Advances and challenges in targeting FGFR signalling in cancer. *Nat. Rev. Cancer* **2017**, *17*, 318–332. [CrossRef]

36. Trueb, B. Biology of FGFRL1, the fifth fibroblast growth factor receptor. *Cell. Mol. Life Sci.* **2011**, *68*, 951–964. [CrossRef]

37. Helsten, T.; Elkin, S.; Arthur, E.; Tomson, B.N.; Carter, J.; Kurzrock, R. The FGFR Landscape in Cancer: Analysis of 4,853 Tumors by Next-Generation Sequencing. *Clin. Cancer Res.* **2016**, *22*, 259–267. [CrossRef]

38. Latko, M.; Czyrek, A.; Porebska, N.; Kucinska, M.; Otlewski, J.; Zakrzewska, M.; Opalinski, L. Cross-Talk between Fibroblast Growth Factor Receptors and Other Cell Surface Proteins. *Cells* **2019**, *8*, 455. [CrossRef]

39. Mulloy, B.; Linhardt, R.J. Order out of complexity – protein structures that interact with heparin. *Curr. Opin. Struct. Boil.* **2001**, *11*, 623–628. [CrossRef]

40. Beenken, A.; Mohammadi, M. The FGF family: Biology, pathophysiology and therapy. *Nat. Rev. Drug Discov.* **2009**, *8*, 235–253. [CrossRef]

41. Venero Galanternik, M.; Kramer, K.L.; Piotrowski, T. Heparan Sulfate Proteoglycans Regulate Fgf Signaling and Cell Polarity during Collective Cell Migration. *Cell Rep.* **2015**, *10*, 414–428. [CrossRef]

42. Goetz, R.; Mohammadi, M. Exploring mechanisms of FGF signalling through the lens of structural biology. *Nat. Rev. Mol. Cell Biol.* **2013**, *14*, 166–180. [CrossRef]

43. Xu, H.; Lee, K.W.; Goldfarb, M. Novel Recognition Motif on Fibroblast Growth Factor Receptor Mediates Direct Association and Activation of SNT Adapter Proteins. *J. Boil. Chem.* **1998**, *273*, 17987–17990. [CrossRef]

44. Hoch, R.V.; Soriano, P. Context-specific requirements for Fgfr1 signaling through Frs2 and Frs3 during mouse development. *Development* **2006**, *133*, 663–673. [CrossRef]

45. Gotoh, N.; Laks, S.; Nakashima, M.; Lax, I.; Schlessinger, J. FRS2 family docking proteins with overlapping roles in activation of MAP kinase have distinct spatial-temporal patterns of expression of their transcripts. *FEBS Lett.* **2004**, *564*, 14–18. [CrossRef]

46. Mohammadi, M.; Olsen, S.K.; Ibrahimi, O.A. Structural basis for fibroblast growth factor receptor activation. *Cytokine Growth Factor Rev.* **2005**, *16*, 107–137. [CrossRef]

47. Zhang, Y.; McKeehan, K.; Lin, Y.; Zhang, J.; Wang, F. Fibroblast growth factor receptor 1 (FGFR1) tyrosine phosphorylation regulates binding of FGFR substrate 2alpha (FRS2alpha) but not FRS2 to the receptor. *Mol. Endocrinol.* **2008**, *22*, 167–175. [CrossRef]

48. Fürthauer, M.; Lin, W.; Ang, S.-L.; Thisse, B.; Thisse, C. Sef is a feedback-induced antagonist of Ras/MAPK-mediated FGF signalling. *Nat. Cell Biol.* **2002**, *4*, 170–174. [CrossRef]

49. Tsang, M.; Friesel, R.; Kudoh, T.; Dawid, I.B. Identification of Sef, a novel modulator of FGF signalling. *Nat. Cell Biol.* **2002**, *4*, 165–169. [CrossRef]

50. Tsang, M.; Dawid, I.B. Promotion and attenuation of FGF signaling through the Ras-MAPK pathway. *Sci. Stke* **2004**, *2004*, pe17. [CrossRef]

51. Dey, J.H.; Bianchi, F.; Voshol, J.; Bonenfant, D.; Oakeley, E.J.; Hynes, N.E. Targeting Fibroblast Growth Factor Receptors Blocks PI3K/AKT Signaling, Induces Apoptosis, and Impairs Mammary Tumor Outgrowth and Metastasis. *Cancer Res.* **2010**, *70*, 4151–4162. [CrossRef]

52. Eswarakumar, V.; Lax, I.; Schlessinger, J.; Eswarakumar, J. Cellular signaling by fibroblast growth factor receptors. *Cytokine Growth Factor Rev.* **2005**, *16*, 139–149. [CrossRef]

53. Turner, N.; Grose, R. Fibroblast growth factor signalling: From development to cancer. *Nat. Rev. Cancer* **2010**, *10*, 116–129. [CrossRef]

54. Kamata, T. Keratinocyte growth factor regulates proliferation and differentiation of hematopoietic cells expressing the receptor gene K-sam. *Exp. Hematol.* **2002**, *30*, 297–305. [CrossRef]

55. Cailliau, K.; Browaeys-Poly, E.; Vilain, J.P. Fibroblast growth factors 1 and 2 differently activate MAP kinase in Xenopus oocytes expressing fibroblast growth factor receptors 1 and 4. *Biochim. Biophys. Acta* **2001**, *1538*, 228–233. [CrossRef]

56. Peters, K.G.; Marie, J.; Wilson, E.; Ives, H.E.; Escobedo, J.; Del Rosario, M.; Mirda, D.; Williams, L.T. Point mutation of an FGF receptor abolishes phosphatidylinositol turnover and Ca^{2+} flux but not mitogenesis. *Nature* **1992**, *358*, 678–681. [CrossRef]

57. Dudka, A.A.; Sweet, S.M.; Heath, J.K. Signal transducers and activators of transcription-3 binding to the fibroblast growth factor receptor is activated by receptor amplification. *Cancer Res.* **2010**, *70*, 3391–3401. [CrossRef]

58. Candi, E.; Rufini, A.; Terrinoni, A.; Giamboi-Miraglia, A.; Lena, A.M.; Mantovani, R.; Knight, R.; Melino, G. DeltaNp63 regulates thymic development through enhanced expression of FgfR2 and Jag2. *Proc. Natl. Acad. Sci. USA* **2007**, *104*, 11999–12004. [CrossRef]

59. Katoh, M.; Katoh, M. Cross-talk of WNT and FGF signaling pathways at GSK3beta to regulate beta-catenin and SNAIL signaling cascades. *Cancer Biol. Ther.* **2006**, *5*, 1059–1064. [CrossRef]

60. Fogarty, M.P.; Emmenegger, B.A.; Grasfeder, L.L.; Oliver, T.G.; Wechsler-Reya, R.J. Fibroblast growth factor blocks Sonic hedgehog signaling in neuronal precursors and tumor cells. *Proc. Natl. Acad. Sci. USA* **2007**, *104*, 2973–2978. [CrossRef]

61. Dudley, A.T.; Godin, R.E.; Robertson, E.J. Interaction between FGF and BMP signaling pathways regulates development of metanephric mesenchyme. *Genes Dev.* **1999**, *13*, 1601–1613. [CrossRef]

62. Peters, K.G.; Werner, S.; Chen, G.; Williams, L.T. Two FGF receptor genes are differentially expressed in epithelial and mesenchymal tissues during limb formation and organogenesis in the mouse. *Development* **1992**, *114*, 233–243.

63. Orr-Urtreger, A.; Bedford, M.T.; Burakova, T.; Arman, E.; Zimmer, Y.; Yayon, A.; Givol, D.; Lonai, P. Developmental Localization of the Splicing Alternatives of Fibroblast Growth Factor Receptor-2 (FGFR2). *Dev. Boil.* **1993**, *158*, 475–486. [CrossRef]

64. Eswarakumar, V.P.; Monsonego-Ornan, E.; Pines, M.; Antonopoulou, I.; Morriss-Kay, G.M.; Lonai, P. The IIIc alternative of Fgfr2 is a positive regulator of bone formation. *Development* **2002**, *129*, 3783–3793.

65. Zhang, X.; Ibrahimi, O.A.; Olsen, S.K.; Umemori, H.; Mohammadi, M.; Ornitz, D.M. Receptor specificity of the fibroblast growth factor family. The complete mammalian FGF family. *J. Boil. Chem.* **2006**, *281*, 15694–15700. [CrossRef]

66. Ornitz, D.M.; Xu, J.; Colvin, J.S.; McEwen, D.G.; MacArthur, C.A.; Coulier, F.; Gao, G.; Goldfarb, M. Receptor specificity of the fibroblast growth factor family. *J. Biol. Chem.* **1996**, *271*, 15292–15297. [CrossRef]

67. Nyeng, P.; Norgaard, G.A.; Kobberup, S.; Jensen, J. FGF10 signaling controls stomach morphogenesis. *Dev.·Boil.* **2007**, *303*, 295–310. [CrossRef]

68. Hattori, Y.; Odagiri, H.; Nakatani, H.; Miyagawa, K.; Naito, K.; Sakamoto, H.; Katoh, O.; Yoshida, T.; Sugimura, T.; Terada, M. K-sam, an amplified gene in stomach cancer, is a member of the heparin-binding growth factor receptor genes. *Proc. Natl. Acad. Sci. USA* **1990**, *87*, 5983–5987. [CrossRef]

69. Nakatani, H.; Sakamoto, H.; Yoshida, T.; Yokota, J.; Tahara, E.; Sugimura, T.; Terada, M. Isolation of an Amplified DNA Sequence in Stomach Cancer. *Jpn. J. Cancer Res.* **1990**, *81*, 707–710. [CrossRef]

70. Cristescu, R.; Lee, J.; Nebozhyn, M.; Kim, K.-M.; Ting, J.C.; Wong, S.S.; Liu, J.; Yue, Y.G.; Wang, J.; Yu, K.; et al. Molecular analysis of gastric cancer identifies subtypes associated with distinct clinical outcomes. *Nat. Med.* **2015**, *21*, 449–456. [CrossRef]

71. Deng, N.; Goh, L.K.; Wang, H.; Das, K.; Tao, J.; Tan, I.B.; Zhang, S.; Lee, M.; Wu, J.; Lim, K.H.; et al. A comprehensive survey of genomic alterations in gastric cancer reveals systematic patterns of molecular exclusivity and co-occurrence among distinct therapeutic targets. *Gut* **2012**, *61*, 673–684. [CrossRef]

72. Jung, E.-J.; Jung, E.-J.; Min, S.Y.; Kim, M.A.; Kim, W.H. Fibroblast growth factor receptor 2 gene amplification status and its clinicopathologic significance in gastric carcinoma. *Hum. Pathol.* **2012**, *43*, 1559–1566. [CrossRef]

73. Matsumoto, K.; Arao, T.; Hamaguchi, T.; Shimada, Y.; Kato, K.; Oda, I.; Taniguchi, H.; Koizumi, F.; Yanagihara, K.; Sasaki, H.; et al. FGFR2 gene amplification and clinicopathological features in gastric cancer. *Br. J. Cancer* **2012**, *106*, 727–732. [CrossRef]

74. Park, Y.S.; Na, Y.-S.; Ryu, M.-H.; Lee, C.-W.; Park, H.J.; Lee, J.-K.; Park, S.R.; Ryoo, B.-Y.; Kang, Y.-K. FGFR2 Assessment in Gastric Cancer Using Quantitative Real-Time Polymerase Chain Reaction, Fluorescent In Situ Hybridization, and Immunohistochemistry. *Am. J. Clin. Pathol.* **2015**, *143*, 865–872. [CrossRef]

75. Das, K.; Gunasegaran, B.; Tan, I.B.; Deng, N.; Lim, K.H.; Tan, P. Mutually exclusive FGFR2, HER2, and KRAS gene amplifications in gastric cancer revealed by multicolour FISH. *Cancer Lett.* **2014**, *353*, 167–175. [CrossRef]

76. Kuboki, Y.; Schatz, C.A.; Koechert, K.; Schubert, S.; Feng, J.; Wittemer-Rump, S.; Ziegelbauer, K.; Krahn, T.; Nagatsuma, A.K.; Ochiai, A. In situ analysis of FGFR2 mRNA and comparison with FGFR2 gene copy number by dual-color in situ hybridization in a large cohort of gastric cancer patients. *Gastric Cancer* **2018**, *21*, 401–412. [CrossRef]

77. Hosoda, K.; Yamashita, K.; Ushiku, H.; Ema, A.; Moriya, H.; Mieno, H.; Washio, M.; Watanabe, M. Prognostic relevance of FGFR2 expression in stage II/III gastric cancer with curative resection and S-1 chemotherapy. *Oncol. Lett.* **2018**, *15*, 1853–1860.

78. Van Cutsem, E.; Bang, Y.J.; Mansoor, W.; Petty, R.D.; Chao, Y.; Cunningham, D.; Ferry, D.R.; Smith, N.R.; Frewer, P.; Ratnayake, J.; et al. A randomized, open-label study of the efficacy and safety of AZD4547 monotherapy versus paclitaxel for the treatment of advanced gastric adenocarcinoma with FGFR2 polysomy or gene amplification. *Ann. Oncol.* **2017**, *28*, 1316–1324. [CrossRef]

79. Nakazawa, K.; Yashiro, M.; Hirakawa, K. Keratinocyte growth factor produced by gastric fibroblasts specifically stimulates proliferation of cancer cells from scirrhous gastric carcinoma. *Cancer Res.* **2003**, *63*, 8848–8852.

80. Huang, T.; Wang, L.; Liu, D.; Li, P.; Xiong, H.; Zhuang, L.; Sun, L.; Yuan, X.; Qiu, H. FGF7/FGFR2 signal promotes invasion and migration in human gastric cancer through upregulation of thrombospondin-1. *Int. J. Oncol.* **2017**, *50*, 1501–1512. [CrossRef]

81. Ren, C.; Chen, H.; Han, C.; Fu, D.; Wang, F.; Wang, D.; Ma, L.; Zhou, L.; Han, D. The anti-apoptotic and prognostic value of fibroblast growth factor 9 in gastric cancer. *Oncotarget* **2016**, *7*, 36655–36665. [CrossRef]

82. Ooi, A.; Oyama, T.; Nakamura, R.; Tajiri, R.; Ikeda, H.; Fushida, S.; Nakamura, H.; Dobashi, Y. Semi-comprehensive analysis of gene amplification in gastric cancers using multiplex ligation-dependent probe amplification and fluorescence in situ hybridization. *Mod. Pathol.* **2015**, *28*, 861–871. [CrossRef]

83. Cerami, E.; Gao, J.; Dogrusoz, U.; Gross, B.E.; Sumer, S.O.; Aksoy, B.A.; Jacobsen, A.; Byrne, C.J.; Heuer, M.L.; Larsson, E.; et al. The cBio cancer genomics portal: An open platform for exploring multidimensional cancer genomics data. *Cancer Discov.* **2012**, *2*, 401–404. [CrossRef]

84. Sun, Q.; Lin, P.; Zhang, J.; Li, X.; Yang, L.; Huang, J.; Zhou, Z.; Liu, P.; Liu, N. Expression of Fibroblast Growth Factor 10 Is Correlated with Poor Prognosis in Gastric Adenocarcinoma. *Tohoku J. Exp. Med.* **2015**, *236*, 311–318. [CrossRef]

85. Zhang, J.; Zhou, Y.; Huang, T.; Wu, F.; Pan, Y.; Dong, Y.; Wang, Y.; Chan, A.K.Y.; Liu, L.; Kwan, J.S.H.; et al. FGF18, a prominent player in FGF signaling, promotes gastric tumorigenesis through autocrine manner and is negatively regulated by miR-590-5p. *Oncogene* **2019**, *38*, 33–46. [CrossRef]

86. Chang, J.; Wang, S.; Zhang, Z.; Liu, X.; Wu, Z.; Geng, R.; Ge, X.; Dai, C.; Liu, R.; Zhang, Q.; et al. Multiple receptor tyrosine kinase activation attenuates therapeutic efficacy of the fibroblast growth factor receptor 2 inhibitor AZD4547 in FGFR2 amplified gastric cancer. *Oncotarget* **2015**, *6*, 2009–2022. [CrossRef]

87. Liu, Y.J.; Shen, D.; Yin, X.; Gavine, P.; Zhang, T.; Su, X.; Zhan, P.; Xu, Y.; Lv, J.; Qian, J.; et al. HER2, MET and FGFR2 oncogenic driver alterations define distinct molecular segments for targeted therapies in gastric carcinoma. *Br. J. Cancer* **2014**, *110*, 1169–1178. [CrossRef]

88. Lau, W.M.; Teng, E.; Huang, K.K.; Tan, J.W.; Das, K.; Zang, Z.; Chia, T.; The, M.; Kono, K.; Yong, W.P.; et al. Acquired Resistance to FGFR Inhibitor in Diffuse-Type Gastric Cancer through an AKT-Independent PKC-Mediated Phosphorylation of GSK3beta. *Mol. Cancer Ther.* **2018**, *17*, 232–242. [CrossRef]

89. Huang, T.; Liu, D.; Wang, Y.; Li, P.; Sun, L.; Xiong, H.; Dai, Y.; Zou, M.; Yuan, X.; Qiu, H. FGFR2 Promotes Gastric Cancer Progression by Inhibiting the Expression of Thrombospondin4 via PI3K-Akt-Mtor Pathway. *Cell. Physiol. Biochem.* **2018**, *50*, 1332–1345. [CrossRef]

90. Zhu, D.-Y.; Guo, Q.-S.; Li, Y.-L.; Cui, B.; Guo, J.; Liu, J.-X.; Li, P. Twist1 correlates with poor differentiation and progression in gastric adenocarcinoma via elevation of FGFR2 expression. *World J. Gastroenterol.* **2014**, *20*, 18306–18315. [CrossRef]

91. Grygielewicz, P.; Dymek, B.; Bujak, A.; Gunerka, P.; Stanczak, A.; Lamparska-Przybysz, M.; Wieczorek, M.; Dzwonek, K.; Zdzalik, D. Epithelial-mesenchymal transition confers resistance to selective FGFR inhibitors in SNU-16 gastric cancer cells. *Gastric Cancer* **2016**, *19*, 53–62. [CrossRef]

92. Hallinan, N.; Finn, S.; Cuffe, S.; Rafee, S.; O'Byrne, K.; Gately, K. Targeting the fibroblast growth factor receptor family in cancer. *Cancer Treat. Rev.* **2016**, *46*, 51–62. [CrossRef]

93. Hui, Q.; Jin, Z.; Li, X.; Liu, C.; Wang, X. FGF Family: From Drug Development to Clinical Application. *Int. J. Mol. Sci.* **2018**, *19*, 1875. [CrossRef]

94. Zhao, W.-M.; Wang, L.; Park, H.; Chhim, S.; Tanphanich, M.; Yashiro, M.; Kim, K.J. Monoclonal Antibodies to Fibroblast Growth Factor Receptor 2 Effectively Inhibit Growth of Gastric Tumor Xenografts. *Clin. Cancer Res.* **2010**, *16*, 5750–5758. [CrossRef]

95. Sommer, A.; Kopitz, C.; Schatz, C.A.; Nising, C.F.; Mahlert, C.; Lerchen, H.G.; Stelte-Ludwig, B.; Hammer, S.; Greven, S.; Schuhmacher, J.; et al. Preclinical Efficacy of the Auristatin-Based Antibody-Drug Conjugate BAY 1187982 for the Treatment of FGFR2-Positive Solid Tumors. *Cancer Res.* **2016**, *76*, 6331–6339. [CrossRef]

96. Katoh, M. Therapeutics Targeting FGF Signaling Network in Human Diseases. *Trends Pharmacol. Sci.* **2016**, *37*, 1081–1096. [CrossRef]

97. Ronca, R.; Giacomini, A.; Di Salle, E.; Coltrini, D.; Pagano, K.; Ragona, L.; Matarazzo, S.; Rezzola, S.; Maiolo, D.; Torella, R.; et al. Long-Pentraxin 3 Derivative as a Small-Molecule FGF Trap for Cancer Therapy. *Cancer Cell* **2015**, *28*, 225–239. [CrossRef]

98. Pierce, K.L.; Deshpande, A.M.; Stohr, B.A.; Gemo, A.T.; Patil, N.S.; Brennan, T.J.; Bellovin, D.I.; Palencia, S.; Giese, T.; Huang, C.; et al. FPA144, a humanized monoclonal antibody for both *FGFR2*-amplified and nonamplified, *FGFR2b*-overexpressing gastric cancer patients. *J. Clin. Oncol.* **2014**, *32*, e15074. [CrossRef]

99. Jiang, X.F.; Dai, Y.; Peng, X.; Shen, Y.Y.; Su, Y.; Wei, M.M.; Liu, W.R.; Ding, Z.B.; Zhang, A.; Shi, Y.H.; et al. SOMCL-085, a novel multi-targeted FGFR inhibitor, displays potent anticancer activity in FGFR-addicted human cancer models. *Acta Pharmacol. Sin.* **2018**, *39*, 243–250. [CrossRef]

100. Hilberg, F.; Tontsch-Grunt, U.; Baum, A.; Le, A.T.; Doebele, R.C.; Lieb, S.; Gianni, D.; Voss, T.; Garin-Chesa, P.; Haslinger, C.; et al. Triple Angiokinase Inhibitor Nintedanib Directly Inhibits Tumor Cell Growth and Induces Tumor Shrinkage via Blocking Oncogenic Receptor Tyrosine Kinases. *J. Pharmacol. Exp. Ther.* **2018**, *364*, 494–503. [CrossRef]

101. Cha, Y.; Kim, H.-P.; Lim, Y.; Han, S.-W.; Song, S.-H.; Kim, T.-Y. FGFR2 amplification is predictive of sensitivity to regorafenib in gastric and colorectal cancers in vitro. *Mol. Oncol.* **2018**, *12*, 993–1003. [CrossRef]

102. Burbridge, M.F.; Bossard, C.J.; Saunier, C.; Fejes, I.; Bruno, A.; Leonce, S.; Ferry, G.; Da Violante, G.; Bouzom, F.; Cattan, V.; et al. S49076 is a novel kinase inhibitor of MET, AXL, and FGFR with strong preclinical activity alone and in association with bevacizumab. *Mol. Cancer Ther.* **2013**, *12*, 1749–1762. [CrossRef]

103. Gozgit, J.M.; Wong, M.J.; Moran, L.; Wardwell, S.; Mohemmad, Q.K.; Narasimhan, N.I.; Shakespeare, W.C.; Wang, F.; Clackson, T.; Rivera, V.M. Ponatinib (AP24534), a multitargeted pan-FGFR inhibitor with activity in multiple FGFR-amplified or mutated cancer models. *Mol. Cancer Ther.* **2012**, *11*, 690–699. [CrossRef]

104. Jang, J.; Kim, H.K.; Bang, H.; Kim, S.T.; Kim, S.Y.; Park, S.H.; Lim, H.Y.; Kang, W.K.; Lee, J.; Kim, K.-M. Antitumor Effect of AZD4547 in a Fibroblast Growth Factor Receptor 2–Amplified Gastric Cancer Patient–Derived Cell Model1. *Transl. Oncol.* **2017**, *10*, 469–475. [CrossRef]

105. Michael, M.; Bang, Y.J.; Park, Y.S.; Kang, Y.K.; Kim, T.M.; Hamid, O.; Thornton, D.; Tate, S.C.; Raddad, E.; Tie, J. A Phase 1 Study of LY2874455, an Oral Selective pan-FGFR Inhibitor, in Patients with Advanced Cancer. *Target. Oncol.* **2017**, *12*, 463–474. [CrossRef]

106. Kim, S.Y.; Ahn, T.; Bang, H.; Ham, J.S.; Kim, J.; Kim, S.T.; Jang, J.; Shim, M.; Kang, S.Y.; Park, S.H.; et al. Acquired resistance to LY2874455 in FGFR2-amplified gastric cancer through an emergence of novel FGFR2-ACSL5 fusion. *Oncotarget* **2017**, *8*, 15014–15022. [CrossRef]

107. Brameld, K.A.; Owens, T.D.; Verner, E.; Venetsanakos, E.; Bradshaw, J.M.; Phan, V.T.; Tam, D.; Leung, K.; Shu, J.; LaStant, J.; et al. Discovery of the Irreversible Covalent FGFR Inhibitor 8-(3-(4-Acryloylpiperazin-1-yl) propyl)-6-(2,6-dichloro-3,5-dimethoxyphenyl)-2-(me thylamino) pyrido [2,3-d] pyrimidin-7 (8 H)-one (PRN1371) for the Treatment of Solid Tumors. *J. Med. Chem.* **2017**, *60*, 6516–6527. [CrossRef]

108. Tsimafeyeu, I.; Daeyaert, F.; Joos, J.B.; Aken, K.V.; Ludes-Meyers, J.; Byakhov, M.; Tjulandin, S. Molecular Modeling, de novo Design and Synthesis of a Novel, Extracellular Binding Fibroblast Growth Factor Receptor 2 Inhibitor Alofanib (RPT835). *Med. Chem.* **2016**, *12*, 303–317. [CrossRef]

109. Hall, T.G.; Yu, Y.; Eathiraj, S.; Wang, Y.; Savage, R.E.; Lapierre, J.M.; Schwartz, B.; Abbadessa, G. Preclinical Activity of ARQ 087, a Novel Inhibitor Targeting FGFR Dysregulation. *PLoS ONE* **2016**, *11*, e0162594. [CrossRef]

110. Schmidt, K.; Moser, C.; Hellerbrand, C.; Zieker, D.; Wagner, C.; Redekopf, J.; Schlitt, H.J.; Geissler, E.K.; Lang, S.A. Targeting Fibroblast Growth Factor Receptor (FGFR) with BGJ398 in a Gastric Cancer Model. *Anticancer Res.* **2015**, *35*, 6655–6665.

111. Inokuchi, M.; Murase, H.; Otsuki, S.; Kawano, T.; Kojima, K. Different clinical significance of FGFR1-4 expression between diffuse-type and intestinal-type gastric cancer. *World J. Surg. Oncol.* **2017**, *15*, 2. [CrossRef]

112. Pearson, A.; Smyth, E.; Babina, I.S.; Herrera-Abreu, M.T.; Tarazona, N.; Peckitt, C.; Kilgour, E.; Smith, N.R.; Geh, C.; Rooney, C.; et al. High-Level Clonal FGFR Amplification and Response to FGFR Inhibition in a Translational Clinical Trial. *Cancer Discov.* **2016**, *6*, 838–851. [CrossRef]

113. Hierro, C.; Alsina, M.; Sanchez, M.; Serra, V.; Rodon, J.; Tabernero, J. Targeting the fibroblast growth factor receptor 2 in gastric cancer: Promise or pitfall? *Ann. Oncol.* **2017**, *28*, 1207–1216. [CrossRef]

8

Membrane-Associated, Not Cytoplasmic or Nuclear, FGFR1 Induces Neuronal Differentiation

Katalin Csanaky [1], Michael W. Hess [2] and Lars Klimaschewski [1,*]

[1]   Division of Neuroanatomy, Medical University of Innsbruck, 6020 Innsbruck, Austria;
      katalin.csanaky@i-med.ac.at
[2]   Division of Histology and Embryology, Medical University of Innsbruck, 6020 Innsbruck, Austria;
      michael.hess@i-med.ac.at
*   Correspondence: lars.klimaschewski@i-med.ac.at

Abstract: The intracellular transport of receptor tyrosine kinases results in the differential activation of various signaling pathways. In this study, optogenetic stimulation of fibroblast growth factor receptor type 1 (FGFR1) was performed to study the effects of subcellular targeting of receptor kinases on signaling and neurite outgrowth. The catalytic domain of FGFR1 fused to the algal light-oxygen-voltage-sensing (LOV) domain was directed to different cellular compartments (plasma membrane, cytoplasm and nucleus) in human embryonic kidney (HEK293) and pheochromocytoma (PC12) cells. Blue light stimulation elevated the pERK and pPLCγ1 levels in membrane-opto-FGFR1-transfected cells similarly to ligand-induced receptor activation; however, no changes in pAKT levels were observed. PC12 cells transfected with membrane-opto-FGFR1 exhibited significantly longer neurites after light stimulation than after growth factor treatment, and significantly more neurites extended from their cell bodies. The activation of cytoplasmic FGFR1 kinase enhanced ERK signaling in HEK293 cells but not in PC12 cells and did not induce neuronal differentiation. The stimulation of FGFR1 kinase in the nucleus also did not result in signaling changes or neurite outgrowth. We conclude that FGFR1 kinase needs to be associated with membranes to induce the differentiation of PC12 cells mainly via ERK activation.

Keywords: optogenetics; FGF2; ERK; AKT; receptor kinase; neurite outgrowth; HEK293; PC12

1. Introduction

The fibroblast growth factor receptor (FGFR) family comprises four closely related receptors (FGFR1-4) consisting of a signal peptide, three extracellular immunoglobulin (Ig)-like domains, an acidic box, a single transmembrane helix, and an intracellular split tyrosine kinase domain [1]. The binding of FGF ligands results in receptor dimerization and autophosphorylation of the cytoplasmic kinase domains. Following the recruitment of various adapter molecules, several intracellular signaling pathways are activated, of which, the Ras/extracellular signal-regulated kinase (ERK) and phosphatidylinositol-3 kinase (PI3K)/AKT signaling cascades play a key role in neuronal differentiation and axon growth [2]. All FGFR subtypes drive the same signaling cascades but with different strengths [3].

Activated FGFR complexes are endocytosed and partly recycled to the plasma membrane or transported to late endosomes/multivesicular bodies followed by degradation in lysosomes [4]. The internalization of activated FGFR1 does not implicate the inactivation of the receptor or attenuation of signaling, but allows ongoing and even stronger signaling activities particularly of the ERK pathway. In neurons, ERK1/2 activation requires receptor internalization, while AKT activation does not [5]. We and others have demonstrated that the extent of the stimulation of the major signaling pathways involved in neuronal differentiation and axon outgrowth is crucially dependent on the amount and localization of activated kinase domains. For example, leupeptin, a tripeptide inhibitor of cysteine and aspartic acid cleaving proteases, prevents the lysosomal degradation of FGFR1 and promotes basic fibroblast growth factor (FGF2)-induced axon regeneration by enhanced receptor recycling [6].

In addition to its membrane localization, FGFR1 is constitutively found in the cytoplasm and nucleus [7]. The nuclear translocation of FGFR1 is importin-dependent [8] and results in the binding of the receptor to transcriptionally active chromatin which drives the gene expression of FGF2 and tyrosine hydroxylase [9]. Furthermore, interaction with RSK1 promotes FGFR1 release from the pre-Golgi compartments to the cytoplasm, increases the mobile population of FGFR1 and facilitates the nuclear accumulation of FGFR1 [10,11]. Nerve growth factor (NGF) has also been suggested to utilize integrative nuclear FGFR1 signaling (INFS) for its gene-activating functions [12]. In addition, INFS is apparently involved in the dendritic outgrowth of sympathetic neurons treated with bone morphogenetic protein [13].

The plethora of FGFR-dependent biological effects is not only explained by differences in subcellular receptor targeting but also by the differential expression of adapter proteins and other signaling components. Since the duration and kinetics of receptor activation play a decisive role in determining the functional and morphological outcome [14], FGFR1 represents an ideal object for optogenetic manipulation.

Recently, light-activatable FGF receptors that are incapable of ligand binding were developed [15]. Chimeric opto-fibroblast growth factor receptors (opto-FGFRs) exhibiting the catalytic domain were fused to the algal light-oxygen-voltage-sensing (LOV) domain for light-induced dimerization. These constructs lack the extracellular receptor domain and are, therefore, insensitive to endogenous or exogenous ligands. Short pulses of blue light induce LOV domain dimerization which results in the transphosphorylation of two receptor kinase molecules. Receptor phosphorylation diminishes rapidly after light stimulation and is followed by the inactivation of the main signaling pathways. Thus, intracellular FGFR activity can be externally controlled by repeated light stimulation. In this study, plasma membrane-targeted (mem-opto-FGFR1), cytoplasmic (cyto-opto-FGFR1) and nuclear FGFR1 (nucl-opto-FGFR1) were produced and studied with regard to signal pathway activation and neurite outgrowth.

2. Materials and Methods

2.1. Plasmid Construction

Mem-opto-FGFR1, which is in the pcDNA3.1(−) plasmid backbone, was described previously [15]. To clone cyto-opto-FGFR1, the myristoylation signal (MYR) of mem-opto-FGFR1 was replaced by the Kozak sequence using reverse PCR. Three nuclear localization sequences (NLSs) were inserted into cyto-opto-FGFR1 to obtain nucl-opto-FGFR1. Fluorescently labeled versions of opto-FGFR1 were produced by inserting mVenus (mV) (Figure 1). All plasmids that were generated were verified by sequencing.

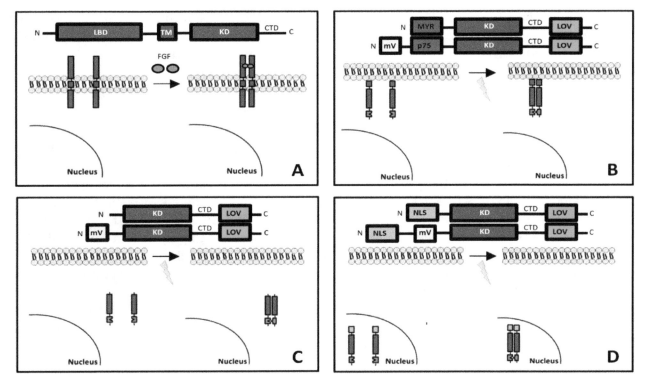

Figure 1. Design of light-controlled opto-fibroblast growth factor receptors (FGFR1s) and their molecular architecture: sequence of the different FGFR1 coding genes, spatial relation of FGFR1s to the surface membrane and nucleus, and their activation/dimerization. (**A**) Naturally occurring full-length FGFR1 consists of the extracellular ligand-binding (LBD), transmembrane (TM), kinase (KD) and C-terminal tail (CTD) domains. (**B**) Artificial mem-opto-FGFR1 is anchored to the plasma membrane with an N-terminal myristoylation signal (MYR) followed by the KD, CTD and LOV domain (mV-mem-opto-FGFR1 is inserted into the plasma membrane by incorporation of the transmembrane domain of p75). Fluorescent mVenus and the light-oxygen-voltage-sensing (LOV) protein are separated to avoid non-specific activation of the LOV domain by mVenus (by Förster resonance energy transfer). mVenus can be detected by excitation with green light (514 nm) that does not activate the LOV domain. (**C**) Cyto-opto-FGFR1 consists of only the KD and the LOV domain. (**D**) Three nuclear localization sequence (NLS) signals are inserted into nucl-opto-FGFR1 with or without mVenus.

2.2. Cell Culture and Transfection

Human embryonic kidney cells (HEK293, ATCC, Manassas, VA, USA) were grown in Dulbecco's Modified Eagle's Medium (DMEM, Sigma, St. Louis, MI, USA) supplemented with 10% fetal bovine serum (FBS; Thermo Fisher Scientific, Waltham, MA, USA, TFS) and 1% antibiotic–antimycotic solution (100 U/mL penicillin, 100 µg/mL streptomycin, 0.25 µg/mL amphotericin, TFS). Rat adrenal pheochromocytoma (PC12) cells (Sigma) were grown in RPMI (Roswell Park Memorial Institute) 1640 (TFS) supplemented with 10% horse serum, 5% FBS, and 1% antibiotic–antimycotic solution on dishes coated with 50 µg/mL collagen type I (Merck, Kenilworth, NJ, USA). Human glioblastoma cells (U251) were cultivated in RPMI 1640 cell culture medium supplemented with 5% FBS and 1% antibiotic–antimycotic solution. All cell lines were grown under standard conditions in humidified atmosphere at 37 °C, with 5% CO_2. Transient transfections were performed with Lipofectamine 2000 (Invitrogen, Carlsbad, CA, USA) according to the manufacturer's protocol.

2.3. Confocal Laser Scanning Microscopy

For the localization of mV-opto-FGFR1s in HEK293 cells, 4×10^5 cells were seeded on #1.5 coverslips and co-transfected with 1 µg mV-opto-FGFR1 and 1 µg LifeAct–mCherry (Addgene #40908). Six hours after transfection, 25 µM importazole (Sigma) was added to the cells overnight to prevent nuclear import of the mV-nucl-opto-FGFR1 construct. Cells were fixed in 4% formaldehyde solution (made from paraformaldehyde; PFA), stained with Hoechst 33258 dye, mounted in ProLong Diamond Antifade Mountant (TFS) and imaged using SP8 confocal laser scanning microscopy (Leica, Wetzlar, Germany).

2.4. Immunogold Electron Microscopy

For immunogold electron microscopy, Tokuyasu cryosection and cryo-based pre-embedding immunogold labelling were applied [16–18]. HEK293 cells were transfected with 1 μg mV-mem-opto-FGFR1 in 10-cm diameter dishes and kept overnight in DMEM supplemented with 10% FBS and 1% antibiotic–antimycotic solution. Cells were fixed with 4% (w/v) buffered formaldehyde solution for >3 days at room temperature (RT). Ultra-thin cryosections were labelled with goat anti-GFP (#600-101-215, 1:500, Rockland, Limerick, PA, USA), followed by rabbit anti-goat IgG 15-nm colloidal gold conjugates (#EM.RAG15, 1:50, British Biocell Intl., Cardiff, UK). U251 cells were cultured on sapphire coverslips (Ø 3 mm), transfected with 0.2 μg mV-mem-opto-FGFR1 and kept overnight in RPMI 1640 supplemented with 5% FBS and 1% antibiotic–antimycotic solution. Samples were then subjected to high-pressure freezing, freeze-substitution, and rehydration, followed by pre-embedding labelling. Primary antibodies were goat anti-GFP (#600-101-215, 1:2000, Rockland), secondary antibodies were Nanogold®-Fab' rabbit anti-goat IgG (H+L) (#2006; 1:700), followed by silver enhancement (SE) with HQ-Silver® (#2012, all from Nanoprobes) and standard plastic embedding. Sections were analyzed by transmission electron microscopy.

2.5. Stimulation and Immunoblot

HEK293 cells were seeded on 6-well dishes (1×10^6 per dish, Corning, NY, USA) and transfected with either 2 μg opto-FGFR1 or 2 μg FGFR1–eGFP [19]. Six hours after transfection, the medium was replaced with reduced serum medium (0.5% FBS) and the cells were incubated overnight. One day after transfection, opto-FGFR1-transfected cells were stimulated for 5 min by 2.5 μW/mm² blue light in a reptile egg incubator (PT2499, Exo Terra, Hagen, Holm, Germany) equipped with 300 light-emitting diodes. The blue light intensity was measured with a digital power meter connected to the Microscope Power Sensor Head (Thorlabs, Newton, NJ, USA). Cells that were kept in the dark (wrapped in aluminum foil) in the same incubator served as controls. FGFR1–eGFP-transfected cells were stimulated by treating them with 100 ng/mL FGF2 (Stemcell, Vancouver, BC, Canada) and 1 μg/mL heparin sulfate (HepS, Sigma) for 5 min.

The lysis of light-stimulated cells was performed in the incubator under blue light. The dark samples were lysed under non-stimulating red light. Cells were harvested in 50 μL lysis buffer on ice, sonicated and centrifuged (13,000 rpm, 20 min, 4 °C). To prevent ERK signaling, 50 μM MAPK kinase inhibitor (PD98059; Sigma) was added to the cells 2 h prior to light activation or FGF2 stimulation. For the measurement of ERK kinetics, the ligand was washed out two times with PBS and lysates made after 5, 15 and 30 min. In the case of opto-FGFR1-transfected cells, the lysates were obtained after 5, 15 and 30 min of dark period. A total of 30 μg protein per lane was separated by SDS-PAGE and transferred to Immobilon-FL PVDF membranes (Merck, Kenilworth, NJ, USA). Blots were incubated with primary antibodies in Odyssey Blocking Buffer (LI-COR) overnight at 4 °C (pFGFR1, #3476, 1:1000; ERK1/2, #9107, 1:1000; pERK1/2, #9101, 1:2000; AKT, #2920, 1:1000; pAKT, #4060, 1:2000; pPLCγ1, #2821, 1:2000, all from Cell Signaling Technology/CST, Danvers, MA, USA; PLCγ1, #16955, 1:1000, Abcam; GAPDH, sc-32233, 1:500, Santa Cruz Biotechnology, Dallas, TX, USA). Secondary antibodies (IRDye 800CW goat anti-rabbit IgG, #926-32231, and IRDye 680RD goat anti-mouse IgG, #926-68070, both 1:20.000, LI-COR, Lincoln, NE, USA) were applied for 1 h at RT. Signals were recorded with an infrared fluorescence detection system (Odyssey Fc Imaging System).

Long-term light and ligand stimulation for neurite outgrowth assay and immunocytochemistry pheochromocytoma (PC12) cells was used for neuronal differentiation experiments. Cell lines enriched with mV-mem-opto-FGFR1-, mV-cyto-opto-FGFR1- or mV-nucl-opto-FGFR1-transfected cells were obtained as follows. Cells with a low passage number (<P10) were seeded on 6-well plates (BD) and

transiently transfected with 1 μg of plasmids. After 24 h, the culture medium was replaced by a medium containing 300 μg/mL Geneticin (G418, TFS), which was changed every other day. When sufficient numbers of resistant cell colonies were observed, the cells were triturated with fire-polished Pasteur pipettes for 10 min and then transferred to collagen-coated culture dishes. For microscopic analysis, cells were plated on 35-mm diameter collagen-coated imaging dishes (IBIDI). PC12 cells were also transiently transfected with 1 μg FGFR1–eGFP. After 6 h, cells were cultured in RPMI 1640 supplemented with 10% horse serum, 5% FBS, and 1% antibiotic–antimycotic solution overnight followed by a starvation medium supplemented with 2 mM L-glutamin, 1% antibiotic–antimycotic solution, and N2 supplement (100×, TFS) for 2 h. FGFR1–eGFP-transfected cells and naive PC12 cells were then treated for 48 h with 100 ng/mL FGF2 + 1 μg/mL HepS and with 100 ng/mL nerve growth factor (NGF), respectively. For light stimulation, the dishes were placed in a humidified atmosphere at 37 °C with 5% CO_2 in an incubator with light-emitting diodes as described above. The blue light intensity was measured with a digital power meter and set to 2.5 μW/mm^2. The control cells were kept in the dark in the same incubator during stimulation. A repetitive 5 min ON and 55 min OFF cycle was used for stimulation over 48 h.

Cells were then fixed with 4% PFA for 15 min and their neurites were visualized by immunocytochemistry. Neuron-specific class III beta-tubulin (1:1000, 1 h at RT; R&D system) and Alexa 568 goat anti-mouse IgG (1:1000, 2 h at RT; TFS) were applied as the primary and secondary antibody, respectively. Inverted fluorescence microscopy (AxioVert, Zeiss, Oberkochen, Germany) at 40× magnification was used for imaging. Neurite outgrowth was defined if cells exhibited at least one process of more than one cell body diameter length.

For immunocytochemistry, PC12 cells enriched with mV-opto-FGFR1-transfected cells were used. Following serum starvation, light stimulation and fixation, cells were incubated overnight with primary antibodies (pERK1/2, #9101, 1:400, CST; pAKT, #4060, 1:400, CST) followed by incubation at RT for 2 h with Alexa 546 goat anti-rabbit IgG (1:1000, TFS). SP8 confocal laser scanning microscopy (Leica) was used for cell imaging. Controls included the omission of primary antibodies (negative) and stimulation of cells with NGF (positive).

2.6. Quantification of Results and Statistical Analysis

The total neurite length (TNL, sum of the length of all neurites extending from the cell body), maximal distance (MD, length of longest neurite) and number of neurites per cell were quantified (Figure S1). Metamorph software version 4.6 (Visitron Systems, Puchheim, Germany) was used for the quantification of neurite outgrowth and immunocytochemistry. Band intensities of the Western blots were densitometrically quantified using Image Studio LiteVer 5.2. GraphPad Prism software was used for statistical analysis (two-way ANOVA with Sidak correction).

3. Results

3.1. Localization of mV-opto-FGFR1s

In HEK293 cells, mV-mem-opto-FGFR1 localized predominantly to the plasma membrane and to endosomes (Figure 2A1–A4), while mV-cyto-opto-FGFR1 diffused freely in the cytoplasm and nucleus (Figure 2B1–B4). The mV-nucl-opto-FGFR1 construct was found exclusively in the nucleus with enrichment of the protein in presumptive nuclear speckle domains (Figure 2C1–C4). Inhibition of the importin β transport receptors with 25 μM importazole resulted in diffuse cytoplasmic yellow fluorescence of mV-nucl-opto-FGFR1, indicating a partial block of the nuclear import (Figure S2). Immunogold electron microscopy of the thawed cryosections revealed mV-mem-opto-FGFR1 in the plasma membrane, in (late) endosomes/multivesicular bodies (MVBs) and lysosomes of HEK293 cells (Figure 2D1–D3). These data were further confirmed by a complementary immunogold labeling method applied to a different cell line (U251 glioblastoma cells) in which the same subcellular distribution of mV-mem-opto-FGFR1 was found (Figure S3).

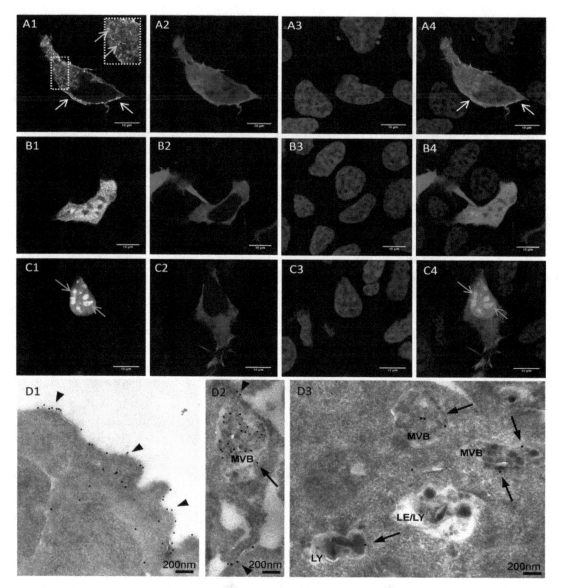

Figure 2. Light and electron microscopic localization of mV-opto-FGFR1s in human embryonic kidney (HEK293) cells. (**A–C**) Immunofluorescence microscopy of cells co-transfected with mV-opto-FGFR1s and LifeAct–mCherry to visualize cell bodies and all cytoplasmic processes. (**A1**) mV-mem-opto-FGFR1 is observed in the plasma membrane (white arrows) and in endosomes (inset; cyan arrows) indicating endosomal receptor internalization. (**B1**) mV-cyto-opto-FGFR1-transfected cells reveal diffuse yellow fluorescence in the cytoplasm and nucleus of transfected cells. (**C1**) mV-nucl-opto-FGFR1 is only located in the nucleus (nuclear speckle domains are indicated by green arrows). Fixed cell nuclei are stained with Hoechst (blue) and A4–C4 represent overlays. Scale bars for all images are 10 μm. (**D**) Immunogold electron microscopy of the thawed cryosections reveals mV-mem-opto-FGFR1 in the plasma membrane (arrowheads in **D1** and **D2**) and in the limiting membrane as well as inside various endocytic compartments (arrows in **D2** and **D3**); MVB = multivesicular body, LE = late endosome, LY = lysosome.

3.2. Opto-FGFR1-Dependent Signaling Pathway Activation in HEK293 Cells

For all quantitative signaling assays, the levels of active FGFR1 (pFGFR1) and the control protein (GAPDH) were determined to normalize for different transfection efficiencies (Figure 3A,B). Immunoblot results showed that serum starvation (naive state) significantly reduced the basal level of pERK but not the basal levels of pAKT or pPLCγ1 in HEK293 cells. The overexpression of FGFR1–eGFP induced ERK and AKT activation which was further increased by adding FGF2 for 5 min. No marked changes in pPLCγ1 levels were observed following FGF2 treatment.

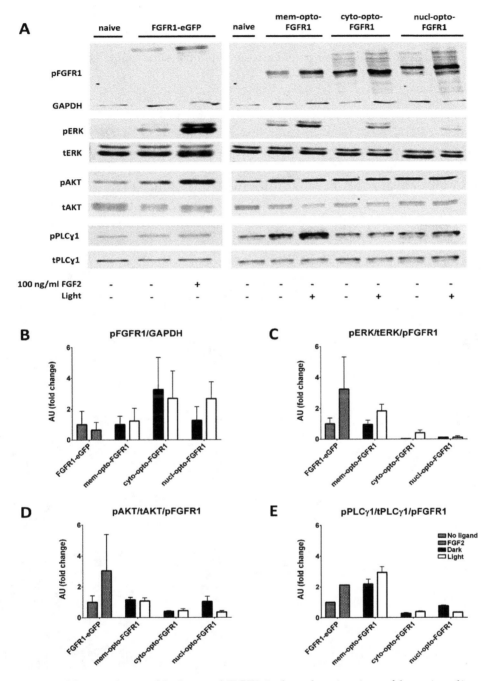

Figure 3. Immunoblot analysis of light- and FGF2-induced activation of key signaling molecules in HEK293 cells. (**A**) Representative examples of Western blots for pFGFR1/GAPDH, pERK/tERK, pAKT/tAKT and pPLCγ1/tPLCγ1 are shown. (**B–E**) Average intensities of ERK, AKT, and PLCγ1 phosphorylation were quantified after normalization for pFGFR1/GAPDH to control for differences in plasmid expression levels. All ratios are relative to FGFR1–eGFP (=1) and calculated from three independent experiments (mean ± SEM).

The overexpression of mem-opto-FGFR1 alone (without light stimulation) resulted in receptor autoactivation, as indicated by increased pERK and pPLCγ1 but unchanged pAKT levels (Figure 3A). Autoactivation was not observed after cyto-opto-FGFR1 or nucl-opto-FGFR1 transfection. Blue light stimulation for 5 min following mem-opto-FGFR1 or cyto-opto-FGFR1 transfection elevated the pERK levels (Figure 3C). Ligand-induced ERK activation lasted longer, whereas light-induced ERK activation was terminated faster (30 min versus 5–15 min, Figure S4). Minimal ERK activation was observed after transfection with nucl-opto-FGFR1. Unexpectedly, the pAKT levels remained unchanged after blue light stimulation of HEK293 cells following transfection with any of the opto-FGFR1 constructs (Figure 3D). On the other hand, the pPLCγ1/tPLCγ1 ratio slightly increased after blue light stimulation of mem-opto-FGFR1- but not of cyto- or nucl-opto-FGFR1-transfected cells (Figure 3E).

3.3. Immunocytochemistry of mV-opto-FGFR1-Transfected PC12 Cells

Immunocytochemistry was performed instead of protein blotting to visualize signaling pathway activation in PC12 cells (due to their lower transfection rate). Similarly to HEK293 cells, mem-opto-FGFR1-transfected pheochromocytoma cells exhibited increased pERK levels, as indicated by intense pERK immunolabeling in the cytoplasm (Figure 4A2/A6,G). As with HEK293 cells, no AKT activation was observed following intermittent blue light stimulation (Figure 4B2/B6,H). Furthermore, the pERK and pAKT levels were unchanged in stimulated mV-cyto-opto-FGFR1 (Figure 4C2/C6,D2/D6) or mV-nucl-opto-FGFR1-expressing PC12 cells (Figure 4E2/E6,F2/F6).

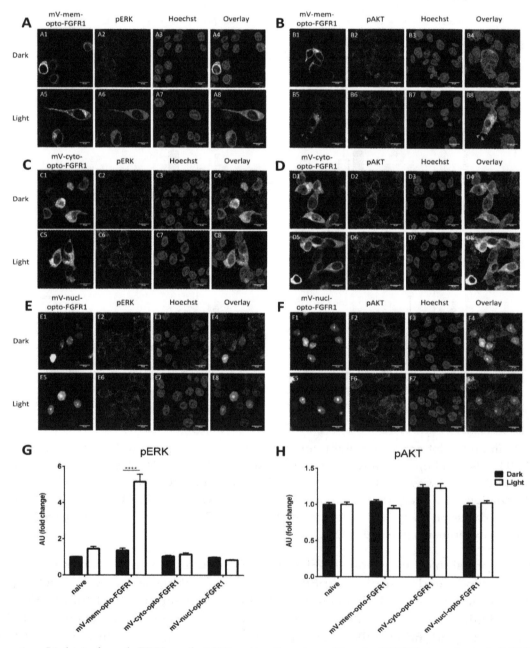

Figure 4. Light-induced ERK and AKT activation in mV-opto-FGFR1-transfected PC12 cells. Although mV-mem-opto-FGFR1-transfected cells exhibit short neurite extensions (sprouts) in the dark

(A1), blue light induces neuronal differentiation with long, slender neurites in transfected cells (A5). (**A**) Significantly increased cytoplasmic pERK levels following light stimulation of mV-mem-opto-FGFR1-transfected cells, while non-transfected cells show no changes. (**C,E**) The pERK level is unchanged after stimulation of mV-cyto- and nucl-opto-FGFR1-transfected cells. (**B,D,F**) The fluorescence intensities of pAKT signals are similar in mV-opto-FGFR1-transfected and non-transfected cells in the dark and no changes are observed after blue light stimulation. Images are acquired using the same laser intensity for both dark and light in each fluorescent channel and presented without adjusting contrast or subtracting background. (**G,H**) Quantification of average fluorescence intensities. Results are calculated from two independent experiments and presented as mean $\pm$ SEM ($30 < n < 60$), **** $p < 0.0001$. Scale bars = 10 μm.

3.4. Neuronal Differentiation of PC12 Cells Induced by Blue Light

PC12 cells exhibited no spontaneous or FGF2-induced neurite outgrowth, suggesting that the clone used in the present study does not express significant levels of endogenous FGF receptors (Figure 5A and Figure S5). In fact, all four FGFR mRNAs are endogenously expressed but the levels are low, particularly for FGFR1 (Figure S5E). Two days after treatment with NGF, neuronal differentiation was observed (Figure 5B; 120 ± 11.9 μm total neurite length, TNL, Figure 5K; 52.7 ± 4 μm of maximal neurite length, MD, Figure 5L; 2.6 ± 0.12 processes extending from the cell body, Figure 5M). Cells transiently transfected with FGFR1–eGFP revealed significantly longer neurites compared to naive cells (Figure 5C) and increased neurite initiation (Figure 5M). FGF2 treatment further enhanced neuronal differentiation with long neurites (Figure 5D). Although the autoactivation of mV-mem-opto-FGFR1 induced mild neurite outgrowth in the dark state (Figure 5E), blue light stimulation resulted in dramatically increased neuronal differentiation (Figure 5F,K) which was significantly inhibited by prior PD98059 treatment (Figure S6). A significant increase in the number of neurites extending from mV-mem-opto-FGFR1-transfected cells after blue light stimulation was observed as well as significantly longer neurites when compared to NGF and FGF2 treatment (Figure 5L,M). Cells expressing either mV-cyto-opto-FGFR1 or mV-nucl-opto-FGFR1 showed flattened, spindle-shaped morphology with short cytoplasmic extensions but failed to grow processes longer than one cell body in diameter (Figure 5G–J).

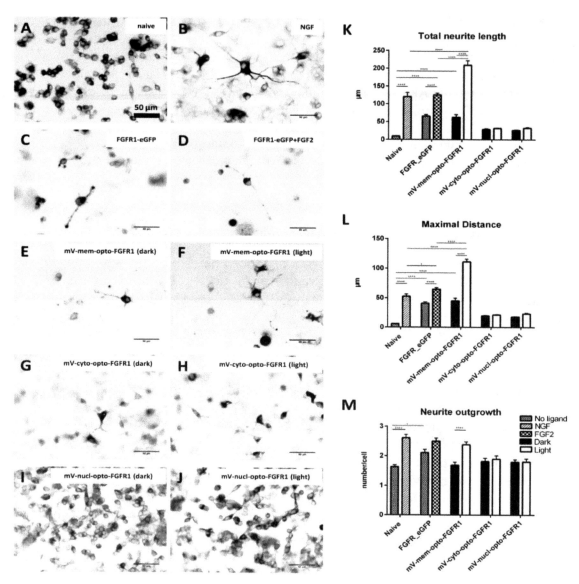

Figure 5. Ligand- and light-induced neurite outgrowth by pheochromocytoma (PC12) cells. (A–J) Inverted immunofluorescence images following neuron-specific class III β-tubulin staining to identify neurites (red nuclei in nucl-opto-FGFR1 cells allow identification of transfected cells in I/J). (K–M) Quantification of morphological parameters (total neurite outgrowth, longest process and number of processes per cell; see Figure S1 for details). Results are calculated from three independent experiments and presented as mean ± SEM (50 < n < 100), * $p < 0.05$, **** $p < 0.0001$. Scale bars = 50 μm.

4. Discussion

Light-sensitive G-protein-coupled receptors (e.g., rhodopsin) occur naturally, whereas light-sensitive receptor tyrosine kinases (RTKs) need to be artificially produced. Recent studies have been aimed at subcellular targeting of opto-TrkA and light-gated adenylate cyclase [20,21]. In addition, various membrane-associated opto-RTK constructs were synthesized, such as opto-TrkB [22] and three different opto-FGFR1 constructs [15,23,24]. One of the light-activated FGFR1 proteins (through the homointeraction of cryptochrome 2) induced cell polarization and directed cell migration through changes in the actin–tubulin cytoskeleton [23]. Furthermore, opto-FGFR1 was applied for light-induced sprouting of human bronchial epithelial cells [15].

The opto-FGFR1 constructs used here were designed for specific targeting of the kinase domain to only the plasma membrane, cytoplasm, and nucleus, respectively, to investigate the possible effects of subcellular FGFR kinase activation on signal pathway induction and neurite outgrowth as a biological read-out. Similarly to full-length FGFR1, immunoelectron microscopy revealed that mV-mem-opto-FGFR1s were anchored to the plasma membrane, internalized and transported to multivesicular bodies (MVBs)/late endosomes and lysosomes [25,26]. Although our construct was expected to only attach to membranes (plasma membrane, endosomal/lysosomal), mV-mem-opto-FGFR1 was also occasionally observed in the cytoplasm and nucleus. It is known that internalized full-length FGFR1 may be released from endosomes and travels to the nucleus through importin β-mediated translocation and that newly synthetized FGFR1 may enter the nucleus directly as well [27–30].

Intranuclear FGFR1 is localized within nuclear matrix-attached speckle domains in the form of large discrete spots [31–33]. In this study, such fluorescence patterns were also observed in mV-nucl-opto-FGFR1-transfected cells exhibiting the split kinase domain of FGFR1 coupled to three NLSs. Biologically active, soluble kinase fragments are also created by cleavage at the transmembrane domain [34,35]. Similarly to these natural cytoplasmic FGFR1 fragments, mV-cyto-opto-FGFR1 constructs lacking specific targeting signals diffuse freely in the cytoplasm.

In this study, the activation of ERK but not of AKT was observed in cells expressing membrane-associated or, to a lesser extent, cytoplasmic opto-FGFR1 after blue light stimulation. The light-induced activation of both pathways has been described in connection with membrane-localized opto-TrkA and -TrkB receptors and with cytoplasmic opto-TrkA [20,22]. Moreover, significant ERK phosphorylation was observed after targeting active kinase domains of FGFR1, FGFR3 and FGFR4 to the plasma membrane [36]. Our findings are consistent with these and earlier studies from our own group demonstrating that FGFR1, in contrast to TrkA, exerts a significantly stronger influence on the ERK than on the AKT pathway [37–39].

Opto-constructs provide additional advantages such as that activation ceases shortly after switching off the light stimulus. In contrast to ligand-dependent RTK signaling, light-induced stimulation is terminated faster. In other studies, ERK nuclear translocation and immunoblot assays indicated similar ERK phosphorylation kinetics following light-induced Raf1 and Ras recruitment, as observed in our experiments [40,41]. However, although PI3K/AKT was expected to be activated after FGFR phosphorylation, we did not observe changes in the pAKT levels after light stimulation of opto-FGFR1s in transfected HEK293 and PC12 cells. These variations in signal pathway activation between different cell lines and different modes of induction may be explained by the differential expression of signaling adapters and possible crosstalk, particularly of the ERK- and AKT-dependent pathways [42,43].

PC12 cells are widely used to study neuronal differentiation. NGF induces biochemical, electrophysiological and morphological changes in these cells, recapitulating many characteristic features of differentiated sympathetic neurons [44,45]. The optogenetic activation of Trk receptors and stimulation of the Raf–MEK–ERK cascade have also been demonstrated to induce neuronal differentiation [41]. Due to the selective activation of ERK in the latter study, the neurites were longer than those of NGF-treated cells which could be explained by multiple downstream pathway activation including PI3K/AKT-mediated process branching. Taken together, we suggest that repetitive ERK stimulation acts as the primary driver of neurite extension, whereas AKT-dependent pathways primarily stimulate the formation of branches [46].

In this study, PC12 cells transfected with mV-mem-opto-FGFR1 exhibited longer neurites than NGF- or FGF2-treated cells and significantly more neurites extended from the cell body. Interestingly, light-induced stimulation of mem-opto-FGFR1 did not increase pAKT levels as compared to dark control cells, suggesting that the effects on neurite outgrowth are mainly, if not exclusively, dependent on the ERK pathway. Our immunocytochemistry results showed intense pERK immunolabeling in the cytoplasm, but only weak fluorescence in the nucleus 55 min after blue light stimulation. In contrast to other cell lines, the activated ERK2 variant (ERK2–MEK1 fusion protein) diffusely localized to the cytoplasm and to extending cellular processes. This localization pattern suggests that ERK2 is involved in promoting neurite formation, not only through its actions on gene transcription but also through effects at the site of neurite extension [47]. Light-induced neuronal differentiation was not observed in mV-cyto- or in nucl-opto-FGFR1-transfected cells, corroborating earlier studies by Donoghue and his coworkers who achieved differentiation from plasma membrane-bound FGFR kinase domains in PC12 cells [36,48]. However, the full-length FGFR1 targeted to the nucleus also induced PC12 cell differentiation [12,49]. As discussed above, the binding of nuclear FGFR1 to CREB-binding protein (CBP) or to ribosomal S6 kinase isoform 1 (RSK1) may be involved in nuclear FGFR1 signaling [49,50]. Nevertheless, the activated kinase domain alone does not appear to be sufficient to modify gene expression, resulting in neuronal differentiation.

In summary, we demonstrated that only membrane-bound opto-FGFR1 constructs are capable of activating the ERK pathway sufficiently to induce neuronal cell differentiation in PC12 cells. We assume that fully functional signaling platforms only form in association with membranes. Active RTKs apparently recruit different adaptors and scaffold proteins in plasma membranes as compared to endosomal and other membranes which exhibit different curvatures and phosphatidylinositol compositions. Finally, the present study provides further evidence that FGFR1-dependent signaling pathways and neurite outgrowth can be controlled and manipulated optogenetically, which allows the study of subcellular receptor activation with spatial and temporal precision.

Author Contributions: Conceptualization, L.K.; methodology and formal analysis, K.C. and M.W.H.; writing—original draft preparation, K.C.; writing—review and editing, M.W.H. and L.K.; supervision, L.K.; funding acquisition, K.C. and L.K.

Acknowledgments: Special thanks go to Markus Reindl for providing HEK293 and U251 cells. K.C. is grateful to Barbara Hausott, Barbara Fogli and Martina Wick for help with several methods used in this study. Maximilian Klein helped in quantification of PC12 neuronal differentiation. Barbara Witting and Karin Gutleben are acknowledged for excellent technical assistance.

References

1.   Lee, P.L.; Johnson, D.E.; Cousens, L.S.; Fried, V.A.; Williams, L.T. Purification and complementary DNA cloning of a receptor for basic fibroblast growth factor. *Science* **1989**, *245*, 57–60. [CrossRef]

2.   Zhou, F.Q.; Snider, W.D. Intracellular control of developmental and regenerative axon growth. *Philos. Trans. R. Soc. Lond. B. Biol. Sci.* **2006**, *361*, 1575–1592. [CrossRef] [PubMed]

3.   Eswarakumar, V.P.; Lax, I.; Schlessinger, J. Cellular signaling by fibroblast growth factor receptors. *Cytokine Growth Factor Rev.* **2005**, *16*, 139–149. [CrossRef] [PubMed]

4.   Platta, H.W.; Stenmark, H. Endocytosis and signaling. *Curr. Opin. Cell Biol.* **2011**, *23*, 393–403. [CrossRef]

5.   MacInnis, B.L.; Campenot, R.B. Retrograde support of neuronal survival without retrograde transport of nerve growth factor. *Science* **2002**, *295*, 1536–1539. [CrossRef] [PubMed]

6.   Hausott, B.; Vallant, N.; Hochfilzer, M.; Mangger, S.; Irschick, R.; Haugsten, E.M.; Klimaschewski, L. Leupeptin enhances cell surface localization of fibroblast growth factor receptor 1 in adult sensory neurons by increased recycling. *Eur. J. Cell Biol.* **2012**, *91*, 129–138. [CrossRef] [PubMed]

7.   Stachowiak, E.K.; Maher, P.A.; Tucholski, J.; Mordechai, E.; Joy, A.; Moffett, J.; Coons, S.; Stachowiak, M.K. Nuclear accumulation of fibroblast growth factor receptors in human glial cells-association with cell proliferation. *Oncogene* **1997**, *14*, 2201–2211. [CrossRef] [PubMed]

8.   Planque, N. Nuclear trafficking of secreted factors and cell-surface receptors: New pathways to regulate cell proliferation and differentiation, and involvement in cancers. *Cell Commun. Signal.* **2006**, *4*, 7. [CrossRef] [PubMed]

9.   Peng, H.; Myers, J.; Fang, X.; Stachowiak, E.K.; Maher, P.A.; Martins, G.G.; Popescu, G.; Berezney, R.; Stachowiak, M.K. Integrative nuclear FGFR1 signaling (INFS) pathway mediates activation of the tyrosine hydroxylase gene by angiotensin II, depolarization and protein kinase C. *J. Neurochem.* **2002**, *81*, 506–524. [CrossRef]

10. Dunham-Ems, S.M.; Pudavar, H.E.; Myers, J.M.; Maher, P.A.; Prasad, P.N.; Stachowiak, M.K. Factors controlling fibroblast growth factor receptor-1's cytoplasmic trafficking and its regulation as revealed by FRAP analysis. *Mol. Biol. Cell* **2006**, *17*, 2223–2235. [CrossRef]

11. Stachowiak, M.K.; Birkaya, B.; Aletta, J.M.; Narla, S.T.; Benson, C.A.; Decker, B.; Stachowiak, E.K. Nuclear FGF receptor-1 and CREB binding protein: An integrative signaling module. *J. Cell. Physiol.* **2015**, *230*, 989–1002. [CrossRef] [PubMed]

12. Lee, Y.W.; Stachowiak, E.K.; Birkaya, B.; Terranova, C.; Capacchietti, M.; Claus, P.; Aletta, J.M.; Stachowiak, M.K. NGF-induced cell differentiation and gene activation is mediated by integrative nuclear FGFR1 signaling (INFS). *PLoS ONE* **2013**, *8*, e68931. [CrossRef] [PubMed]

13. Horbinski, C.; Stachowiak, E.K.; Chandrasekaran, V.; Miuzukoshi, E.; Higgins, D.; Stachowiak, M.K. Bone morphogenetic protein-7 stimulates initial dendritic growth in sympathetic neurons through an intracellular fibroblast growth factor signaling pathway. *J. Neurochem.* **2002**, *80*, 54–63. [CrossRef]

14. Li, X.; Wang, C.; Xiao, J.; McKeehan, W.L.; Wang, F. Fibroblast growth factors, old kids on the new block. *Semin. Cell Dev. Biol.* **2016**, *53*, 155–167. [CrossRef] [PubMed]

15. Grusch, M.; Schelch, K.; Riedler, R.; Reichhart, E.; Differ, C.; Berger, W.; Inglés-Prieto, Á.; Janovjak, H. Spatio-temporally precise activation of engineered receptor tyrosine kinases by light. *EMBO J.* **2014**, *33*, 1713–1726. [CrossRef]

16. Tokuyasu, K.T. A technique for ultracryotomy of cell suspensions and tissues. *J. Cell Biol.* **1973**, *57*, 551–565. [CrossRef]

17. Liou, W.; Geuze, H.J.; Slot, J.W. Improving structural integrity of cryosections for immunogold labeling. *Histochem. Cell Biol.* **1996**, *106*, 41–58. [CrossRef]

18. Hess, M.W.; Vogel, G.F.; Yordanov, T.E.; Witting, B.; Gutleben, K.; Ebner, H.L.; de Araujo, M.E.G.; Filipek, P.A.; Huber, L.A. Combining high-pressure freezing with pre-embedding immunogold electron microscopy and tomography. *Traffic* **2018**, *19*, 639–649. [CrossRef] [PubMed]

19. Jin, Y.; Pasumarthi, K.B.; Bock, M.E.; Lytras, A.; Kardami, E.; Cattini, P.A. Cloning and expression of fibroblast growth factor receptor-1 isoforms in the mouse heart: Evidence for isoform switching during heart development. *J. Mol. Cell. Cardiol.* **1994**, *26*, 1449–1459. [CrossRef] [PubMed]

20. Duan, L.; Hope, J.M.; Guo, S.; Ong, Q.; François, A.; Kaplan, L.; Scherrer, G.; Cui, B. Optical Activation of TrkA Signaling. *ACS Synth. Biol.* **2018**, *7*, 1685–1693. [CrossRef] [PubMed]

21. Tsvetanova, N.G.; von Zastrow, M. Spatial encoding of cyclic AMP signaling specificity by GPCR endocytosis. *Nat. Chem. Biol.* **2014**, *10*, 1061–1065. [CrossRef]

22. Chang, K.Y.; Woo, D.; Jung, H.; Lee, S.; Kim, S.; Won, J.; Kyung, T.; Park, H.; Kim, N.; Yang, H.W.; et al. Light-inducible receptor tyrosine kinases that regulate neurotrophin signalling. *Nat. Commun.* **2014**, *5*, 4057. [CrossRef] [PubMed]

23. Kim, N.; Kim, J.M.; Lee, M.; Kim, C.Y.; Chang, K.Y.; Heo, W.D. Spatiotemporal control of fibroblast growth factor receptor signals by blue light. *Chem. Biol.* **2014**, *21*, 903–912. [CrossRef]

24. Reichhart, E.; Ingles-Prieto, A.; Tichy, A.M.; McKenzie, C.; Janovjak, H. A Phytochrome Sensory Domain Permits Receptor Activation by Red Light. *Angew. Chem. Int. Ed. Engl.* **2016**, *55*, 6339–6342. [CrossRef] [PubMed]

25. Haugsten, E.M.; Sørensen, V.; Brech, A.; Olsnes, S.; Wesche, J. Different intracellular trafficking of FGF1 endocytosed by the four homologous FGF receptors. *J. Cell Sci.* **2005**, *118*, 3869–3881. [CrossRef]

26. Irschick, R.; Trost, T.; Karp, G.; Hausott, B.; Auer, M.; Claus, P.; Klimaschewski, L. Sorting of the FGF receptor 1 in a human glioma cell line. *Histochem. Cell Biol.* **2013**, *139*, 135–148. [CrossRef] [PubMed]

27. Maher, P.A. Nuclear Translocation of fibroblast growth factor (FGF) receptors in response to FGF-2. *J. Cell Biol.* **1996**, *134*, 529–536. [CrossRef]

28. Małecki, J.; Wiedłocha, A.; Wesche, J.; Olsnes, S. Vesicle transmembrane potential is required for translocation to the cytosol of externally added FGF-1. *EMBO J.* **2002**, *21*, 4480–4490. [CrossRef]

29. Reilly, J.F.; Maher, P.A. Importin beta-mediated nuclear import of fibroblast growth factor receptor: Role in cell proliferation. *J. Cell Biol.* **2001**, *152*, 1307–1312. [CrossRef]

30. Stachowiak, M.K.; Fang, X.; Myers, J.M.; Dunham, S.M.; Berezney, R.; Maher, P.A.; Stachowiak, E.K. Integrative nuclear FGFR1 signaling (INFS) as a part of a universal "feed-forward-and-gate" signaling module that controls cell growth and differentiation. *J. Cell. Biochem.* **2003**, *90*, 662–691. [CrossRef]

31. Somanathan, S.; Stachowiak, E.K.; Siegel, A.J.; Stachowiak, M.K.; Berezney, R. Nuclear matrix bound fibroblast growth factor receptor is associated with splicing factor rich and transcriptionally active nuclear speckles. *J. Cell. Biochem.* **2003**, *90*, 856–869. [CrossRef] [PubMed]

32. Stachowiak, M.K.; Maher, P.A.; Joy, A.; Mordechai, E.; Stachowiak, E.K. Nuclear localization of functional FGF receptor 1 in human astrocytes suggests a novel mechanism for growth factor action. *Brain. Res. Mol. Brain. Res.* **1996**, *38*, 161–165. [CrossRef]

33. Stachowiak, M.K.; Maher, P.A.; Joy, A.; Mordechai, E.; Stachowiak, E.K. Nuclear accumulation of fibroblast growth factor receptors is regulated by multiple signals in adrenal medullary cells. *Mol. Biol. Cell* **1996**, *7*, 1299–1317. [CrossRef] [PubMed]

34. Chioni, A.M.; Grose, R. FGFR1 cleavage and nuclear translocation regulates breast cancer cell behavior. *J. Cell. Biol.* **2012**, *197*, 801–817. [CrossRef] [PubMed]

35. Coleman, S.J.; Bruce, C.; Chioni, A.M.; Kocher, H.M.; Grose, R.P. The ins and outs of fibroblast growth factor receptor signalling. *Clin. Sci. (Lond)* **2014**, *127*, 217–231. [CrossRef] [PubMed]

36. Hart, K.C.; Robertson, S.C.; Kanemitsu, M.Y.; Meyer, A.N.; Tynan, J.A.; Donoghue, D.J. Transformation and Stat activation by derivatives of FGFR1, FGFR3, and FGFR4. *Oncogene* **2000**, *19*, 3309–3320. [CrossRef]

37. Klimaschewski, L.; Nindl, W.; Feurle, J.; Kavakebi, P.; Kostron, H. Basic fibroblast growth factor isoforms promote axonal elongation and branching of adult sensory neurons in vitro. *Neuroscience* **2004**, *126*, 347–353. [CrossRef]

38. Hausott, B.; Schlick, B.; Vallant, N.; Dorn, R.; Klimaschewski, L. Promotion of neurite outgrowth by fibroblast growth factor receptor 1 overexpression and lysosomal inhibition of receptor degradation in pheochromocytoma cells and adult sensory neurons. *Neuroscience* **2008**, *153*, 461–473. [CrossRef]

39. Hausott, B.; Rietzler, A.; Vallant, N.; Auer, M.; Haller, I.; Perkhofer, S.; Klimaschewski, L. Inhibition of fibroblast growth factor receptor 1 endocytosis promotes axonal branching of adult sensory neurons. *Neuroscience* **2011**, *188*, 13–22. [CrossRef]

40. Toettcher, J.E.; Weiner, O.D.; Lim, W.A. Using optogenetics to interrogate the dynamic control of signal transmission by the Ras/Erk module. *Cell* **2013**, *155*, 1422–1434. [CrossRef]

41. Zhang, K.; Duan, L.; Ong, Q.; Lin, Z.; Varman, P.M.; Sung, K.; Cui, B. Light-mediated kinetic control reveals the temporal effect of the Raf/MEK/ERK pathway in PC12 cell neurite outgrowth. *PLoS ONE* **2014**, *9*, e92917. [CrossRef]

42. Hayashi, H.; Tsuchiya, Y.; Nakayama, K.; Satoh, T.; Nishida, E. Down-regulation of the PI3-kinase/Akt pathway by ERK MAP kinase in growth factor signaling. *Genes Cells* **2008**, *13*, 941–947. [CrossRef] [PubMed]

43. Hensel, N.; Baskal, S.; Walter, L.M.; Brinkmann, H.; Gernert, M.; Claus, P. ERK and ROCK functionally interact in a signaling network that is compensationally upregulated in Spinal Muscular Atrophy. *Neurobiol. Dis.* **2017**, *108*, 352–361. [CrossRef] [PubMed]

44. Angelastro, J.M.; Klimaschewski, L.; Tang, S.; Vitolo, O.V.; Weissman, T.A.; Donlin, L.T.; Shelanski, M.L.; Greene, L.A. Identification of diverse nerve growth factor-regulated genes by serial analysis of gene expression (SAGE) profiling. *Proc. Natl. Acad. Sci. USA* **2000**, *97*, 10424–10429. [CrossRef]

45. Greene, L.A.; Tischler, A.S. Establishment of a noradrenergic clonal line of rat adrenal pheochromocytoma cells which respond to nerve growth factor. *Proc. Natl. Acad. Sci. USA* **1976**, *73*, 2424–2428. [CrossRef] [PubMed]

46. Hausott, B.; Klimaschewski, L. Membrane turnover and receptor trafficking in regenerating axons. *Eur. J. Neurosci.* **2016**, *43*, 309–317. [CrossRef]

47. Robinson, M.J.; Stippec, S.A.; Goldsmith, E.; White, M.A.; Cobb, M.H. A constitutively active and nuclear form of the MAP kinase ERK2 is sufficient for neurite outgrowth and cell transformation. *Curr. Biol.* **1998**, *8*, 1141–1150. [CrossRef]

48. Webster, M.K.; Donoghue, D.J. Enhanced signaling and morphological transformation by a membrane-localized derivative of the fibroblast growth factor receptor 3 kinase domain. *Mol. Cell. Biol.* **1997**, *17*, 5739–5747. [CrossRef]

49. Peng, H.; Moffett, J.; Myers, J.; Fang, X.; Stachowiak, E.K.; Maher, P.; Kratz, E.; Hines, J.; Fluharty, S.J.; Mizukoshi, E.; et al. Novel nuclear signaling pathway mediates activation of fibroblast growth factor-2 gene by type 1 and type 2 angiotensin II receptors. *Mol. Biol. Cell* **2001**, *12*, 449–462. [CrossRef]
50. Hu, Y.; Fang, X.; Dunham, S.M.; Prada, C.; Stachowiak, E.K.; Stachowiak, M.K. 90-kDa ribosomal S6 kinase is a direct target for the nuclear fibroblast growth factor receptor 1 (FGFR1): Role in FGFR1 signaling. *J. Biol. Chem.* **2004**, *279*, 29325–29335. [CrossRef]

9

Fibroblast Growth Factor Receptor Functions in Glioblastoma

Ana Jimenez-Pascual and Florian A. Siebzehnrubl *

European Cancer Stem Cell Research Institute, Cardiff University School of Biosciences, Cardiff CF24 4HQ, UK
* Correspondence: fas@cardiff.ac.uk

Abstract: Glioblastoma is the most lethal brain cancer in adults, with no known cure. This cancer is characterized by a pronounced genetic heterogeneity, but aberrant activation of receptor tyrosine kinase signaling is among the most frequent molecular alterations in glioblastoma. Somatic mutations of fibroblast growth factor receptors (*FGFRs*) are rare in these cancers, but many studies have documented that signaling through FGFRs impacts glioblastoma progression and patient survival. Small-molecule inhibitors of FGFR tyrosine kinases are currently being trialed, underlining the therapeutic potential of blocking this signaling pathway. Nevertheless, a comprehensive overview of the state of the art of the literature on FGFRs in glioblastoma is lacking. Here, we review the evidence for the biological functions of FGFRs in glioblastoma, as well as pharmacological approaches to targeting these receptors.

Keywords: FGFR; review; malignant glioma; brain cancer; astrocytoma; fibroblast growth factor

1. Introduction

Fibroblast growth factors (FGFs) were first isolated from bovine brain extracts in 1939 and characterized by their ability to induce proliferation of fibroblasts [1]. It took another 50 years to discover and clone the first of their cognate receptors [2]. FGFRs control many biological functions, including cell proliferation, survival, and cytoskeletal regulation (for review, see [3]). FGFR signaling is important during embryonal development of the CNS, and as a survival mechanism for adult neurons and astrocytes [4–6]. Furthermore, FGFR signaling was found to promote self-renewal and fate specification of neural stem cells [7].

In many cancers, *FGFR* aberrations have been implicated in tumor development and progression [8,9], and include *FGFR* overexpression, amplification, mutations, splicing isoform variations, and *FGFR* translocations [10,11]. While *FGFR* genomic alterations have been identified in many solid tissue cancers, such events remain rare in glioblastoma (GBM) (Table 1) [12]. Nonetheless, FGFR expression changes in astrocytes can lead to malignant transformation and GBM progression due to the activation of mitogenic, migratory, and antiapoptotic responses [13–15]. Of note, fusions between *FGFR* and *TACC* (transforming acidic coiled-coil containing proteins) genes were shown to be oncogenic in GBM [16], and occur in about 3% of GBM patients [12]. Whole-genome analyses of patient samples have revealed that the number of *FGFR* mutations and amplifications are generally very low in GBM (*FGFR1*: 51/3068 samples, *FGFR2*: 12/2662; *FGFR3*: 16/2887; *FGFR4*: 9/2456; cancer.sanger.ac.uk; [17]). Not only are oncogenic mutations in *FGFRs* rare in GBM, the lack of *FGFR* passenger mutations (i.e., mutations not providing a survival benefit) suggests that these are selected against, and that the maintenance of dynamic FGFR signaling is important for the development and/or progression of GBM.

Table 1. Common *FGFR* genomic aberration in solid tumors. FGF signaling deregulation is involved in the development of many different human cancers. Four FGFR genomic alterations are represented in this table: gene amplification, point mutations, chromosomal translocations, and FGFR splicing isoforms. Each FGFR alteration is linked with the most significant cancers that contain those alterations. The role of the majority of the discovered point mutations in FGFR is unknown in cancer. Adapted from [10,18].

Gene	Gene Amplifications	Point Mutations	Chromosomal Translocations	Splice Variants
FGFR1	Breast, ovarian, bladder, and lung cancer	Majority of cancers. Example: Melanoma	Stem cell leukemia/lymphoma (SCLL), GBM	IIIc: small cell lung carcinoma Iβ: breast cancer and GBM
FGFR2	Breast, gastric, lung cancer	Majority of cancers. Example: Endometrial carcinoma		IIIb: breast, endometrial, cervical, lung, pancreatic and colorectal cancer IIIc: prostate cancers
FGFR3	Bladder cancer	Majority of cancers. Example: bladder cancer	GBM, T-cell lymphoma and bladder	IIIc: bladder cancer
FGFR4	Colorectal cancer	Majority of cancers. Example: metastatic breast cancer and rhabdomyosarcoma		

We hypothesize that the neurodevelopmental and cell survival functions of this signaling pathway are at least partly conserved in GBM. Thus, dynamic FGFR signaling needs to be maintained for the survival of GBM cells, and therefore evolutionary pressure selects against both activating and inactivating mutations. Here, we review the current literature on FGFRs in GBM, and the evidence for differential functions of individual FGFRs in brain tumor progression.

2. FGFR Structure

There are four known FGFRs, FGFR1–4, which are membrane-bound receptor tyrosine kinases (RTKs). A fifth member of the FGFR family, FGFRL1, is lacking a transmembrane domain and is therefore soluble. FGFRL1 acts as an antagonist to FGFR signaling [19,20]. Structurally, FGFR1–4 consist of three different domains: an extracellular ligand binding domain, a transmembrane domain, and an intracellular domain that interacts with cytoplasmic molecules and transduces FGFR signaling [8,21,22] (Figure 1).

The extracellular domain can bind FGF ligands, heparan sulfate (HS), and extracellular matrix molecules, which can act as a scaffold to enable receptor binding of specific FGFs. It is divided into three immunoglobulin-like (Ig) loops: Ig-I, Ig-II, and Ig-III (also called D1, D2, and D3) [23]. Ig-I is linked to Ig-II by a stretch of 30 acidic residues called the acid box, a unique region of FGFRs [10,24]. Ig-I and the acid box have receptor auto-inhibitory functions [25–27] while the Ig-II and Ig-III subdomains form the ligand binding site of the receptor [3,15,18,28,29]. Ig-II contains the heparin/HS binding region and FGF binding activity site, while the junction between Ig-II and Ig-III controls heparin and FGF affinity [21,30–33] (Figure 1).

Multiple FGFR isoforms are generated by alternative splicing of the region encoding for the extracellular domain. This modifies the affinity and sensitivity of the receptors for different FGF ligands [34,35]. Thus, an array of FGFR isoforms is created that can fine-tune the response of cells to the large number of potential FGF ligands available in their specific environment. FGF sensitivity is further modified by co-receptors, such as Klotho family members, which are required for binding of endocrine FGFs [36–38].

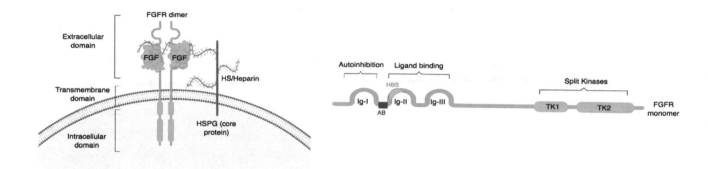

Figure 1. Domain structure of FGFRs: an extracellular domain containing ligand binding site is followed by a single transmembrane domain, and an intracellular domain containing split tyrosine kinases. Left panel: organization of the FGF–FGFR complex at the cell surface. The FGF–FGFR complex is stabilized by a heparin/HS chain of the HS proteoglycan (HSPG). Right panel: The extracellular domain of the receptor is composed of three Ig-like domains: Ig-I, Ig-II, and Ig-III. Ig-I has autoinhibitory capacity while Ig-II and Ig-III form the ligand binding domain. Ig-II contains the heparin/HS binding site (HBS) and is separated from Ig-I by an acid box (AB). The cytoplasmic domain is formed by two tyrosine kinases: tyrosine kinase 1 (TK1) and tyrosine kinase 2 (TK2). Image created with biorender.com.

Alternatively-spliced β isoforms of FGFR1 or FGFR2 are produced by the exclusion of the Ig-I domain, which is encoded by exon 3. Due to the auto-inhibitory function of the Ig-I domain, the β isoform has considerably higher affinity for FGFs, and is oncogenic [39,40]. Retention of the Ig-I domain creates FGFR α isoforms [28].

Additionally, alternative splicing generates two isoforms of the Ig-III domain, known as Ig-IIIb and Ig-IIIc, in FGFR1–3, but not FGFR4 [10,41–43]. Ig-IIIb and Ig-IIIc are generated by exon skipping, and are encoded by exons 8 and 9, respectively (Figure 2). By contrast, exon 7, encoding Ig-IIIa, is present in all splice variants. Different splice-regulatory proteins have been identified that control the splicing of Ig-IIIb, such as regulatory RNA-binding protein (RBP), and the epithelial splicing regulatory proteins (ESRP1/2) [32,40,44].

The expression of FGFR splice variants is tissue-dependent. For example, the Ig-IIIb isoform is more prevalent in epithelial tissues, while Ig-IIIc is preferentially expressed in mesenchymal ones [22,24,32]. Switching of epithelial and mesenchymal isoforms occurs during epithelial–mesenchymal transition, which is known as the IIIb/IIIc switch. Hence, FGFR isoform expression is also related to tissue plasticity, and changes during tissue growth, proliferation, and remodeling [45].

The FGFR transmembrane domain is crucial for transferring the signal from the extracellular to the intracellular domain, the latter consisting of a juxtamembrane domain, two tyrosine kinase domains, and the C-terminal tail [10,32,46,47]. FGFR domains are highly conserved among receptors, and the tyrosine kinase domain shares the highest homology. The Ig-III domain is also highly conserved, especially between FGFR1 and FGFR2 [48].

The binding of FGF ligands to HSPGs causes FGFR dimerization and activation in the -COOH receptor tail of the cytoplasmic tyrosine residues by phosphorylation [49,50]. For instance, autophosphorylation of FGFR1 tyrosine (Y) residues occurs in three steps. Firstly, phosphorylation of Y653 leads to a 50–100-fold increase of the catalytic core activation of the intracellular domain. Secondly, Y583, Y463, Y766, and Y585 sites are consequently phosphorylated, and finally, the second tyrosine kinase domain phosphorylation increases the tyrosine kinase activity 10-fold. This sequence of autophosphorylation follows a specific and controlled order that, if deregulated, can induce malfunction of the pathway [49].

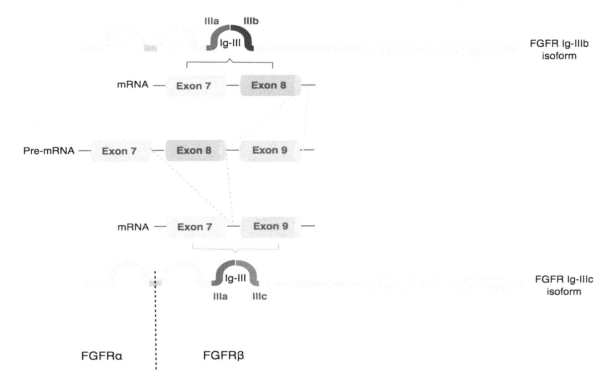

Figure 2. Schematic representation of FGFR splice isoforms. The Ig-III domain of FGFR1–3 is encoded by exons 7–9. Exon 7 encodes Ig-IIIa, which consists of the N-terminal half of the Ig-III loop. The C-terminal half is formed by the IIIb or IIIc sequence, which is generated by the selective inclusion of exons 8 or 9, respectively. Truncation of the Ig-I loop creates FGFRβ isoforms (dotted line), while the full-length receptor is termed FGFRα. Image created with biorender.com.

Different models have been proposed for FGFR dimerization depending on the ligand–heparin–receptor complex, specificity between the ligand and the receptor, and the heparin length required for the binding [21,24]. In the first model, HS increases the association between the receptor and the high affinity binding site of the ligand, forming a ternary complex (1:1:1 FGF–HS–FGFR) that then interacts with a second receptor, inducing FGFR dimerization through the FGF low affinity binding site (2:1:1 FGFR–HS–FGF). On the other hand, the symmetrical model suggests that two individual ternary complexes are formed. The FGFR dimerization will then occur by FGFR–FGFR direct interaction, FGF ligand interaction, or by HS–HS link (2:2:2 FGFR–HS–FGF). In this model, HS enhances FGFR dimerization, but it is not crucial. Finally, according to the asymmetric model, HS attaches to two FGF–FGFR complexes, binding both FGFs but only one of the receptors (2:1:2 FGFR–HS–FGF) [21]. Therefore, although different models have been suggested, more research is needed to clarify the stoichiometry of FGFR dimerization [21,24,51].

3. FGFR Signaling Cascade

FGF–FGFR stimulates cell signaling pathways related to cell proliferation, survival, cytoskeletal regulation, and FGFR degradation [10]. Cell proliferation is mainly induced by RAC/JNK and RAS–MAPK signaling pathways [3]. RAC kinases can be activated by the transient phosphorylation of CRK, which simultaneously stimulates RAC phosphorylation trough DOCK1 or SOS/RAS [10]. RAC kinases promote proliferation by the activation of JNK and p38. Alternatively, the RAS/RAF/MEK/ERK signaling pathway can be activated by the FRS2–GRB2–SOS–SHP2 complex assembly or byPKC activation through PLC phosphorylation [18,22].

Cell survival is mainly promoted by phosphorylation of PI3K/AKT signaling through the FRS2–GRB2–GAB1 complex. Finally, FGFRs are also implicated in cytoskeletal regulation, as PLC phosphorylation leads to the hydrolysis of PIP_2 into IP3, inducing calcium release [18] (Figure 3).

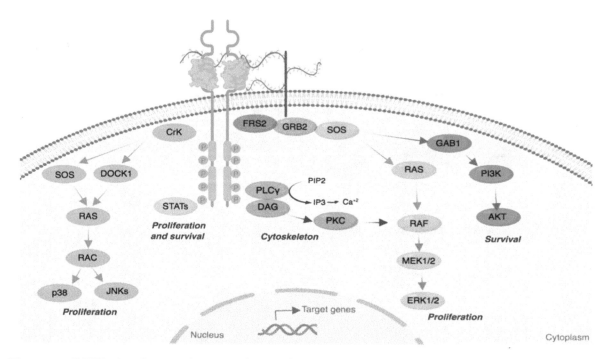

Figure 3. FGFR signaling pathway. After ligand binding, FGFRs dimerize and activate multiple signal transduction pathways. Each pathway induces the expression of specific target genes related to cell proliferation (STATs, RAS/p38/JNKs, and RAS/MAPK/ERK), survival (STATs and PI3K/AKT), and cytoskeleton regulation (PLC/Ca^{2+}). Kinases are color-coded according to their specific signaling pathway. Image created with biorender.com.

Because FGFR signaling acts upon many biological functions, a regulatory system that controls its timing, spread, and balances its activation is required. This is important as the activation of the signaling cascade depends on FGFR expression and localization to the cell membrane. Therefore, receptor availability depends on the balance between its recycling and degradation rate, which differ among receptors. One of these regulatory systems is FGFR internalization or constitutive endocytosis. FGFR synthesis occurs at a higher level than its internalization. However, after ligand-binding, FGFR internalization from the plasma membrane accelerates [52]. FGFR internalization is primarily mediated by clathrin-dependent endocytosis and requires the SRC–FRS2 complex [53]. The internalization rate depends on the receptor type—FGFR1 has the highest internalization rate and FGFR3 the lowest. Endocytosis of activated FGFRs involves detachment from the SRC complex [54]. FGFRs can then re-translocate to the cytosol, mitochondria, nucleus (to directly regulate gene expression), or to the endosomal compartment for receptor degradation [52]. The latter requires interaction between the FRS2–GRB complex and CBL, and is receptor-independent [22] (Figure 4). Indeed, FGFR1 has more ubiquitination sites than FGFR4, so its degradation rate is likely higher [10].

Other regulatory systems of FGFR signaling are the negative regulators SEF, SPRY1/SPRY4, and MKP1/MKP3. The activation of cell proliferation is counterbalanced by SEF, which negatively regulates ERK and AKT activation [18]. Similarly, SPRY1/SPRY4 reduce proliferation by directly interacting with RAS/RAF kinases or by blocking the FRS2–GRB2–SOS–SHP2 complex. MKP1 and MKP3 also attenuate FGFR signaling by dephosphorylating MAPK and ERK [10] (Figure 4).

FGF signaling is also negatively regulated by the autoinhibitory (Ig-I) domain of the receptors. This is controlled by the electrostatic interactions between the negatively charged acid box with the highly basic heparin binding site in Ig-II [25]. This complex blocks the heparin–FGF binding, minimizing FGFR activation. The auto-inhibitory capacity is crucial for the modulation of the pathway, as the high amount of HSPGs from the cell surface and the extracellular matrix increases the probabilities of FGF–heparin binding and activation of the RTK cascade [24].

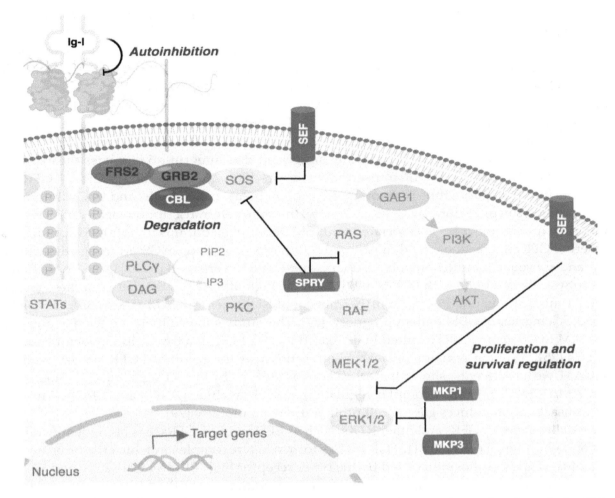

Figure 4. FGFR signaling pathway regulation. FGFR signaling is negatively regulated, partly by CBL (inducing FGFR degradation after receptor internalization), by SEF, SPRY, MKP1, and MPK3 (which negatively regulate proliferation and survival related pathways). FGFRs can also regulate their own activation due to the autoinhibitory function of Ig-I. Image created with biorender.com.

Other factors involved in FGFR pathway regulation are the ligand affinity for the receptor and the ligand amount and availability. Extracellular FGFs are protected and stored by HS proteoglycans. Heparanases are directly involved in FGF signaling regulation, as they cleave the HS chain, thus releasing FGFs in the vicinity of cells. Depending on the cell type and the growth factor released, heparanases are therefore involved in cell growth, differentiation, or stemness maintenance [55]. Likewise, sufficient amounts of ligand and heparin/HSPGs are necessary for stabilizing FGFR dimerization. The necessary ligand concentration is dependent on the ligand-binding affinity of the FGFRs, which depends on the FGFR splice isoforms [3,8,42]. For example, FGF2 activates both FGFR1 IIIb and IIIc isoforms, while it has a higher affinity for the isoform IIIc in FGFR2 and FGFR3 [42].

4. Crosstalk between FGFRs and Other Cell Surface Molecules

FGFRs can be modulated independently of their ligands by integral cell membrane proteins, such as G-protein-coupled receptors (GPCRs), cell adhesion molecules (CAMs), and other RTKs, which play a crucial role in the induction of specific cell responses and fate during development and cancer [56]. GPCRs can transactivate FGFRs by promoting the activation of matrix metalloproteinases, resulting in cleavage of FGFs, or by directly interacting with FGFRs [57]. GPCR-mediated FGFR1 transactivation is associated with neuronal differentiation, neurite growth, and synaptic plasticity [56]. FGFR1 modulation by GPCRs (e.g., CB1A, 5-HT1A, and mAChR) involves the activation of the SRC-ERK1/2 pathway [57–59]. In C6 glioma cells, crosstalk between the mu-Opioid receptor and

FGFR1 was shown to activate this signaling cascade, but the specific mechanism is not yet completely understood [60].

FGFR activity is also modulated by CAMs, cell surface proteins that regulate cell–cell interactions and motility. Importantly, the FGFR acid box region is required for the CAM/FGFR interaction [61], hence the FGFR β isoforms cannot be transactivated by CAMs. CAMs of the integrin, cadherin, and immunoglobulin (e.g., NCAM and L1-CAM) superfamilies signal through FGFRs to induce neurite outgrowth, cell survival, and oncogenesis [61–66]. Integrins signaling can activate cell proliferation, survival, and invasion [67], and integrin α6 [68] and α7 [69] have been linked to GBM cancer stem cells (GSCs). A recent study suggested that integrin α6 regulates the expression of FGFR1 through ZEB1 and YAP transcription factors [70]. N-cadherin stabilizes FGFR1 and decreases its internalization, thus promoting invasion in breast cancer cells, and N-Cadherin/FGFR crosstalk promotes neurite outgrowth [71]. Of note, the stem cell transcription factor ZEB1 regulates N-Cadherin expression, which is associated with EMT and invasion. It is tempting to speculate that N-Cadherin/FGFR1 interactions could constitute a positive feedback loop in GSCs through the activation of ZEB1 and subsequent induction of N-Cadherin and FGFR1 expression [70,72] (see also Section 5.1). NCAMs physically associate with FGFRs and inhibit the high-affinity binding between these receptors and their canonical ligands [56,73]. Furthermore, polysialic acid-NCAM (PSA-NCAM) has been described as a marker of GBM patient prognosis [74]. This study showed that a targeted expression of PSA-NCAM in C6 glioma cells resulted in increased levels of Olig2, a transcription factor associated with GSCs [75]. While it remains unclear whether this was the result of FGFR transactivation by PSA-NCAM, we have recently shown that OLIG2 can be induced by FGFR1 signaling [72]. Furthermore, the L1-CAM/FGFR1/Anosmin-1 complex regulates neurite branching [76–78] and L1-CAM-mediated FGFR1 transactivation induces glioma cell proliferation and motility [79].

Crosstalk between FGFRs and other RTKs, such as EPHs and PDGFRs, has been identified in Y2H screens and endothelial cells [80,81]. FGFR/RTKs form a heterocomplex in which the tyrosine kinase domain of the FGFR is phosphorylated by the other receptor [56]. EPHA4 transactivated FGFR1 in the U251 glioma cell line, promoting cell growth and migration, and EPHA4 expression is increased in glioma [82]. Less is known about potential crosstalk between other RTKs and FGFRs in GBM. In summary, FGFR activity can be modulated non-canonically by other cell surface proteins, resulting in the activation of intracellular signaling pathways and cell responses associated with FGFR signaling.

5. Expression and Functions of FGFRs in Glioblastoma

Gene expression analysis of TCGA data (GBM 540) revealed profound heterogeneity of *FGFR1–4* expression across GBM patients [72]. Below, we discuss the evidence for the functions of individual FGFRs in glioma.

5.1. FGFR1

Yamaguchi et al. found that expression of FGFR1 increases with WHO grade in astrocytomas [39], and increased FGFR1 levels in GBM are not due to amplification of the *FGFR1* gene [83].

In addition to the increased expression of FGFR1 in malignant gliomas, the ratio of alternatively spliced FGFR1 α/β isoforms changes with progression to more aggressive brain cancers. While FGFR1 α is the predominant isoform in normal brain and low-grade gliomas, high-grade gliomas show a shift towards the expression of FGFR1 β [13,39]. Loss of the FGFR1 α exon increases the receptor–ligand affinity [28], thus, changes in alternative splicing may contribute to GBM malignancy by increasing the sensitivity of tumor cells to FGFs present in their environment.

Functionally, FGFR1 expression in malignant glioma has been associated with increased migration of cancer cells [82]. In this study, a high expression of EPHA4 in glioma cells was found to potentiate FGF2–FGFR1 signaling and promoted cell growth and migration through the AKT/MAPK and RAC1/CDC42 pathways, respectively. Data from our lab support that FGFR1 loss results in reduced tumor invasion in vivo (Jimenez-Pascual and Siebzehnrubl, unpublished observation).

Loilome and colleagues identified FGFR1 as a potential transducer of FGF2 effects on glioma cell proliferation [84], but whether other FGFRs also contribute was not directly tested. Nevertheless, the pharmacological inhibition of FGFR signaling significantly reduced tumor cell growth in a range of established and patient-derived glioma lines.

The malignancy-promoting effects of FGFR1 were further demonstrated in a study that found FGFR1 signaling promoting radioresistance in glioma cell lines through PLC1γ and HIF1α [85]. FGFR1 expression is regulated by the stem-cell associated transcription factor ZEB1 [70], suggesting that FGFR1 may be associated with GBM cancer stem cells. We recently performed a comprehensive analysis of the functions of FGFR1-3 in GBM and found that FGFR1 indeed is preferentially expressed on GSCs, where it regulates the expression of the critical stem cell transcription factors SOX2, OLIG2, and ZEB1, thereby promoting tumorigenicity in vivo [72]. In summary, FGFR1 is a key regulator of tumor growth, invasion, therapy resistance, and cancer stemness in malignant glioma.

5.2. FGFR2

While FGFR1 is mainly expressed on neurons [6], FGFR2 is the primary FGFR on astrocytes [5]. In contrast to FGFR1, FGFR2 expression decreases with glioma grade [43]. Reduced expression of FGFR2, as well as its IIIb and IIIc isoforms, is associated with a higher tumor grade and poorer survival in glioma patients [43]. Tumors with a higher expression of FGFR2 showed significantly less proliferation, as identified by Ki-67 staining, but whether there is a direct link between FGFR2 signaling and slowing or exiting the cell cycle remains unclear. By contrast, experimental tumors derived from in vivo implantation of C6 glioma cells exhibited decreased tumor growth after inhibition of FGFR2 signaling by a dominant negative construct [86].

Our recent analysis of cell-surface FGFR expression patterns in GBM stem cell lines indicates that FGFR2 is nevertheless highly prevalent on GBM cells in vitro [72], but it remains to be tested whether FGFR2 loss results in increased proliferation and/or tumorigenicity. Loss of FGFR2 is associated with a loss of Chr. 10q, which in and of itself carries an unfavorable prognosis [87]. It is therefore conceivable that FGFR2 loss is not causally linked to reduced patient survival, and further work is needed to clarify the functional relevance of FGFR2 signaling in GBM.

5.3. FGFR3

In a small subset of GBM patients, fusion of the *FGFR3* and *TACC3* genes generates an oncogenic *FGFR3* form [16]. In rare cases, fusion between *FGFR1* and *TACC1* can occur as well [88]. In *FGFR3–TACC3*, the FGFR tyrosine kinase domain is fused to the TACC coiled-coil domain, resulting in constitutive activation of the fused receptor. Small-molecule FGFR inhibitors were effective at blocking tumor growth where *FGFR–TACC* fusion occurred, indicating that the fused receptor is causally linked to tumor development. Overall, *FGFR–TACC* fusions are found in ~3% of gliomas and are mutually exclusive with *EGFR* amplifications. Recently, it has been shown that *FGFR–TACC* fusions affect cell metabolism, activating oxidative phosphorylation and mitochondrial activity [89]. Of note, we have recently found that GSCs preferentially utilize oxidative phosphorylation and mitochondrial respiration [90], and therefore it would be interesting to investigate whether *FGFR–TACC* fusions also affect stemness pathways in GBM.

Whether FGFR3 has specific functions that differ from FGFR1 or -2 in GBM, and/or whether signaling through FGFR3 activates specific downstream signaling pathways remains unclear. Of note, global transcriptomic analysis of TCGA and CGGA datasets found increased expression of *FGFR3* in the classical and neural subtypes of GBM [91]. Gene ontology analysis showed an association of FGFR3 expression with biological processes of cell differentiation in this study.

A recent study investigating gene expression using single-cell RNA-Seq in GBM patients found that *FGFR3* expression is five-fold higher in invasive GBM cells compared to the tumor core [92]. Indeed, *FGFR3* was the second highest differentially expressed gene between invasive and tumor core GBM cells. While this suggests that FGFR3 may be functionally associated with tumor invasion, whether FGFR3 signaling is driving GBM invasion remains to be shown.

5.4. FGFR4

Very little evidence exists of the expression of FGFR4 in GBM. An early study found increased expression of the FGFR4 protein, but not mRNA, with increasing grade in astrocytoma [93]. Another study demonstrated the expression of FGFR4 across different GBM cell lines [84]. We recently investigated FGFR protein expression in GBM cells [72]. In our study, FGFR4 was not detectable by western blot in primary patient-derived GBM cells, and analysis of GBM patient data in the TCGA dataset showed heterogeneous, but overall low expression of *FGFR4*. Moreover, we could not find differences in survival when stratifying patients for *FGFR4* high or low expression. More research is needed to fully characterize whether FGFR4 is expressed on subsets of GBM cells, and whether it is functional in these cancers.

6. FGFRs as Therapeutic Targets in GBM

Oncogenic FGFR signaling promotes malignancy in many cancers, including CNS malignancies. Thus, pharmacological targeting of FGFRs may be therapeutically beneficial. A number of RTK inhibitors have been developed that show selectivity of FGFRs over other RTKs [18]. While some small-molecule inhibitors also target non-FGFR RTKs (e.g., dovitinib, levatinib, brivanib), others are selective for FGFR1–3 (e.g., PD173074, BGJ398, AZ4547, JNJ-493). To date, no small-molecule inhibitors exist with good selectivity for individual FGFR subtypes or isoforms [18].

Recently, a study identified FGFR signaling as a potential therapeutic target in pediatric glioma using a large-scale shRNA screen [94]. In this study, FGFR inhibitors (AZ4547, dovitanib, PD173074, ponatinib) were more effective in reducing the growth of pediatric glioma cells in vitro than the first-line chemotherapeutic agent Temozolomide.

The selective FGFR inhibitors AZ4547 and BGJ398 have been tested in clinical phase I/II (NCT028224133) and phase II trials (NCT01975701), respectively. AZ4547 was trialed in patients with recurrent IDH wild-type gliomas with *FGFR1–TACC1* or *FGFR3–TACC3* fusions, but this trial was suspended after analysis of the data from the first 12 patients. A trial of BGJ398 in malignant glioma patients with *FGFR1–TACC1* or *FGFR3–TACC3* fusion, and/or activating mutation in *FGFR1, -2, or -3*, was completed, but so far, no results have been published.

A phase I/II trial of the irreversible FGFR inhibitor TAS-120 (NCT02052778) is currently recruiting patients with advanced solid tumors, including brain tumors. As with the AZ4547 and BGJ398 trials, the focus of this trial is on patients with *FGFR* gene fusions or activating mutations.

Due to the prevalence of FGFRs on many CNS cells, and the importance of FGFR signaling for CNS cell survival, as well as the apparent intratumoral and intertumoral heterogeneity of FGFR expression in GBM, it will be interesting to see whether FGFR inhibitors are successful as monotherapy. Based on the evidence implicating FGFR1 as a GSC regulator, it is further tempting to speculate whether the targeted inhibition of FGFR1 in combination with conventional chemo/radiotherapy could prevent or delay recurrence in GBM.

7. Conclusions

Gene expression profiling and whole-genome sequencing data indicate that all four FGFRs are expressed to varying degrees in GBM, underlining the heterogeneity of this disease. Several studies have documented that high-grade gliomas show an increased expression of *FGFR1*, and decreased expression of *FGFR2* [13,39,43,72,83], but in almost all cases, *FGFR* expression was detected at a global level, and limited or no effort was made to further identify *FGFR* splice isoforms. A recent report investigating gene expression at the single-cell level [92] found that *FGFR3* was expressed the second-highest in invasive GBM cells. This illustrates that much more research is needed to unravel the functions of individual FGFRs and their splice isoforms in brain cancers in general, and GBM in particular.

To what extent signaling through individual FGFRs contributes to disease progression, and whether individual FGFRs and/or isoforms activate specific pathways linked to different pathobiological aspects of these cancers (e.g., invasion, tumor initiation, therapy resistance) remains largely unknown. Yet, the fact that FGFR subtypes are differentially expressed on cellular subpopulations within the same tumor suggests that individual FGFRs may have divergent functions in GBM [72,92].

The strongest evidence by far indicates that FGFR1 is an important contributor to poor outcome in GBM, and FGFR1 signaling is linked to cancer stemness, invasion, and radioresistance [70,72,85]. However, recent evidence from TCGA datasets highlights that *FGFR1–4* are expressed to varying degrees and in different combinations in patient samples [72]. This calls for a more detailed analysis of FGFR distribution across GBM patients and within individual tumors. It is conceivable that different combinations of FGFR subtypes and splice isoforms mediate and/or modulate different aspects of FGF signaling in GBM cells. Only a comprehensive analysis of the cell-surface expression of FGFRs at the single-cell level will dissect the intratumoral heterogeneity of these receptors, and thus provide the foundation for new, targeted approaches to blocking FGFR signaling for glioma therapy.

Author Contributions: Conceptualization, A.J.-P. and F.A.S.; writing—original draft preparation, A.J.-P. and F.A.S.; writing—review and editing, A.J.-P. and F.A.S.; visualization, A.J.-P.; supervision, F.A.S.; project administration, F.A.S.; funding acquisition, F.A.S.

References

1. Trowell, O.A.; Willmer, E.N. Studies on the growth of tissues in vitro VI. The effects of some tissue extracts on the growth of periosteal fibroblasts. *J. Exp. Biol.* **1939**, *16*, 60–70.

2. Lee, P.L.; Johnson, D.E.; Cousens, L.S.; Fried, V.A.; Williams, L.T. Purification and complementary DNA cloning of a receptor for basic fibroblast growth factor. *Science* **1989**, *245*, 57–60. [CrossRef] [PubMed]

3. Beenken, A.; Mohammadi, M. The FGF family: Biology, pathophysiology and therapy. *Nat. Rev. Drug. Discov.* **2009**, *8*, 235–253. [CrossRef]

4. Guillemot, F.; Zimmer, C. From cradle to grave: The multiple roles of fibroblast growth factors in neural development. *Neuron* **2011**, *71*, 574–588. [CrossRef] [PubMed]

5. Miyake, A.; Hattori, Y.; Ohta, M.; Itoh, N. Rat oligodendrocytes and astrocytes preferentially express fibroblast growth factor receptor-2 and -3 mRNAs. *J. Neurosci. Res.* **1996**, *45*, 534–541. [CrossRef]

6. Gonzalez, A.M.; Berry, M.; Maher, P.A.; Logan, A.; Baird, A. A comprehensive analysis of the distribution of FGF-2 and FGFR1 in the rat brain. *Brain Res.* **1995**, *701*, 201–226. [CrossRef]

7. Frinchi, M.; Bonomo, A.; Trovato-Salinaro, A.; Condorelli, D.F.; Fuxe, K.; Spampinato, M.G.; Mudo, G. Fibroblast growth factor-2 and its receptor expression in proliferating precursor cells of the subventricular zone in the adult rat brain. *Neurosci. Lett.* **2008**, *447*, 20–25. [CrossRef] [PubMed]

8. Haugsten, E.M.; Wiedlocha, A.; Olsnes, S.; Wesche, J. Roles of fibroblast growth factor receptors in carcinogenesis. *Mol. Cancer Res.* **2010**, *8*, 1439–1452. [CrossRef]

9. Greulich, H.; Pollock, P.M. Targeting mutant fibroblast growth factor receptors in cancer. *Trends Mol. Med.* **2011**, *17*, 283–292. [CrossRef]

10. Tiong, K.H.; Mah, L.Y.; Leong, C.O. Functional roles of fibroblast growth factor receptors (FGFRs) signaling in human cancers. *Apoptosis* **2013**, *18*, 1447–1468. [CrossRef]

11. Costa, R.; Carneiro, B.A.; Taxter, T.; Tavora, F.A.; Kalyan, A.; Pai, S.A.; Chae, Y.K.; Giles, F.J. FGFR3-TACC3 fusion in solid tumors: Mini review. *Oncotarget* **2016**, *7*, 55924–55938. [CrossRef] [PubMed]

12. Lasorella, A.; Sanson, M.; Iavarone, A. FGFR-TACC gene fusions in human glioma. *Neuro Oncol.* **2017**, *19*, 475–483. [CrossRef] [PubMed]

13. Morrison, R.S.; Yamaguchi, F.; Saya, H.; Bruner, J.M.; Yahanda, A.M.; Donehower, L.A.; Berger, M. Basic fibroblast growth factor and fibroblast growth factor receptor I are implicated in the growth of human astrocytomas. *J. Neuro Oncol.* **1994**, *18*, 207–216. [CrossRef]

14. Yamada, S.M.; Yamaguchi, F.; Brown, R.; Berger, M.S.; Morrison, R.S. Suppression of glioblastoma cell growth following antisense oligonucleotide-mediated inhibition of fibroblast growth factor receptor expression. *Glia* **1999**, *28*, 66–76. [CrossRef]

15. Dienstmann, R.; Rodon, J.; Prat, A.; Perez-Garcia, J.; Adamo, B.; Felip, E.; Cortes, J.; Iafrate, A.J.; Nuciforo, P.; Tabernero, J. Genomic aberrations in the FGFR pathway: Opportunities for targeted therapies in solid tumors. *Ann. Oncol.* **2014**, *25*, 552–563. [CrossRef] [PubMed]

16. Singh, D.; Chan, J.M.; Zoppoli, P.; Niola, F.; Sullivan, R.; Castano, A.; Liu, E.M.; Reichel, J.; Porrati, P.; Pellegatta, S.; et al. Transforming fusions of FGFR and TACC genes in human glioblastoma. *Science* **2012**, *337*, 1231–1235. [CrossRef]

17. Forbes, S.A.; Beare, D.; Boutselakis, H.; Bamford, S.; Bindal, N.; Tate, J.; Cole, C.G.; Ward, S.; Dawson, E.; Ponting, L.; et al. COSMIC: Somatic cancer genetics at high-resolution. *Nucleic Acids Res.* **2017**, *45*, D777–D783. [CrossRef]

18. Dieci, M.V.; Arnedos, M.; Andre, F.; Soria, J.C. Fibroblast growth factor receptor inhibitors as a cancer treatment: From a biologic rationale to medical perspectives. *Cancer Discov.* **2013**, *3*, 264–279. [CrossRef]

19. Wiedemann, M.; Trueb, B. Characterization of a novel protein (FGFRL1) from human cartilage related to FGF receptors. *Genomics* **2000**, *69*, 275–279. [CrossRef]

20. Steinberg, F.; Zhuang, L.; Beyeler, M.; Kalin, R.E.; Mullis, P.E.; Brandli, A.W.; Trueb, B. The FGFRL1 receptor is shed from cell membranes, binds fibroblast growth factors (FGFs), and antagonizes FGF signaling in Xenopus embryos. *J. Biol. Chem.* **2010**, *285*, 2193–2202. [CrossRef]

21. Harmer, N.J.; Ilag, L.L.; Mulloy, B.; Pellegrini, L.; Robinson, C.V.; Blundell, T.L. Towards a resolution of the stoichiometry of the fibroblast growth factor (FGF)-FGF receptor-heparin complex. *J. Mol. Biol.* **2004**, *339*, 821–834. [CrossRef] [PubMed]

22. Ahmad, I.; Iwata, T.; Leung, H.Y. Mechanisms of FGFR-mediated carcinogenesis. *Biochim. Biophys. Acta* **2012**, *1823*, 850–860. [CrossRef] [PubMed]

23. Ornitz, D.M.; Marie, P.J. Fibroblast growth factor signaling in skeletal development and disease. *Genes Dev.* **2015**, *29*, 1463–1486. [CrossRef] [PubMed]

24. Mohammadi, M.; Olsen, S.K.; Ibrahimi, O.A. Structural basis for fibroblast growth factor receptor activation. *Cytokine Growth Factor Rev.* **2005**, *16*, 107–137. [CrossRef] [PubMed]

25. Olsen, S.K.; Ibrahimi, O.A.; Raucci, A.; Zhang, F.; Eliseenkova, A.V.; Yayon, A.; Basilico, C.; Linhardt, R.J.; Schlessinger, J.; Mohammadi, M. Insights into the molecular basis for fibroblast growth factor receptor autoinhibition and ligand-binding promiscuity. *Proc. Natl. Acad. Sci. USA* **2004**, *101*, 935–940. [CrossRef] [PubMed]

26. Kiselyov, V.V.; Kochoyan, A.; Poulsen, F.M.; Bock, E.; Berezin, V. Elucidation of the mechanism of the regulatory function of the Ig1 module of the fibroblast growth factor receptor 1. *Protein Sci.* **2006**, *15*, 2318–2322. [CrossRef]

27. Kalinina, J.; Dutta, K.; Ilghari, D.; Beenken, A.; Goetz, R.; Eliseenkova, A.V.; Cowburn, D.; Mohammadi, M. The alternatively spliced acid box region plays a key role in FGF receptor autoinhibition. *Structure* **2012**, *20*, 77–88. [CrossRef] [PubMed]

28. Wang, F.; Kan, M.; Yan, G.; Xu, J.; McKeehan, W.L. Alternately spliced NH2-terminal immunoglobulin-like Loop I in the ectodomain of the fibroblast growth factor (FGF) receptor 1 lowers affinity for both heparin and FGF-1. *J. Biol. Chem.* **1995**, *270*, 10231–10235. [CrossRef] [PubMed]

29. Yeh, B.K.; Igarashi, M.; Eliseenkova, A.V.; Plotnikov, A.N.; Sher, I.; Ron, D.; Aaronson, S.A.; Mohammadi, M. Structural basis by which alternative splicing confers specificity in fibroblast growth factor receptors. *Proc. Natl. Acad. Sci. USA* **2003**, *100*, 2266–2271. [CrossRef]

30. Rapraeger, A.C.; Krufka, A.; Olwin, B.B. Requirement of heparan sulfate for bFGF-mediated fibroblast growth and myoblast differentiation. *Science* **1991**, *252*, 1705–1708. [CrossRef]

31. Yayon, A.; Klagsbrun, M.; Esko, J.D.; Leder, P.; Ornitz, D.M. Cell surface, heparin-like molecules are required for binding of basic fibroblast growth factor to its high affinity receptor. *Cell* **1991**, *64*, 841–848. [CrossRef]

32. Gong, S.G. Isoforms of receptors of fibroblast growth factors. *J. Cell Physiol.* **2014**, *229*, 1887–1895. [CrossRef]

33. Schlessinger, J.; Plotnikov, A.N.; Ibrahimi, O.A.; Eliseenkova, A.V.; Yeh, B.K.; Yayon, A.; Linhardt, R.J.; Mohammadi, M. Crystal structure of a ternary FGF-FGFR-heparin complex reveals a dual role for heparin in FGFR binding and dimerization. *Mol. Cell* **2000**, *6*, 743–750. [CrossRef]

34. Miki, T.; Bottaro, D.P.; Fleming, T.P.; Smith, C.L.; Burgess, W.H.; Chan, A.M.; Aaronson, S.A. Determination of ligand-binding specificity by alternative splicing: Two distinct growth factor receptors encoded by a single gene. *Proc. Natl. Acad. Sci. USA* **1992**, *89*, 246–250. [CrossRef] [PubMed]

35. Chellaiah, A.T.; McEwen, D.G.; Werner, S.; Xu, J.; Ornitz, D.M. Fibroblast growth factor receptor (FGFR) 3. Alternative splicing in immunoglobulin-like domain III creates a receptor highly specific for acidic FGF/FGF-1. *J. Biol. Chem.* **1994**, *269*, 11620–11627. [PubMed]

36. Kurosu, H.; Ogawa, Y.; Miyoshi, M.; Yamamoto, M.; Nandi, A.; Rosenblatt, K.P.; Baum, M.G.; Schiavi, S.; Hu, M.C.; Moe, O.W.; et al. Regulation of fibroblast growth factor-23 signaling by klotho. *J. Biol. Chem.* **2006**, *281*, 6120–6123. [CrossRef] [PubMed]

37. Urakawa, I.; Yamazaki, Y.; Shimada, T.; Iijima, K.; Hasegawa, H.; Okawa, K.; Fujita, T.; Fukumoto, S.; Yamashita, T. Klotho converts canonical FGF receptor into a specific receptor for FGF23. *Nature* **2006**, *444*, 770–774. [CrossRef] [PubMed]

38. Adams, A.C.; Cheng, C.C.; Coskun, T.; Kharitonenkov, A. FGF21 requires betaklotho to act in vivo. *PLoS ONE* **2012**, *7*, e49977. [CrossRef] [PubMed]

39. Yamaguchi, F.; Saya, H.; Bruner, J.M.; Morrison, R.S. Differential expression of two fibroblast growth factor-receptor genes is associated with malignant progression in human astrocytomas. *Proc. Natl. Acad. Sci. USA* **1994**, *91*, 484–488. [CrossRef]

40. Tomlinson, D.C.; Knowles, M.A. Altered splicing of FGFR1 is associated with high tumor grade and stage and leads to increased sensitivity to FGF1 in bladder cancer. *Am. J. Pathol.* **2010**, *177*, 2379–2386. [CrossRef]

41. Eswarakumar, V.P.; Lax, I.; Schlessinger, J. Cellular signaling by fibroblast growth factor receptors. *Cytokine Growth Factor Rev.* **2005**, *16*, 139–149. [CrossRef] [PubMed]

42. Holzmann, K.; Grunt, T.; Heinzle, C.; Sampl, S.; Steinhoff, H.; Reichmann, N.; Kleiter, M.; Hauck, M.; Marian, B. Alternative Splicing of Fibroblast Growth Factor Receptor IgIII Loops in Cancer. *J. Nucleic Acids* **2012**, *2012*, 950508. [CrossRef] [PubMed]

43. Ohashi, R.; Matsuda, Y.; Ishiwata, T.; Naito, Z. Downregulation of fibroblast growth factor receptor 2 and its isoforms correlates with a high proliferation rate and poor prognosis in high-grade glioma. *Oncol. Rep.* **2014**, *32*, 1163–1169. [CrossRef] [PubMed]

44. Matlin, A.J.; Clark, F.; Smith, C.W. Understanding alternative splicing: Towards a cellular code. *Nat. Rev. Mol. Cell Biol.* **2005**, *6*, 386–398. [CrossRef] [PubMed]

45. Yamaguchi, A.; Ishii, H.; Morita, I.; Oota, I.; Takeda, H. mRNA expression of fibroblast growth factors and hepatocyte growth factor in rat plantaris muscle following denervation and compensatory overload. *Pflugers Arch.* **2004**, *448*, 539–546. [CrossRef]

46. Burgar, H.R.; Burns, H.D.; Elsden, J.L.; Lalioti, M.D.; Heath, J.K. Association of the signaling adaptor FRS2 with fibroblast growth factor receptor 1 (Fgfr1) is mediated by alternative splicing of the juxtamembrane domain. *J. Biol. Chem.* **2002**, *277*, 4018–4023. [CrossRef] [PubMed]

47. Sarabipour, S.; Hristova, K. FGFR3 unliganded dimer stabilization by the juxtamembrane domain. *J. Mol. Biol.* **2015**, *427*, 1705–1714. [CrossRef] [PubMed]

48. Zhang, Y.; Gorry, M.C.; Post, J.C.; Ehrlich, G.D. Genomic organization of the human fibroblast growth factor receptor 2 (FGFR2) gene and comparative analysis of the human FGFR gene family. *Gene* **1999**, *230*, 69–79. [CrossRef]

49. Lew, E.D.; Furdui, C.M.; Anderson, K.S.; Schlessinger, J. The precise sequence of FGF receptor autophosphorylation is kinetically driven and is disrupted by oncogenic mutations. *Sci. Signal.* **2009**, *2*, ra6. [CrossRef]

50. Touat, M.; Ileana, E.; Postel-Vinay, S.; Andre, F.; Soria, J.C. Targeting FGFR Signaling in Cancer. *Clin. Cancer Res.* **2015**, *21*, 2684–2694. [CrossRef]

51. Ori, A.; Wilkinson, M.C.; Fernig, D.G. The heparanome and regulation of cell function: Structures, functions and challenges. *Front. Biosci.* **2008**, *13*, 4309–4338. [CrossRef] [PubMed]

52. Opalinski, L.; Sokolowska-Wedzina, A.; Szczepara, M.; Zakrzewska, M.; Otlewski, J. Antibody-induced dimerization of FGFR1 promotes receptor endocytosis independently of its kinase activity. *Sci. Rep.* **2017**, *7*, 7121. [CrossRef] [PubMed]

53. Auciello, G.; Cunningham, D.L.; Tatar, T.; Heath, J.K.; Rappoport, J.Z. Regulation of fibroblast growth factor receptor signalling and trafficking by Src and Eps8. *J. Cell Sci.* **2013**, *126*, 613–624. [CrossRef] [PubMed]

54. Sandilands, E.; Akbarzadeh, S.; Vecchione, A.; McEwan, D.G.; Frame, M.C.; Heath, J.K. Src kinase modulates the activation, transport and signalling dynamics of fibroblast growth factor receptors. *EMBO Rep.* **2007**, *8*, 1162–1169. [CrossRef] [PubMed]

55. Kundu, S.; Xiong, A.; Spyrou, A.; Wicher, G.; Marinescu, V.D.; Edqvist, P.D.; Zhang, L.; Essand, M.; Dimberg, A.; Smits, A.; et al. Heparanase promotes glioma progression and is inversely correlated with patient survival. *Mol. Cancer Res.* **2016**. [CrossRef] [PubMed]

56. Latko, M.; Czyrek, A.; Porebska, N.; Kucinska, M.; Otlewski, J.; Zakrzewska, M.; Opalinski, L. Cross-talk between fibroblast growth factor receptors and other cell surface proteins. *Cells* **2019**, *8*, 455. [CrossRef] [PubMed]

57. Di Liberto, V.; Mudo, G.; Belluardo, N. Crosstalk between receptor tyrosine kinases (RTKs) and G protein-coupled receptors (GPCR) in the brain: Focus on heteroreceptor complexes and related functional neurotrophic effects. *Neuropharmacology* **2019**, *152*, 67–77. [CrossRef]

58. Asimaki, O.; Leondaritis, G.; Lois, G.; Sakellaridis, N.; Mangoura, D. Cannabinoid 1 receptor-dependent transactivation of fibroblast growth factor receptor 1 emanates from lipid rafts and amplifies extracellular signal-regulated kinase 1/2 activation in embryonic cortical neurons. *J. Neurochem.* **2011**, *116*, 866–873. [CrossRef]

59. Borroto-Escuela, D.O.; Carlsson, J.; Ambrogini, P.; Narvaez, M.; Wydra, K.; Tarakanov, A.O.; Li, X.; Millon, C.; Ferraro, L.; Cuppini, R.; et al. Understanding the role of GPCR heteroreceptor complexes in modulating the brain networks in health and disease. *Front. Cell Neurosci.* **2017**, *11*, 37. [CrossRef]

60. Belcheva, M.M.; Haas, P.D.; Tan, Y.; Heaton, V.M.; Coscia, C.J. The fibroblast growth factor receptor is at the site of convergence between mu-opioid receptor and growth factor signaling pathways in rat C6 glioma cells. *J. Pharmacol. Exp. Ther.* **2002**, *303*, 909–918. [CrossRef]

61. Sanchez-Heras, E.; Howell, F.V.; Williams, G.; Doherty, P. The fibroblast growth factor receptor acid box is essential for interactions with N-cadherin and all of the major isoforms of neural cell adhesion molecule. *J. Biol. Chem.* **2006**, *281*, 35208–35216. [CrossRef] [PubMed]

62. Williams, E.J.; Furness, J.; Walsh, F.S.; Doherty, P. Activation of the FGF receptor underlies neurite outgrowth stimulated by L1, N-CAM, and N-cadherin. *Neuron* **1994**, *13*, 583–594. [CrossRef]

63. Condic, M.L.; Letourneau, P.C. Ligand-induced changes in integrin expression regulate neuronal adhesion and neurite outgrowth. *Nature* **1997**, *389*, 852–856. [CrossRef] [PubMed]

64. Boscher, C.; Mege, R.M. Cadherin-11 interacts with the FGF receptor and induces neurite outgrowth through associated downstream signalling. *Cell Signal.* **2008**, *20*, 1061–1072. [CrossRef] [PubMed]

65. Meiri, K.F.; Saffell, J.L.; Walsh, F.S.; Doherty, P. Neurite outgrowth stimulated by neural cell adhesion molecules requires growth-associated protein-43 (GAP-43) function and is associated with GAP-43 phosphorylation in growth cones. *J. Neurosci.* **1998**, *18*, 10429–10437. [CrossRef] [PubMed]

66. Doherty, P.; Cohen, J.; Walsh, F.S. Neurite outgrowth in response to transfected N-CAM changes during development and is modulated by polysialic acid. *Neuron* **1990**, *5*, 209–219. [CrossRef]

67. Bianconi, D.; Unseld, M.; Prager, G.W. Integrins in the spotlight of cancer. *Int. J. Mol. Sci.* **2016**, *17*, 2037. [CrossRef] [PubMed]

68. Lathia, J.D.; Gallagher, J.; Heddleston, J.M.; Wang, J.; Eyler, C.E.; Macswords, J.; Wu, Q.; Vasanji, A.; McLendon, R.E.; Hjelmeland, A.B.; et al. Integrin alpha 6 regulates glioblastoma stem cells. *Cell Stem Cell* **2010**, *6*, 421–432. [CrossRef]

69. Haas, T.L.; Sciuto, M.R.; Brunetto, L.; Valvo, C.; Signore, M.; Fiori, M.E.; di Martino, S.; Giannetti, S.; Morgante, L.; Boe, A.; et al. Integrin alpha7 is a functional marker and potential therapeutic target in glioblastoma. *Cell Stem Cell* **2017**, *21*, 35–50. [CrossRef]

70. Kowalski-Chauvel, A.; Gouaze-Andersson, V.; Baricault, L.; Martin, E.; Delmas, C.; Toulas, C.; Cohen-Jonathan-Moyal, E.; Seva, C. Alpha6-integrin regulates FGFR1 expression through the ZEB1/YAP1 transcription complex in glioblastoma stem cells resulting in enhanced proliferation and stemness. *Cancers* **2019**, *11*, 406. [CrossRef]

71. Nguyen, T.; Mege, R.M. N-cadherin and fibroblast growth factor receptors crosstalk in the control of developmental and cancer cell migrations. *Eur. J. Cell Biol.* **2016**, *95*, 415–426. [CrossRef] [PubMed]

72. Hale, J.S.; Jimenez-Pascual, A.; Kordowski, A.; Pugh, J.; Rao, S.; Silver, D.J.; Alban, T.; Watson, D.B.; Chen, R.; McIntyre, T.M.; et al. ADAMDEC1 maintains a novel growth factor signaling loop in cancer stem cells. *bioRxiv* **2019**. [CrossRef]

73. Francavilla, C.; Loeffler, S.; Piccini, D.; Kren, A.; Christofori, G.; Cavallaro, U. Neural cell adhesion molecule regulates the cellular response to fibroblast growth factor. *J. Cell Sci.* **2007**, *120*, 4388–4394. [CrossRef] [PubMed]

74. Amoureux, M.C.; Coulibaly, B.; Chinot, O.; Loundou, A.; Metellus, P.; Rougon, G.; Figarella-Branger, D. Polysialic acid neural cell adhesion molecule (PSA-NCAM) is an adverse prognosis factor in glioblastoma, and regulates olig2 expression in glioma cell lines. *BMC Cancer* **2010**, *10*, 91. [CrossRef] [PubMed]

75. Ligon, K.L.; Huillard, E.; Mehta, S.; Kesari, S.; Liu, H.; Alberta, J.A.; Bachoo, R.M.; Kane, M.; Louis, D.N.; Depinho, R.A.; et al. Olig2-regulated lineage-restricted pathway controls replication competence in neural stem cells and malignant glioma. *Neuron* **2007**, *53*, 503–517. [CrossRef] [PubMed]

76. Bribian, A.; Barallobre, M.J.; Soussi-Yanicostas, N.; de Castro, F. Anosmin-1 modulates the FGF-2-dependent migration of oligodendrocyte precursors in the developing optic nerve. *Mol. Cell Neurosci.* **2006**, *33*, 2–14. [CrossRef]

77. Garcia-Gonzalez, D.; Clemente, D.; Coelho, M.; Esteban, P.F.; Soussi-Yanicostas, N.; de Castro, F. Dynamic roles of FGF-2 and Anosmin-1 in the migration of neuronal precursors from the subventricular zone during pre- and postnatal development. *Exp. Neurol.* **2010**, *222*, 285–295. [CrossRef]

78. Murcia-Belmonte, V.; Esteban, P.F.; Garcia-Gonzalez, D.; De Castro, F. Biochemical dissection of Anosmin-1 interaction with FGFR1 and components of the extracellular matrix. *J. Neurochem.* **2010**, *115*, 1256–1265. [CrossRef]

79. Mohanan, V.; Temburni, M.K.; Kappes, J.C.; Galileo, D.S. L1CAM stimulates glioma cell motility and proliferation through the fibroblast growth factor receptor. *Clin. Exp. Metastasis* **2013**, *30*, 507–520. [CrossRef]

80. Chen, P.Y.; Simons, M.; Friesel, R. FRS2 via fibroblast growth factor receptor 1 is required for platelet-derived growth factor receptor beta-mediated regulation of vascular smooth muscle marker gene expression. *J. Biol. Chem.* **2009**, *284*, 15980–15992. [CrossRef]

81. Yokote, H.; Fujita, K.; Jing, X.; Sawada, T.; Liang, S.; Yao, L.; Yan, X.; Zhang, Y.; Schlessinger, J.; Sakaguchi, K. Trans-activation of EphA4 and FGF receptors mediated by direct interactions between their cytoplasmic domains. *Proc. Natl. Acad. Sci. USA* **2005**, *102*, 18866–18871. [CrossRef] [PubMed]

82. Fukai, J.; Yokote, H.; Yamanaka, R.; Arao, T.; Nishio, K.; Itakura, T. EphA4 promotes cell proliferation and migration through a novel EphA4-FGFR1 signaling pathway in the human glioma U251 cell line. *Mol. Cancer Ther.* **2008**, *7*, 2768–2778. [CrossRef] [PubMed]

83. Morrison, R.S.; Yamaguchi, F.; Bruner, J.M.; Tang, M.; McKeehan, W.; Berger, M.S. Fibroblast growth factor receptor gene expression and immunoreactivity are elevated in human glioblastoma multiforme. *Cancer Res.* **1994**, *54*, 2794–2799. [PubMed]

84. Loilome, W.; Joshi, A.D.; ap Rhys, C.M.; Piccirillo, S.; Vescovi, A.L.; Gallia, G.L.; Riggins, G.J. Glioblastoma cell growth is suppressed by disruption of fibroblast growth factor pathway signaling. *J. Neuro Oncol.* **2009**, *94*, 359–366. [CrossRef] [PubMed]

85. Gouaze-Andersson, V.; Delmas, C.; Taurand, M.; Martinez-Gala, J.; Evrard, S.; Mazoyer, S.; Toulas, C.; Cohen-Jonathan-Moyal, E. FGFR1 induces glioblastoma radioresistance through the PLCgamma/Hif1alpha pathway. *Cancer Res.* **2016**, *76*, 3036–3044. [CrossRef] [PubMed]

86. Miraux, S.; Lemiere, S.; Pineau, R.; Pluderi, M.; Canioni, P.; Franconi, J.M.; Thiaudiere, E.; Bello, L.; Bikfalvi, A.; Auguste, P. Inhibition of FGF receptor activity in glioma implanted into the mouse brain using the tetracyclin-regulated expression system. *Angiogenesis* **2004**, *7*, 105–113. [CrossRef]

87. Daido, S.; Takao, S.; Tamiya, T.; Ono, Y.; Terada, K.; Ito, S.; Ouchida, M.; Date, I.; Ohmoto, T.; Shimizu, K. Loss of heterozygosity on chromosome 10q associated with malignancy and prognosis in astrocytic tumors, and discovery of novel loss regions. *Oncol. Rep.* **2004**, *12*, 789–795. [CrossRef]

88. Di Stefano, A.L.; Fucci, A.; Frattini, V.; Labussiere, M.; Mokhtari, K.; Zoppoli, P.; Marie, Y.; Bruno, A.; Boisselier, B.; Giry, M.; et al. Detection, characterization, and inhibition of FGFR-TACC fusions in IDH wild-type glioma. *Clin. Cancer Res.* **2015**, *21*, 3307–3317. [CrossRef]

89. Frattini, V.; Pagnotta, S.M.; Fan, J.J.; Russo, M.V.; Lee, S.B.; Garofano, L.; Zhang, J.; Shi, P.; Lewis, G.; Zhang, J.; et al. A metabolic function of FGFR3-TACC3 gene fusions in cancer. *Nature* **2018**, *553*, 222–227. [CrossRef]

90. Hoang-Minh, L.B.; Siebzehnrubl, F.A.; Yang, C.; Suzuki-Hatano, S.; Dajac, K.; Loche, T.; Andrews, N.; Schmoll Massari, M.; Patel, J.; Amin, K.; et al. Infiltrative and drug-resistant slow-cycling cells support metabolic heterogeneity in glioblastoma. *EMBO J.* **2018**, *37*, e98772. [CrossRef]

91. Wang, Z.; Zhang, C.; Sun, L.; Liang, J.; Liu, X.; Li, G.; Yao, K.; Zhang, W.; Jiang, T. FGFR3, as a receptor tyrosine kinase, is associated with differentiated biological functions and improved survival of glioma patients. *Oncotarget* **2016**, *7*, 84587–84593. [CrossRef] [PubMed]

92. Darmanis, S.; Sloan, S.A.; Croote, D.; Mignardi, M.; Chernikova, S.; Samghababi, P.; Zhang, Y.; Neff, N.; Kowarsky, M.; Caneda, C.; et al. Single-cell RNA-Seq analysis of infiltrating neoplastic cells at the migrating front of human glioblastoma. *Cell Rep.* **2017**, *21*, 1399–1410. [CrossRef] [PubMed]

93. Yamada, S.M.; Yamada, S.; Hayashi, Y.; Takahashi, H.; Teramoto, A.; Matsumoto, K. Fibroblast growth factor receptor (FGFR) 4 correlated with the malignancy of human astrocytomas. *Neurol. Res.* **2002**, *24*, 244–248. [CrossRef] [PubMed]

94. Schramm, K.; Iskar, M.; Statz, B.; Jager, N.; Haag, D.; Slabicki, M.; Pfister, S.M.; Zapatka, M.; Gronych, J.; Jones, D.T.W.; et al. DECIPHER pooled shRNA library screen identifies PP2A and FGFR signaling as potential therapeutic targets for DIPGs. *Neuro Oncol.* **2019**. [CrossRef] [PubMed]

10

Fibroblast Growth Factor Receptor 4 Targeting in Cancer: New Insights into Mechanisms and Therapeutic Strategies[†]

Liwei Lang [1] and Yong Teng [1,2,3,*]

[1]   Department of Oral Biology and Diagnostic Sciences, Dental College of Georgia, Augusta University, Augusta, GA 30912, USA; llang@augusta.edu
[2]   Georgia Cancer Center, Department of Biochemistry and Molecular Biology, Medical College of Georgia, Augusta University, Augusta, GA 30912, USA
[3]   Department of Medical Laboratory, Imaging and Radiologic Sciences, College of Allied Health, Augusta University, Augusta, GA 30912, USA
*   Correspondence: yteng@augusta.edu
†   Running Title: Targeting FGFR4 for cancer therapy.

Abstract: Fibroblast growth factor receptor 4 (FGFR4), a tyrosine kinase receptor for FGFs, is involved in diverse cellular processes, including the regulation of cell proliferation, differentiation, migration, metabolism, and bile acid biosynthesis. High activation of FGFR4 is strongly associated with the amplification of its specific ligand FGF19 in many types of solid tumors and hematologic malignancies, where it acts as an oncogene driving the cancer development and progression. Currently, the development and therapeutic evaluation of FGFR4-specific inhibitors, such as BLU9931 and H3B-6527, in animal models and cancer patients, are paving the way to suppress hyperactive FGFR4 signaling in cancer. This comprehensive review not only covers the recent discoveries in understanding FGFR4 regulation and function in cancer, but also reveals the therapeutic implications and applications regarding emerging anti-FGFR4 agents. Our aim is to pinpoint the potential of FGFR4 as a therapeutic target and identify new avenues for advancing future research in the field.

Keywords: FGFR4; FGF19; gene regulation; cancer signaling; anticancer

1. Introduction

Fibroblast growth factor receptors (FGFRs) have been found to play a vital role in tumorigenesis and cancer progression through increased cell proliferation, metastasis, and survival [1,2]. Compared with the other three FGFR family members, the signaling pathways and mechanisms of FGFR4 involved in cancer development are less characterized. The expression of FGFR4 is strictly regulated in human adult organs and tissues after fetal development, suggesting it perhaps has a particular relevance to tissue functions. Recently, elevated FGFR4 has been tightly correlated with cancer development and progression, making it an attractive target to develop novel and effective anticancer therapeutics. More efforts have been focused on developing selective inhibitors to target FGFR4, which show particular promise as an anticancer monotherapy or an adjunct treatment.

2. Molecular Characters of FGFR4 and Its Ligands

2.1. The Molecular Structure of FGFR4

FGFR4 is one of four family members harboring tyrosine kinase (TK) domains. The human *FGFR4* gene is located on the long arm of chromosome 5 (5q 35.1). The *FGFR4* gene consists of 18 exons and has five transcript variants with three of them encoding the FGFR4 isoform 1 (Figure 1A) [3]. The 802 amino acid (aa) core region in the FGFR4 protein contains four parts, signal peptide (1–21 aa), extracellular region (22–369 aa), transmembrane region (70–390 aa), and the intracellular region (391–802 aa) (Figure 1B). Similar to the other three FGFR members, the extracellular region of FGFR4 consists of three immunoglobulin-like domains (IgI, IgII, and IgIII), which are essential for specific ligand-binding. IgI is located in 50–107 aa with a length of 97 aa. IgII and IgIII are located in order in 157–241 aa and 264–351 aa. Compared with the other three family members, FGFR4 does not have a splice variant on the IgIII [4]. Several ligand binding sites have been identified, such as 273, 278–280, 309–310, 316, and 337 aa. The TK domains locate in the C terminal from 454–767 aa with several tyrosine (Y) for autocatalysis, such as Y642, Y643, and Y764 (Figure 2).

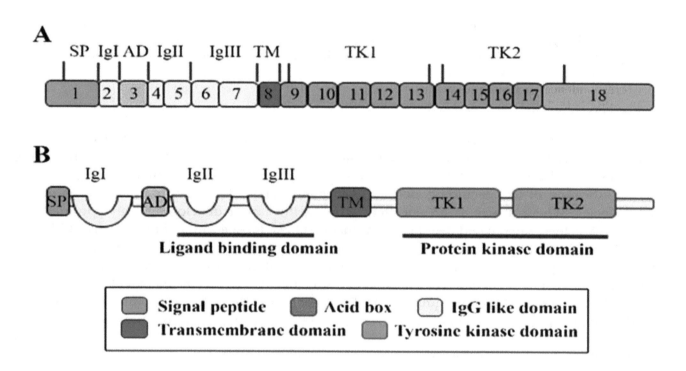

Figure 1. The molecular structure of FGFR4. (**A**) The illustration of FGFR4 with mRNA structure. The transcript variant 1 of FGFR4 contains 18 exons and encodes isoform 1 of FGFR4 protein with the main function domains. (**B**) The main domains of FGFR4 with the corresponding function.

2.2. The Ligands of FGFR4

FGFs are a family of 22 different proteins in vertebrates and are classified into seven subfamilies including FGF1, FGF4, FGF7, FGF8, FGF9, FGF19 ligand subfamily, and FGF11 subfamily [5]. The members of FGF11 subfamily are not ligands of FGFRs and are known as FGF homologous factors [5], while all other six subfamilies work as ligands to bind with FGFR4 (Figure 2) [6]. In other words, ten canonical FGF subfamily members (FGF1, FGF2, FGF4, FGF6, FGF7, FGF8, FGF9, FGF16, FGF17, and FGF18) and three FGF19 subfamily members (FGF19, FGF21, and FGF23) have the potential to bind FGFR4 (Figure 2). Canonical FGFs bind to and activate FGFR4 with heparin/heparin sulfate (HS) [7], while FGF19 subfamily members need β-klotho (KLB) as a co-receptor to bind with FGFR4. FGF1, FGF4, and FGF8, have a higher affinity to bind FGFR4 than other canonical FGFs. Most importantly, FGF19, as an endocrine ligand, has a more specific selective affinity to FGFR4 than other FGFR members [8,9].

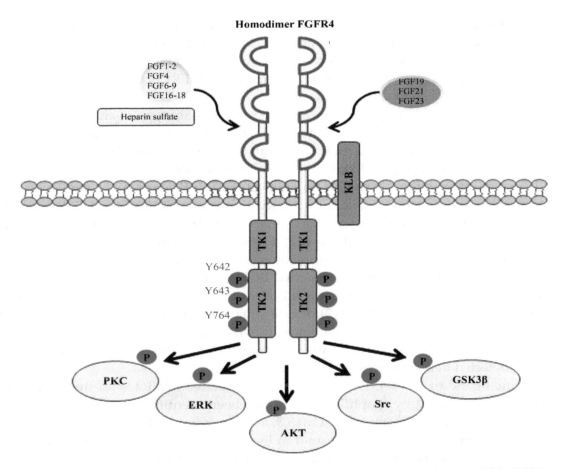

Figure 2. The FGF/FGFR4 signal axis. The signal transduction mediated by the FGF/FGFR4 axis is extremely complex, which includes PKC, ERK1/2, AKT, Src, and GSK3β signaling cascades. The homodimer of FGFR4 forms when binding to either canonical FGF subfamily members (FGF1, FGF2, FGF4, FGF6, FGF7, FGF8, FGF9, FGF16, FGF17, and FGF18) or FGF19 subfamily members (FGF19, FGF21, and FGF23). Heparin or heparin sulfate is required for the binding of canonical FGF subfamily members to FGFR4, whereas KLB acts as a co-receptor of FGFR4 to facilitate FGFR4 interacting with FGF19 subfamily members. When FGFR4 forms protein complexes with FGFs, it can be phosphorylated on three main tyrosine residues: Y642, Y643, and Y764.

2.3. The Physiologic Functions of FGFR4

As an important mediator of homeostasis in the liver, FGFR4 function is required for the maintenance of both lipid and glucose metabolism under normal dietary conditions, in addition to its established role in cholesterol [10]. Particularly, FGFR4 activated by endocrine FGF19 represses the gluconeogenesis and stimulates of glycogen and protein synthesis in hepatocytes [11]. The liver-protective effect of FGFR4 becomes even clearer in the model of carbon tetrachloride-induced liver damage, where more significant liver fibrosis was observed in FGFR4 knock-out compared with wild-type (wt) mice [12]. The importance of FGFR4 in controlling bile acids was also established. It has been reported that bile acids secretion and cholesterol metabolism are regulated by FGF19 through binding to FGFR4 in physical activity [13]. It is worth mentioning that the FGF6/FGFR4 pathway plays important role in myoblast differentiation and myotube regeneration [14,15].

3. The Genetic Alterations of FGFR4 Gene in Cancer

The high expression levels of *FGFR4* can be detected during fetal human and mouse embryonic development. However, deletion of *FGFR4* does not lead to developmental abnormalities in adult mice only with changed cholesterol metabolism and elevated bile acids [16,17]. The expression of FGFR4 is dramatically decreased although it still consistently expresses in several organs, especially in the liver. Gene alterations of FGFRs, including amplification, translocation, and mutation of gain of function, have been linked to tumorigenesis and cancer progression in solid and hematological malignancies. Recently, one study was conducted to evaluate the alterations of FGFR genes in a variety of cancer types [18], which showed that gene alterations of FGFRs occurred in 7.1% of 4853 solid tumors, with the majority being gene amplification (66% of the aberrations), followed by mutations (26%) and translocations (8%). Amplification was the predominant type of alteration for the *FGFR4* gene, accounting for 78% of all FGFR4 gene alterations. Interestingly, the amplified FGFR4 gene was identified in 10% of breast cancer, which more frequently harbors estrogen- and progesterone-receptor with lymph-node metastases [11]. Unlike FGFR1, the translocation of FGFR4 is very rare in human cancers [18]. Two point mutations in the TK domains of the *FGFR4* gene, K535 and E550, have been identified in rhabdomyosarcoma [19]. Another activating point mutation in *FGFR4* gene (Y367C) inducing constitutive FGFR4 dimerization, has been found in MDA-MB-453 breast cancer cells [20]. Although the gene alteration is relatively low, FGFR4 overexpression has been reported in many types of cancer. Increased FGFR4 mRNA expression has been detected in one-third of hepatocellular carcinoma (HCC) [21]. In another study, elevated FGFR4 mRNA levels were detected in 32% of breast cancer samples [22]. FGFR4 overexpression is also observed in 64% (153/238) of oropharyngeal squamous cell carcinoma and 41% (87/212) of oral squamous cell carcinoma [23]. Overexpressed FGFR4 has also been found in pancreatic carcinomas and derived cell lines, which are mediated by an intronic enhancer activated by hepatic nuclear factor 1 alpha [24]. Additionally, highly FGFR4 expression was detected in rhabdomyosarcoma [19].

As a specific ligand of FGFR4, FGF19 can bind and active FGFR4 with the co-receptor KLB. FGFR4 consistently activated by amplified FGF19 has been identified in several types of cancer. The *FGF19* gene is located on chromosome 11q13.3, a region commonly amplified in human cancer. The amplification of the *FGF19* gene was found in liver cancer, breast cancer, lung cancer, bladder cancer, head and neck squamous cell carcinoma (HNSCC), and esophageal cancer [25–29]. For example, the frequency of the amplified *FGF19* gene is as high as 15% in HCC [26]. Moreover, compared with adjacent normal liver tissues, HCC tissues have significantly elevated mRNA levels of FGF19 [30], suggesting the increased mRNA expression is tightly associated with its amplification. A similar tendency was also identified in HNSCC where FGF19 amplification corresponds with an increased dependency upon FGF19–FGFR4 autocrine signaling [31].

4. Mechanisms and Functions of FGFR4 in Cancer Development and Treatment

Accumulating observations indicate that the FGFR4 plays vital roles for cancer development, especially for those harboring *FGF19* amplification. Unlike other family members, the mechanisms and functions of FGFR4 are still poorly characterized at the molecular level in cancer development and progression. Here, the novel observations of mechanisms and functions about oncogenic FGFR4 signaling in cancer development and progression have been summarized and discussed.

4.1. FGFR4-Mediated AKT and ERK Signaling Cascades Promote Cancer Development

The MAPK-ERK and PI3K-AKT signaling are two main pathways regulated by the FGF/FGFR protein complexes. After binding FGFs with HS or KLB, FGFR4 will be activated through autophosphorylation and forms a homodimer (Figure 2). FGFR4 also has the potential to form a heterodimer receptor with other family members, especially with FGFR3 [32,33]. Mechanistic studies showed that phosphorylated FGFR4 recruits and phosphorylates two important intracellular targets, phospholipase γ (PLCγ) and FGFR substrate 2 (FRS2) [4]. MAPK then can be stimulated by activated protein kinase C (PKC) through PLCγ. Meanwhile, the MAPK and PI3K-AKT pathway can be triggered by activated FRS2 through recruitment of growth factor receptor bound 2 (GRB2) (Figure 3) [4]. Upregulated activity of AKT and ERK1/2 leads to enhanced cell proliferation and survival in HCC upon the activation of FGF19/FGFR4 signaling (Figure 3) [34–36].

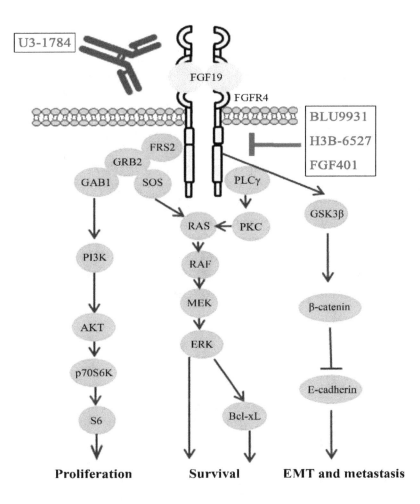

Figure 3. The signal transduction cascades of FGF19/FGFR4 in cancer development and progression. In cancer cells, once FGFR4 receives the extracellular signal from FGF19, it activates many downstream pathways, including PI3K-AKT, MEK-ERK, and GSK3β-β-catenin, leading to increased tumor-promoting activities. FGFR4 activation can be blocked by two non-genetic strategies, using either monoclonal antibodies (e.g., U3-1798) or selective small-molecule inhibitors (e.g., BLU9931, H3B-6527 and FGF401).

FGF19 is more highly expressed in the breast cancer tissue than the adjacent normal tissue [37], and co-expression of *FGFR4* and *FGF19* accounts over 28% primary breast cancer [38]. AKT phosphorylation is strongly associated with co-expression of *FGF19* and *FGFR4*, which can be blocked with FGF19 antibody (1A6) or siRNA-mediated silencing of *FGF19* in breast cancer cells [27]. Our recent findings reveal that FGFR4-mediated hyperactivation of AKT increases breast cancer cells proliferation, but not metastasis [37]. Inactivation of FGFR4 by its inhibitor BLU9931 significantly attenuates FGF19-induced tumor-promoting activity, suggesting interruption of FGFR4 function is sufficient to affect FGF19-driven breast cancer [37].

Our recent study demonstrates that *FGF19* amplification and overexpression are associated with a poorer overall survival rate for HNSCC patients, provoking FGFR4-dependent ERK/AKT-p70S6K-S6 signaling activation to increase HNSCC cell proliferation [31]. Blocking activation of FGFR4 by small hairpin RNAs (shRNAs) or BLU9931, not only attenuates FGF19-induced ERK1/2 and AKT activation, but also abrogates its ability to induce cell proliferation [31]. FGFR4-induced activation of ERK1/2 and AKT pathways was also correlated with increased cell proliferation and survival in colorectal cancer (CRC) [39].

4.2. The FGF19–FGFR4 Axis Promotes Epithelial-Mesenchymal Transition (EMT) to Accelerate Metastasis

The FGF19–FGFR4 axis has been linked to metastasis and poor survival [26]. FGFR4 is predominantly expressed in the liver and responsive for FGF19 stimulation to regulate cholesterol metabolism. There is no doubt that elevated FGF19–FGFR4 signaling is associated with HCC progression, especially for metastasis [26,40]. Our research team has demonstrated that the FGF19/FGFR4 axis facilitates HCC cell EMT through upregulating GSK3β-β-catenin signaling and consequently increases HCC metastasis (Figure 3) [30]. Recently, a vital role of FGFR4 was found in CRC metastasis. Activated FGFR4 phosphorylates AKT, ERK1/2, and Src, leading to increased CRC cell invasion. Silencing FGFR4 reduces adhesion, migration, and invasion of CRC cells [39]. Further study shows that depletion FGFR4 by CRISPR-Cas9 results in the morphological changes and reduced metastasis ability, accompanied by upregulation of E-cadherin and downregulation of Snail and other EMT mediators [39]. Moreover, FGFR4-GSK3β-β-catenin is also elucidated in CRC metastasis. Elevated expression of *Forkhead box C1* (*FOXC1*) is tightly correlated with metastasis of CRC, and FGFR4 is the main target of this gene [41]. BLU9931, the specific inhibitor of FGFR4, can inhibit the activation of GSK3β/β-catenin induced by *FOXC1* overexpression in vitro and metastatic colonization of CRC in vivo [41].

4.3. FGFR4-Associated Chemotherapy Resistance in Cancer

Cancer cells may develop a mechanism that inactivates the drug which represents the main obstacle for cancer treatment. It can be achieved by cancer cells through different mechanisms, such as drug inactivation, drug target alteration, drug efflux, DNA damage repair, cell death inhibition, and EMT [42]. The FGF19–FGFR4 axis has participated in chemotherapy resistance in several types of cancers. The expression levels of FGFR4 are significantly increased in doxorubicin-resistant breast cancer clones [43]. Moreover, FGFR4 overexpression has been detected in those insensitive breast cancer cell lines to doxorubicin [43]. Silencing *FGFR4* with small interfering RNA (siRNA) in chemo-resistant clones increases their sensitivity to doxorubicin. Furthermore, inhibition of FGFR4 with an antagonistic antibody also enhances the sensitivity of endogenously FGFR4-expressing cell lines to doxorubicin. Inhibition of apoptosis by FGFR4 is the main mechanism of doxorubicin resistance in breast cancer [43]. Bcl-xL, an anti-apoptotic protein, is upregulated by FGFR4 via MAPK cascade and responsive for the increased resistance to doxorubicin [43]. Other studies indicate that upregulation of FGF19–FGFR4 signaling increases drug resistance to doxorubicin in basal-like breast cancer [27]. Inactivation of FGFR4 signaling by an anti-FGF19 antibody or siRNA-mediated *FGF19* gene silencing, can sensitize FGFR4$^+$/FGF19$^+$ breast cancer cells to doxorubicin treatment [27]. Increased sensitivity to 5-fluorouracil (5-FU) or oxaliplatin treatment has also been observed after *FGFR4* silencing in CRC cells [44].

As a multiple TKI, sorafenib is an efficient target therapy agent to treat HCC. However, sorafenib is always restricted to continuous administration by occurring drug resistance with unknown mechanisms [45,46]. Recently, our study shows that activation of the FGF19–FGFR4 axis is one of the main mechanisms for sorafenib resistance in the treatment of HCC [47]. The outbalanced oxidative stress induced by reactive oxygen species (ROS) plays a pivotal role in apoptosis [48]. Mechanistically, sorafenib induces ROS-associated apoptosis, but this can be suppressed by *FGF19* overexpression in HCC cells. The FGF19–FGFR4 axis has the potential to assist HCC cancer cells to escape apoptosis in sorafenib treatment through suppression of ROS. *FGFR4* knockout increases the sensitivity to sorafenib treatment in HCC cells, accompanied by enhanced cell apoptosis [47]. Silencing the *FGF19* gene or inactivating FGFR4 with the FGFR-pan inhibitor ponatinib, significantly increases the sensitivity to sorafenib in sorafenib-resistant HCC cells, with induced apoptosis and accumulated ROS generation [47]. Additionally, a similar phenomenon is also observed in young adult mouse colonic epithelial cells. Upregulating FGF19–FGFR4 signaling significantly reduces ROS-mediated apoptosis caused by H_2O_2 through blocking the caspase-3 pathway [49], which also prevents prostate cancer cells from apoptosis in TNFα treatment [50].

5. Develop Specific FGFR4 Inhibitors Targeting Cancer Harboring Elevated FGF19/FGFR4 Signaling

As a promising target, FGFR4 attracts intensive pharmaceutical and academic attention to develop novel target therapy against cancers driven by FGFR4. Three strategies have been developed to target FGFR4, including neutral antibodies, antisense oligonucleotides, and small molecule inhibitors. Two monoclonal neutralizing antibodies of FGFR4, LD-1, and U3-1784, have been developed to competitively targeting extracellular Ig domains of FGFR4. The therapeutic efficacy of U3-1784 is currently being evaluated in Phase I clinical trials for the treatment of HCC and other advanced solid tumors [51] (Table 1 and Figure 3). As an antisense oligonucleotide targeting FGFR4 mRNA, ISIS-FGFR4RX has entered a Phase I clinical trial for obesity (NCT02476019). However, the potential anticancer activity of ISIS-FGFR4RX has not been reported.

Comparing two strategies above, targeting FGFR4 using small-molecule inhibitors is more feasible and can be developed through structure-guided drug design. Not surprisingly, multi-targeted tyrosine kinase inhibitors (mTKIs) can be used to inactivate FGFR4 by disrupting ATP binding in its TK domains. The anticancer activity of many mTKIs, including lenvatinib and ponatinib, have been tested on FGFR-driven solid tumors in animals or in clinical trials (Table 1) [52–54]. However, the limited selective activity of mTKIs on FGFRs induces less efficiency and increases side effects in these treatments. Therefore, pan-FGFR inhibitors are developed and are being evaluated in clinical trials to treat cancers driven by abnormal FGFR pathways. Most of these inhibitors target ATP binding pocket in the TK domains of FGFRs through reversible or covalent bonds. For example, ponatinib can impede the autophosphorylation activity of FGFRs by binding to the hinge region of FGFRs and block the ATP-binding cassette motif [5]. As such, ponatinib has the great potential to inhibit the enzyme activity of FGFRs which are always hyperactive in cancer cells. Other inhibitors in this category include ATP-competitive inhibitors NVP-BGJ398 [55] and AZD4547 [56], ATP-binding pocket inhibitor LY2874455 [57], and FGFRs-FIIN-3 which generates a covalent bond with a conserved cysteine located in the ATP binding site (Table 1) [58]. However, the low specificity of these pan-FGFR inhibitors to FGFR4 cannot sufficiently suppress the oncogenic FGFR4 signaling. For example, the IC_{50} of AZD4547 on FGFR4 is over 100-fold higher than other FGFR members [59]. Moreover, inevitable on-target toxicities and off-target activity resulting from the use of nonspecific FGFR inhibitors lead to several adverse effects such as soft-tissue mineralization and hyperphosphatemia [60]. Such disadvantages eventually limit their usage in cancer patients.

Compared with other FGFR family members, FGFR4 is more specifically expressed in the liver and several other organs for bile acid secretion and cholesterol metabolism. Therefore, the generation of more specific inhibitors which only abolish FGFR4 can improve FGFR4 sensitivity and overcome the drawbacks of pan-FGFR inhibitors. BLU9931 is the first selective FGFR4 inhibitor for the treatment of HCC with an activated FGFR4 signaling pathway [61,62]. As a novel irreversible kinase inhibitor, BLU9931 creates a covalent bond with Cysteine 552 near the ATP-binding site that is only present in FGFR4 among FGFRs [62]. BLU9931 can effectively inhibit HCC tumor harboring elevated FGF19–FGFR4 axis in vivo. Moreover, BLU9931 also displays the potent anticancer ability in breast cancer, CRC, and HNSCC with upregulated FGF19–FGFR4 signaling [37,41]. BLU554 was derived from BLU9931 with improved pharmaceutical properties, which is now in Phase I clinical trial to treat HCC with elevated FGF19–FGFR4 axis (NCT02508467) (Table 1) [63]. H3B-6527 is another selective FGFR4 inhibitor currently in clinical trials for HCC treatment (Table 1). H3B-6527 also targets Cysteine 552 through forming a covalent bond near the ATP binding site of FGFR4, and it exhibits an inhibitory effect on FGFR4 activation in FGF19-driven HCC in vitro and in vivo [64]. By studying a panel of 40 HCC cell lines and 30 HCC patient-derived xenograft models, the expression levels of FGF19 are implicated as a predictive biomarker for H3B-6527 response [64]. Moreover, the combination of H3B-6527 with the CDK4/6 inhibitor palbociclib has a superior effect on the repression of tumors in a xenograft model of HCC [64]. FGF401 is another selective FGFR4 inhibitor, which is under investigation in a phase I/II study to treat HCC with FGFR4 and KLB expression (NCT02325739) [65]

(Table 1). FGF401 is evaluated to treat HCC as the single use or combined with a humanized anti-PD1 IgG4 antibody PDR001 [65]. These novel FGFR4-targeting therapies provide a novel and promising approach which could potentially be developed into a therapeutic strategy to combat cancer.

Table 1. Clinical trials of FGFR4 inhibitors for cancer treatment.

Drug	Structure	Target(s)	Cancer Type	Clinical Trial Number/Phase
Ponatinib		Multiple RTKs, including FGFRs	Advanced solid tumors with activating mutations of FGFRs	NCT02272998/II
AZD4547		Pan-FGFRs	Recurrent malignant glioma expressing FGFR-TACC gene fusion	NCT02824133/I/II
LY2874455		Pan-FGFRs	Advanced cancer	NCT01212107/I
NVP-BGJ398		Pan-FGFRs	Solid tumors and hematologic malignancies with FGFR genetic alterations	NCT02160041/II
BLU554		FGFR4	Hepatocellular carcinoma	NCT02508467/I
FGF401		FGFR4	Hepatocellular carcinoma and other solid tumors	NCT02325739/I/II
H3B-6527		FGFR4	Advanced hepatocellular carcinoma and intrahepatic cholangiocarcinoma	NCT02834780/I
U3-1784	Monoclonal antibody	FGFR4	Hepatocellular cancer and other advanced solid tumors	NCT02690350/I

6. Conclusions

Increasing evidence indicates that upregulation of FGF19–FGFR4 signaling plays an essential role in tumorigenesis and cancer progression. FGFR4 has been proven as an attracted target to develop a novel therapy for the subgroup of cancers associated with the FGF19–FGFR4 pathway. Given that overexpression of FGFR4 significantly correlates with EpCAM, a marker of hepatic cancer stem cells, within the fatty liver-steatosis-cirrhosis-HCC sequence [66], FGFR4 may have the ability to regulate cancer stem cells and lead to chemoresistance in HCC or other cancers. Gaining these insights will improve our comprehensive understanding of the role of FGFR4 in cancer development and treatment. Recently, more specific inhibitors targeting FGFR4 have been developed and evaluated, which are demonstrating promise as a single agent therapy or in combination with other anticancer agents. Thus, there is no doubt that FGFR4-targeting inhibitors offer the most immediate prospects of reducing cancer mortality rate. Perhaps, we can design stapled peptides [67] to incorporate the hydrophobic staple at the interface of FGF19–FGFR4 binding sites, which would increase the specificity of signal targeting compared to the FGFR4 inhibitors that are commercially available. Although the critical role of FGFR4 in metastasis has been demonstrated in animal models of several cancers, prospective studies are warranted to provide evidence regarding the therapeutic efficacy of FGFR4 inhibitors in clinical metastatic cancers.

Abbreviation

aa	amino acid
CRC	colorectal cancer
EpCAM	Epithelial cell adhesion molecule
FGF	fibroblast growth factor
FGFR	fibroblast growth factor receptor
FOXC1	Forkhead box C1
FRS2	FGFR substrate 2
GRB2	growth factor receptor bound 2
HCC	hepatocellular carcinoma
HNSCC	head and neck squamous cell carcinoma
HS	heparin sulfate
mTKI	multi-targeted tyrosine kinase inhibitor
PLCγ	phospholipase γ
p70S6K	p70 ribosomal protein S6 kinase
ROS	reactive oxygen species
shRNA	small hairpin RNA
siRNA	small interfering RNA
TK	tyrosine kinase

Author Contributions: Writing—L.L., Y.T.; Review and editing—Y.T.

References

1.   Babina, I.S.; Turner, N.C. Advances and challenges in targeting FGFR signalling in cancer. *Nat. Rev. Cancer* **2017**, *17*, 318–332. [CrossRef]

2.   Porta, R.; Borea, R.; Coelho, A.; Khan, S.; Araújo, A.; Reclusa, P.; Franchina, T.; Van Der Steen, N.; Van Dam, P.; Ferri, J. FGFR a promising druggable target in cancer: Molecular biology and new drugs. *Crit. Rev. Oncol./Hematol.* **2017**, *113*, 256–267. [CrossRef]

3.   Heinzle, C.; Erdem, Z.; Paur, J.; Grasl-Kraupp, B.; Holzmann, K.; Grusch, M.; Berger, W.; Marian, B. Is fibroblast growth factor receptor 4 a suitable target of cancer therapy? *Curr. Pharm. Des.* **2014**, *20*, 2881–2898. [CrossRef]

4.   Touat, M.; Ileana, E.; Postel-Vinay, S.; André, F.; Soria, J.C. Targeting FGFR signaling in cancer. *Clin. Cancer Res.* **2015**, *21*, 2684–2694. [CrossRef]

5.   Prieto-Dominguez, N.; Shull, A.Y.; Teng, Y. Making way for suppressing the FGF19/FGFR4 axis in cancer. *Future Med. Chem.* **2018**, *10*, 2457–2469. [CrossRef]

6.   Helsten, T.; Schwaederle, M.; Kurzrock, R. Fibroblast growth factor receptor signaling in hereditary and neoplastic disease: Biologic and clinical implications. *Cancer Metastasis Rev.* **2015**, *34*, 479–496. [CrossRef]

7.   Lin, B.C.; Wang, M.; Blackmore, C.; Desnoyers, L.R. Liver-specific activities of FGF19 require Klotho beta. *J. Biol. Chem.* **2007**, *282*, 27277–27284. [CrossRef]

8.   Ornitz, D.M.; Xu, J.; Colvin, J.S.; McEwen, D.G.; MacArthur, C.A.; Coulier, F.; Gao, G.; Goldfarb, M. Receptor specificity of the fibroblast growth factor family. *J. Biol. Chem.* **1996**, *271*, 15292–15297. [CrossRef]

9.   Zhang, X.; Ibrahimi, O.A.; Olsen, S.K.; Umemori, H.; Mohammadi, M.; Ornitz, D.M. Receptor specificity of the fibroblast growth factor family, part II. *J. Biol. Chem.* **2006**, *281*, 15694–15700. [CrossRef]

10. Huang, X.; Yang, C.; Luo, Y.; Jin, C.; Wang, F.; McKeehan, W.L. FGFR4 Prevents Hyperlipidemia and Insulin Resistance but Underlies High Fat Diet-Induced Fatty Liver. *Diabetes* **2007**, *56*, 2501–2510. [CrossRef]

11. Kir, S.; Beddow, S.A.; Samuel, V.T.; Miller, P.; Previs, S.F.; Suino-Powell, K.; Xu, H.E.; Shulman, G.I.; Kliewer, S.A.; Mangelsdorf, D.J. FGF19 as a postprandial, insulin-independent activator of hepatic protein and glycogen synthesis. *Science* **2011**, *331*, 1621–1624. [CrossRef] [PubMed]

12. Yu, C.; Wang, F.; Jin, C.; Wu, X.; Chan, W.-K.; McKeehan, W.L. Increased carbon tetrachloride-induced liver injury and fibrosis in FGFR4-deficient mice. *Am. J. Pathol.* **2002**, *161*, 2003–2010. [CrossRef]

13. Wu, A.L.; Coulter, S.; Liddle, C.; Wong, A.; Eastham-Anderson, J.; French, D.M.; Peterson, A.S.; Sonoda, J. FGF19 regulates cell proliferation, glucose and bile acid metabolism via FGFR4-dependent and independent pathways. *PLoS ONE* **2011**, *6*, e17868. [CrossRef] [PubMed]

14. Floss, T.; Arnold, H.-H.; Braun, T. A role for FGF-6 in skeletal muscle regeneration. *Genes Dev.* **1997**, *11*, 2040–2051. [CrossRef] [PubMed]

15. Zhao, P.; Hoffman, E.P. Embryonic myogenesis pathways in muscle regeneration. *Dev. Dyn.* **2004**, *229*, 380–392. [CrossRef] [PubMed]

16. Partanen, J.; Mäkelä, T.; Eerola, E.; Korhonen, J.; Hirvonen, H.; Claesson-Welsh, L.; Alitalo, K. FGFR-4, a novel acidic fibroblast growth factor receptor with a distinct expression pattern. *EMBO J.* **1991**, *10*, 1347–1354. [CrossRef] [PubMed]

17. Weinstein, M.; Xu, X.; Ohyama, K.; Deng, C.-X. FGFR-3 and FGFR-4 function cooperatively to direct alveogenesis in the murine lung. *Development* **1998**, *125*, 3615–3623. [PubMed]

18. Helsten, T.; Elkin, S.; Arthur, E.; Tomson, B.N.; Carter, J.; Kurzrock, R. The FGFR landscape in cancer: Analysis of 4,853 tumors by next-generation sequencing. *Clin. Cancer Res.* **2016**, *22*, 259–267. [CrossRef] [PubMed]

19. Taylor, J.G.; Cheuk, A.T.; Tsang, P.S.; Chung, J.Y.; Song, Y.K.; Desai, K.; Yu, Y.; Chen, Q.R.; Shah, K.; Youngblood, V. Identification of FGFR4-activating mutations in human rhabdomyosarcomas that promote metastasis in xenotransplanted models. *J. Clin. Investig.* **2009**, *119*, 3395–3407.

20. Roidl, A.; Foo, P.; Wong, W.; Mann, C.; Bechtold, S.; Berger, H.; Streit, S.; Ruhe, J.; Hart, S.; Ullrich, A. The FGFR4 Y367C mutant is a dominant oncogene in MDA-MB453 breast cancer cells. *Oncogene* **2010**, *29*, 1543–1552. [CrossRef]

21. Ho, H.K.; Pok, S.; Streit, S.; Ruhe, J.E.; Hart, S.; Lim, K.S.; Loo, H.L.; Aung, M.O.; Lim, S.G.; Ullrich, A. Fibroblast growth factor receptor 4 regulates proliferation, anti-apoptosis and alpha-fetoprotein secretion during hepatocellular carcinoma progression and represents a potential target for therapeutic intervention. *J. Hepatol.* **2009**, *50*, 118–127. [CrossRef] [PubMed]

22. Penault-Llorca, F.; Bertucci, F.; Adélaïde, J.; Parc, P.; Coulier, F.; Jacquemier, J.; Birnbaum, D.; Delapeyrière, O. Expression of FGF and FGF receptor genes in human breast cancer. *Int. J. Cancer* **1995**, *61*, 170–176. [CrossRef] [PubMed]

23. Koole, K.; Van Kempen, P.M.; Van Bockel, L.W.; Smets, T.; Van Der Klooster, Z.; Dutman, A.C.; Peeters, T.; Koole, R.; Van Diest, P.; Van Es, R.J. FGFR4 is a potential predictive biomarker in oral and oropharyngeal squamous cell carcinoma. *Pathobiology* **2015**, *82*, 280–289. [CrossRef] [PubMed]

24. Shah, R.N.; Ibbitt, J.C.; Alitalo, K.; Hurst, H.C. FGFR4 overexpression in pancreatic cancer is mediated by an intronic enhancer activated by HNF1α. *Oncogene* **2002**, *21*, 8251–8261. [CrossRef] [PubMed]

25. Huang, X.; Gollin, S.M.; Raja, S.; Godfrey, T.E. High-resolution mapping of the 11q13 amplicon and identification of a gene, TAOS1, that is amplified and overexpressed in oral cancer cells. *Proc. Natl. Acad. Sci. USA* **2002**, *99*, 11369–11374. [CrossRef] [PubMed]

26. Sawey, E.T.; Chanrion, M.; Cai, C.; Wu, G.; Zhang, J.; Zender, L.; Zhao, A.; Busuttil, R.W.; Yee, H.; Stein, L. Identification of a therapeutic strategy targeting amplified FGF19 in liver cancer by Oncogenomic screening. *Cancer Cell* **2011**, *19*, 347–358. [CrossRef] [PubMed]

27. Tiong, K.H.; Tan, B.S.; Choo, H.L.; Chung, F.F.; Hii, L.W.; Tan, S.H.; Khor, N.T.; Wong, S.F.; See, S.J.; Tan, Y.F.; et al. Fibroblast growth factor receptor 4 (FGFR4) and fibroblast growth factor 19 (FGF19) autocrine enhance breast cancer cells survival. *Oncotarget* **2016**, *7*, 57633. [CrossRef]

28. Zhang, X.; Kong, M.; Zhang, Z.; Xu, S.; Yan, F.; Wei, L.; Zhou, J. FGF 19 genetic amplification as a potential therapeutic target in lung squamous cell carcinomas. *Thorac. Cancer* **2017**, *8*, 655–665. [CrossRef]

29. Hoover, H.; Li, J.; Marchese, J.; Rothwell, C.; Borawoski, J.; Jeffery, D.A.; Gaither, L.A.; Finkel, N. Quantitative proteomic verification of membrane proteins as potential therapeutic targets located in the 11q13 amplicon in cancers. *J. Proteome Res.* **2015**, *14*, 3670–3679. [CrossRef]

30. Zhao, H.; Lv, F.; Liang, G.; Huang, X.; Wu, G.; Zhang, W.; Yu, L.; Shi, L.; Teng, Y. FGF19 promotes epithelial-mesenchymal transition in hepatocellular carcinoma cells by modulating the GSK3beta/beta-catenin signaling cascade via FGFR4 activation. *Oncotarget* **2016**, *7*, 13575–13586. [CrossRef]

31. Gao, L.; Lang, L.; Zhao, X.; Shay, C.; Shull, A.Y.; Teng, Y. FGF19 Amplification Reveals an Oncogenic Dependency upon Autocrine FGF19/FGFR4 Signaling in Head and Neck Squamous Cell Carcinoma. *Oncogene* **2018**, in press. [CrossRef] [PubMed]

32. Paur, J.; Nika, L.; Maier, C.; Moscu-Gregor, A.; Kostka, J.; Huber, D.; Mohr, T.; Heffeter, P.; Schrottmaier, W.C.; Kappel, S. Fibroblast growth factor receptor 3 isoforms: Novel therapeutic targets for hepatocellular carcinoma? *Hepatology* **2015**, *62*, 1767–1778. [CrossRef] [PubMed]

33. Del Piccolo, N.; Sarabipour, S.; Hristova, K. A new method to study heterodimerization of membrane proteins and its application to fibroblast growth factor receptors. *J. Biol. Chem.* **2017**, *292*, 1288–1301. [CrossRef] [PubMed]

34. Manning, B.D.; Toker, A. AKT/PKB signaling: Navigating the network. *Cell* **2017**, *169*, 381–405. [CrossRef] [PubMed]

35. Degirolamo, C.; Sabba, C.; Moschetta, A. Therapeutic potential of the endocrine fibroblast growth factors FGF19, FGF21 and FGF23. *Nat. Rev. Drug Discov.* **2016**, *15*, 51–69. [CrossRef] [PubMed]

36. Panera, N.; Ceccarelli, S.; Nobili, V.; Alisi, A. Targeting FGF19 binding to its receptor system: A novel therapeutic approach for hepatocellular carcinoma. *Hepatology* **2015**, *62*, 1324. [CrossRef] [PubMed]

37. Zhao, X.; Xu, F.; Dominguez, N.P.; Xiong, Y.; Xiong, Z.; Peng, H.; Shay, C.; Teng, Y. FGFR4 provides the conduit to facilitate FGF19 signaling in breast cancer progression. *Mol. Carcinog.* **2018**, *57*, 1616–1625. [CrossRef] [PubMed]

38. Dallol, A.; Buhmeida, A.; Merdad, A.; Al-Maghrabi, J.; Gari, M.A.; Abu-Elmagd, M.M.; Elaimi, A.; Assidi, M.; Chaudhary, A.G.; Abuzenadah, A.M. Frequent methylation of the KLOTHO gene and overexpression of the FGFR4 receptor in invasive ductal carcinoma of the breast. *Tumor Biol.* **2015**, *36*, 9677–9683. [CrossRef]

39. Peláez-García, A.; Barderas, R.; Torres, S.; Hernández-Varas, P.; Teixidó, J.; Bonilla, F.; de Herreros, A.G.; Casal, J.I. FGFR4 role in epithelial-mesenchymal transition and its therapeutic value in colorectal cancer. *PLoS ONE* **2013**, *8*, e63695. [CrossRef]

40. Miura, S.; Mitsuhashi, N.; Shimizu, H.; Kimura, F.; Yoshidome, H.; Otsuka, M.; Kato, A.; Shida, T.; Okamura, D.; Miyazaki, M. Fibroblast growth factor 19 expression correlates with tumor progression and poorer prognosis of hepatocellular carcinoma. *BMC Cancer* **2012**, *12*, 56. [CrossRef]

41. Liu, J.; Zhang, Z.; Li, X.; Chen, J.; Wang, G.; Tian, Z.; Qian, M.; Chen, Z.; Guo, H.; Tang, G. Forkhead box C1 promotes colorectal cancer metastasis through transactivating ITGA7 and FGFR4 expression. *Oncogene* **2018**, *37*, 5477–5491. [CrossRef] [PubMed]

42. Housman, G.; Byler, S.; Heerboth, S.; Lapinska, K.; Longacre, M.; Snyder, N.; Sarkar, S. Drug resistance in cancer: An overview. *Cancers* **2014**, *6*, 1769–1792. [CrossRef] [PubMed]

43. Roidl, A.; Berger, H.-J.; Kumar, S.; Bange, J.; Knyazev, P.; Ullrich, A. Resistance to chemotherapy is associated with fibroblast growth factor receptor 4 up-regulation. *Clin. Cancer Res.* **2009**, *15*, 2058–2066. [CrossRef] [PubMed]

44. Turkington, R.; Longley, D.; Allen, W.; Stevenson, L.; McLaughlin, K.; Dunne, P.; Blayney, J.; Salto-Tellez, M.; Van Schaeybroeck, S.; Johnston, P. Fibroblast growth factor receptor 4 (FGFR4): A targetable regulator of drug resistance in colorectal cancer. *Cell. Death Disease* **2014**, *5*, e1046. [CrossRef] [PubMed]

45. Van Malenstein, H.; Dekervel, J.; Verslype, C.; Van Cutsem, E.; Windmolders, P.; Nevens, F.; van Pelt, J. Long-term exposure to sorafenib of liver cancer cells induces resistance with epithelial-to-mesenchymal transition, increased invasion and risk of rebound growth. *Cancer Lett.* **2013**, *329*, 74–83. [CrossRef] [PubMed]

46. Villanueva, A.; Llovet, J.M. Second-line therapies in hepatocellular carcinoma: Emergence of resistance to sorafenib. *Clin. Cancer Res.* **2012**, *18*, 1824–1826. [CrossRef] [PubMed]

47. Gao, L.; Wang, X.; Tang, Y.; Huang, S.; Hu, C.-A.A.; Teng, Y. FGF19/FGFR4 signaling contributes to the resistance of hepatocellular carcinoma to sorafenib. *J. Exp. Clin. Cancer Res.* **2017**, *36*, 1–10. [CrossRef]

48. Octavia, Y.; Brunner-La Rocca, H.P.; Moens, A.L. NADPH oxidase-dependent oxidative stress in the failing heart: From pathogenic roles to therapeutic approach. *Free Radic. Biol. Med.* **2012**, *52*, 291–297. [CrossRef]

49. Valastyan, S.; Weinberg, R.A. Tumor metastasis: Molecular insights and evolving paradigms. *Cell* **2011**, *147*, 275–292. [CrossRef]

50. Hu, L.; Cong, L. Fibroblast growth factor 19 is correlated with an unfavorable prognosis and promotes progression by activating fibroblast growth factor receptor 4 in advanced-stage serous ovarian cancer. *Oncol. Rep.* **2015**, *34*, 2683–2691. [CrossRef]

51. Bartz, R.; Fukuchi, K.; Lange, T.; Gruner, K.; Ohtsuka, T.; Watanabe, I.; Hayashi, S.; Redondo-Müller, M.; Takahashi, M.; Agatsuma, T. U3-1784, a human anti-FGFR4 antibody for the treatment of cancer. *Cancer Res.* **2016**, *76*, 3852. [CrossRef]

52. Dienstmann, R.; Rodon, J.; Prat, A.; Perez-Garcia, J.; Adamo, B.; Felip, E.; Cortes, J.; Iafrate, A.; Nuciforo, P.; Tabernero, J. Genomic aberrations in the FGFR pathway: Opportunities for targeted therapies in solid tumors. *Ann. Oncol.* **2014**, *25*, 552–563. [CrossRef] [PubMed]

53. Li, S.Q.; Cheuk, A.T.; Shern, J.F.; Song, Y.K.; Hurd, L.; Liao, H.; Wei, J.S.; Khan, J. Targeting wild-type and mutationally activated FGFR4 in rhabdomyosarcoma with the inhibitor ponatinib (AP24534). *PLoS ONE* **2013**, *8*, e76551. [CrossRef]

54. Cabanillas, M.E.; Schlumberger, M.; Jarzab, B.; Martins, R.G.; Pacini, F.; Robinson, B.; McCaffrey, J.C.; Shah, M.H.; Bodenner, D.L.; Topliss, D. A phase 2 trial of lenvatinib (E7080) in advanced, progressive, radioiodine-refractory, differentiated thyroid cancer: A clinical outcomes and biomarker assessment. *Cancer* **2015**, *121*, 2749–2756. [CrossRef] [PubMed]

55. Wolf, J.; LoRusso, P.M.; Camidge, R.D.; Perez, J.M.; Tabernero, J.; Hidalgo, M.; Schuler, M.; Tian, G.G.; Soria, J.C.; Delord, J.P. Abstract LB-122: A phase I dose escalation study of NVP-BGJ398, a selective pan FGFR inhibitor in genetically preselected advanced solid tumors. *Cancer Res.* **2012**, *72*, LB-122. [CrossRef]

56. Gavine, P.R.; Mooney, L.; Kilgour, E.; Thomas, A.P.; Al-Kadhimi, K.; Beck, S.; Rooney, C.; Coleman, T.; Baker, D.; Mellor, M.J. AZD4547: An orally bioavailable, potent, and selective inhibitor of the fibroblast growth factor receptor tyrosine kinase family. *Cancer Res.* **2012**, *72*, 2045–2056. [CrossRef]

57. Tie, J.; Bang, Y.-J.; Park, Y.S.; Kang, Y.-K.; Monteith, D.; Hartsock, K.; Thornton, D.E.; Michael, M. Abstract CT215: A phase I trial of LY2874455, a fibroblast growth factor receptor inhibitor, in patients with advanced cance. *Cancer Res.* **2014**, *74*, CT215. [CrossRef]

58. Tan, L.; Wang, J.; Tanizaki, J.; Huang, Z.; Aref, A.R.; Rusan, M.; Zhu, S.-J.; Zhang, Y.; Ercan, D.; Liao, R.G. Development of covalent inhibitors that can overcome resistance to first-generation FGFR kinase inhibitors. *Proc. Natl. Acad. Sci. USA* **2014**, *111*, e4869–e4877. [CrossRef]

59. Repana, D.; Ross, P. Targeting FGF19/FGFR4 Pathway: A Novel Therapeutic Strategy for Hepatocellular Carcinoma. *Diseases* **2015**, *3*, 294–305. [CrossRef]

60. Dieci, M.V.; Arnedos, M.; Andre, F.; Soria, J.C. Fibroblast growth factor receptor inhibitors as a cancer treatment: From a biologic rationale to medical perspectives. *Cancer Discov.* **2013**, *3*, 264–279. [CrossRef]

61. Gao, L.; Shay, C.; Lv, F.; Wang, X.; Teng, Y. Implications of FGF19 on sorafenib-mediated nitric oxide production in hepatocellular carcinoma cells-a short report. *Cell. Oncol.* **2018**, *41*, 85–91. [CrossRef] [PubMed]

62. Hagel, M.; Miduturu, C.; Sheets, M.; Rubin, N.; Weng, W.; Stransky, N.; Bifulco, N.; Kim, J.L.; Hodous, B.; Brooijmans, N. First selective small molecule inhibitor of FGFR4 for the treatment of hepatocellular carcinomas with an activated FGFR4 signaling pathway. *Cancer Discov.* **2015**, *5*, 1–14. [CrossRef] [PubMed]

63. Kim, R.; Sharma, S.; Meyer, T.; Sarker, D.; Macarulla, T.; Sung, M.; Choo, S.; Shi, H.; Schmidt-Kittler, O.; Clifford, C. First-in-human study of BLU-554, a potent, highly-selective FGFR4 inhibitor designed for hepatocellular carcinoma (HCC) with FGFR4 pathway activation. *Eur. J. Cancer* **2016**, *69*, S41. [CrossRef]

64. Joshi, J.J.; Coffey, H.; Corcoran, E.; Tsai, J.; Huang, C.-L.; Ichikawa, K.; Prajapati, S.; Hao, M.-H.; Bailey, S.; Wu, J. H3B-6527 Is a Potent and Selective Inhibitor of FGFR4 in FGF19-Driven Hepatocellular Carcinoma. *Cancer Res.* **2017**, *77*, 6999–7013. [CrossRef] [PubMed]

65. Chan, S.L.; Yen, C.-J.; Schuler, M.; Lin, C.-C.; Choo, S.P.; Weiss, K.-H.; Geier, A.; Okusaka, T.; Lim, H.Y.; Macarulla, T. Abstract CT106: Ph I/II study of FGF401 in adult pts with HCC or solid tumors characterized by FGFR4/KLB expression. *Cancer Res.* **2017**, *77*, CT106. [CrossRef]

66. Li, Y.; Zhang, W.; Doughtie, A.; Cui, G.; Li, X.; Pandit, H.; Yang, Y.; Li, S.; Martin, R. Up-regulation of fibroblast growth factor 19 and its receptor associates with progression from fatty liver to hepatocellular carcinoma. *Oncotarget* **2016**, *7*, 52329–52339. [CrossRef] [PubMed]

67. Xie, X.; Gao, L.; Shull, A.Y.; Teng, Y. Stapled peptides: Providing the best of both worlds in drug development. *Future Med. Chem.* **2016**, *8*, 1969–1980. [CrossRef] [PubMed]

11

Preclinical Evaluation of the Pan-FGFR Inhibitor LY2874455 in FRS2-Amplified Liposarcoma

Robert Hanes [1,2], Else Munthe [1], Iwona Grad [1], Jianhua Han [3], Ida Karlsen [3,4], Emmet McCormack [3,5], Leonardo A. Meza-Zepeda [1,2,6], Eva Wessel Stratford [1] and Ola Myklebost [1,2,7,*]

[1]   Department of Tumor Biology, Institute of Cancer Research, the Norwegian Radium Hospital, Oslo University Hospital, 0379 Oslo, Norway; Robert.Hanes@rr-research.no (R.H.); Else.Munthe@rr-research.no (E.M.); Iwona.Grad@rr-research.no (I.G.); Leonardo.A.Meza-Zepeda@rr-research.no (L.A.M.-Z.); eva.wessel.stratford@rr-research.no (E.W.S.)

[2]   Norwegian Cancer Genomics Consortium, 0379 Oslo, Norway

[3]   Centre for Cancer Biomarkers (CCBIO), Department of Clinical Sciences, University of Bergen, 5021 Bergen, Norway; Jianhua.Han@uib.no (J.H.); idakarlsenemail@gmail.com (I.K.); emmet.mc.cormack@med.uib.no (E.M.)

[4]   KinN Therapeutics AS, 5021 Bergen, Norway

[5]   Department of Internal Medicine, Hematology Section, Haukeland University Hospital, 5021 Bergen, Norway

[6]   Genomics Core Facility, Department of Core Facilities, Institute of Cancer Research, the Norwegian Radium Hospital, Oslo University Hospital, 0379 Oslo, Norway

[7]   Department of Clinical Science, University of Bergen, 5020 Bergen, Norway

*   Correspondence: ola.myklebost@uib.no

Abstract: Background: FGFR inhibition has been proposed as treatment for dedifferentiated liposarcoma (DDLPS) with amplified *FRS2*, but we previously only demonstrated transient cytostatic effects when treating *FRS2*-amplified DDLPS cells with NVP-BGJ398. **Methods:** Effects of the more potent FGFR inhibitor LY2874455 were investigated in three DDLPS cell lines by measuring effects on cell growth and apoptosis *in vitro* and also testing efficacy *in vivo*. Genome, transcriptome and protein analyses were performed to characterize the signaling components in the FGFR pathway. **Results:** LY2874455 induced a stronger, longer-lasting growth inhibitory effect and moderate level of apoptosis for two cell lines. The third cell line, did not respond to FGFR inhibition, suggesting that *FRS2* amplification alone is not sufficient to predict response. Importantly, efficacy of LY2874455 was confirmed *in vivo*, using an independent *FRS2*-amplified DDLPS xenograft model. Expression of *FRS2* was similar in the responding and non-responding cell lines and we could not find any major difference in downstream FGFR signaling. The only FGF expressed by unstimulated non-responding cells was the intracellular ligand FGF11, whereas the responding cell lines expressed extracellular ligand FGF2. **Conclusion:** Our study supports LY2874455 as a better therapy than NVP-BGJ398 for *FRS2*-amplified liposarcoma, and a clinical trial is warranted.

Keywords: FRS2; FGFR; NVP-BGJ398; LY2874455; sarcoma

1. Introduction

Sarcomas are rare cancers of mesenchymal origin, accounting for approximately 1% of all solid cancers, and can be classified into more than 50 distinct histological subtypes [1]. Liposarcomas (LPS), which resemble adipose tissue, are further classified into three main subtypes, well-differentiated/dedifferentiated liposarcoma (WD/DDLPS), myxoid/round cell liposarcoma, and pleomorphic liposarcoma [2]. The heterogeneity makes clinical research and trials challenging.

However, all together rare cancers comprise one of the largest patient groups, which is in great need of new therapeutic approaches. Therefore, a deeper mechanistic understanding is needed to be able to identify and validate new potential targets. In a previous study, we identified amplifications of multiple genes in the 12q14.1-q15 region in the DDLPS cell line NRH-LS1 and investigated several of these as therapeutic targets [3]. One of these, *FRS2*, is generally co-amplified with *MDM2* in WD/DDLPS [4]. *FRS2* codes for an important component of the FGF receptor (FGFR) signaling pathway, which plays crucial roles in multiple biological processes, such as cell growth, survival and differentiation, as well as tumor development and progression [5,6]. FRS2-dependent FGFR signaling is induced through FGFR activation by FGF ligands, and consecutive phosphorylation of FRS2 triggers an intracellular signaling cascade involving RAS/MAPK/ERK and PI3K/AKT [7], leading to oncogenic pro-survival and anti-apoptotic properties and increased proliferation and migration. To date, there are no drugs available that can target FRS2 directly, but attenuating the signal from FGFR, upstream of FRS2, with FGFR inhibitors has been shown to be growth inhibitory in such cells [3,8].

NVP-BGJ398, which is in phase II clinical trials, has been shown to be a potent and selective FGFR inhibitor in a wide panel of cancer cell lines [9]. NVP-BGJ398 has been reported to selectively inhibit FGFR1, −2 and −3 with IC50s of 0.9 nM, 1.4 nM and 1.0 nM, respectively, whereas the IC50 for FGFR4 is 60 nM [10]. Another pan-FGFR inhibitor, LY2874455, recently completed a phase I clinical trial [11], and has been reported to selectively inhibit FGFR1, −2, −3, and −4 with IC50s of 2.8 nM, 2.6 nM, 6.4 nM and 6 nM, respectively [12].

In this study, we have investigated the therapeutic potential of LY2874455 with the aim to improve efficacy for *FRS2*-amplified DDLPS.

2. Materials and Methods

2.1. Cell Line and Culture Conditions

The DDLPS cell lines NRH-LS1, established from a patient-derived xenograft as previously described [3] and LPS510 and LPS853, kindly provided by Dr. Jonathan Fletcher, were cultured in RPMI-1640 medium (Sigma-Aldrich, St. Louis, MO, USA) supplemented with 10% FBS (Sigma-Aldrich), 1% L-alanyl-L-glutamine (Sigma-Aldrich) and 1% penicillin-streptomycin (Sigma-Aldrich) and grown at 37 °C, 5% CO_2. Short tandem repeat DNA profiling was performed on all cell lines to confirm identity. Cells were negative for mycoplasma using the VenorGeM Mycoplasma Detection Kit (Minerva Biolabs, Berlin, Germany).

2.2. Drugs

LY2874455 (#S7057) (Selleck Chemicals, Munich, Germany) was dissolved in DMSO (Sigma-Aldrich) according to the manufacturer's recommendation. For each experiment the appropriate control (referred to as untreated) was used, with a DMSO concentration corresponding to that used with the highest drug concentration. The concentration of DMSO in the control for 100 nM LY2874455 and NVP-BGJ398 was 0.01% and for 1 μM LY2874455 and NVP-BGJ398 0.1%.

2.3. Drug Treatment and Cell Proliferation Assay

The cellular proliferation rate was measured using a live-cell imaging system, IncuCyte ZOOM (Essen Bioscience, Birmingham, UK) with the corresponding software application (version 2013BRev1), (Essen Bioscience, Birmingham, UK). Cells were seeded into 96-well plates, and drug treatment initiated after 16h, in triplicates. Each drug treatment was performed over a period of 1–2 weeks and done at least three times. Proliferation rate was measured as cell confluence over time every third hour.

2.4. Apoptosis

Apoptosis assays were done with the IncuCyte ZOOM. To measure apoptosis, the CellPlayer 96-Well Kinetic Caspase-3/7 reagent containing DEVD-NucViewTM488 (Essen Bioscience) at a concentration of 2 μM was added at the same time as the drug. Total numbers of apoptotic cells were counted in the green channel (488 nm). After 96 h, cells were incubated for 30 minutes with 4 μM of Nuclear-ID Red DNA stain (Enzo Life Sciences, Farmingdale, NY, USA), and total cell count was measured in the red channel (566 nm). The percentage of apoptotic cells per well was calculated as the number of apoptotic cells relative to the total number of nuclei.

2.5. Viability Assay for Dose-Response Curve and IC50 Calculations

Cell viability was measured using the CellTiter-Glo Luminescent Cell Viability Assay (Promega, Madison, WI, USA). 5×10^3 cells per well were seeded onto a 96-well flat and clear bottom polystyrene treated microplate (Corning, Corning, NY, USA). The drug treatment was initiated 16 h after seeding and was applied at concentrations ranging from 0.1 nM to 1000 nM. After 72, 96 and 120 h, ATP levels were used as a measure of viability. Relative IC50 for 120 h was calculated using a four-parameter logistic function [13] based on non-linear regression analysis using SigmaPlot (Systat Software Inc, San Jose, CA, USA) version 12.5.0.38.

2.6. In Vivo Assay

Animal experiments were performed according to protocols approved by the National Animal Research Authority (Mattilsynet) in compliance with the European Convention of the Protection of Vertebrates Used for Scientific Purposes (ID 10175). The LS70x xenograft (established directly from a DDLPS tumor) was implanted into the flank of immunodeficient NOD-scid IL2rγnull (NSG) mice [14]. When tumors reached 150 mm^3, animals were randomized into two groups, each of six mice, and treated twice per day with either 3 mg/kg LY2874455 or vehicle only (2% (v/v) DMSO, 30% (v/v) PEG 300, 5% (v/v) Tween 80 in sterile water), administered by oral gavage. Treatment was performed for 28 days, or until the tumor reached a size of 1 cm^3. Tumor growth was measured by caliper measurements twice per week for the duration of treatment. Unpaired two-tailed t-test was performed to detect significant differences in tumor volumes ($p \leq 0.05$ was considered significant).

2.7. Western Blots

Cells were treated for 24 h with either 100 nM LY2874455, 100 nM NVP-BGJ398 or control-treated with the corresponding concentration of DMSO and the last 15 min with or without 15 ng/ml of recombinant human FGF1 [15] and 10 U/mL of Heparin. In vitro cells were washed with PBS and dissolved in SDS lysis buffer. Xenografts were cut into smaller pieces and snap frozen. Proteins were extracted with T-Per lysis buffer (Thermo Fisher Scientific, Waltham, MA, USA), supplemented with protease and phosphatase inhibitors (both from Thermo Fischer Scientific), using the TissueLyser LT (QIAGEN, Venlo, Netherlands). DTT was added to the lysates before boiling. Proteins were separated in a 4–12% Novex PAGE gel in MOPS running buffer, and transferred to PVDF membranes (Thermo Fisher Scientific). The following antibodies were used: pFRS2-TYR436 (#3861), AKT (#9272), pAKT-SER473 (#9271), ERK (#9102), pERK-T202/Y204 (#4370), PLCγ1 (#5690), pPLCγ1-TYR783 (#2821) (all from Cell Signaling Technology, Danvers, MA, USA), FRS2 (#SC8318) (Santa Cruz Biotechnology, Dallas, TX, USA) and α-Tubulin (#CP06) (Merck KGaA, Darmstadt, Germany). All antibodies were diluted 1:1000, except FRS2 (1:500) and α-Tubulin (1:2000). Secondary antibodies were rabbit anti-mouse immunoglobulins/HRP (#P0260) and goat anti-rabbit immunoglobulins/HRP (#P0448) (Dako, Glostrup, Denmark) at a concentration of 1.3 g/L and 0.25 g/L respectively. The Western blots were developed using the Supersignal Western Dura substrate (Thermo Fisher Scientific), and detected and quantified on a Syngene G-Box (Synoptics Group, Cambridge, UK) with the GeneSnap (version 7.12, Synoptics Group) and the GeneTools (version 4.3.7.0, Synoptics Group) programs, respectively.

2.8. Quantitative Real-Time PCR-Based Copy Number Assay

DNA was isolated from cells using the AllPrep DNA/RNA Mini Kit (QIAGEN) according to the manufacturer's protocol. Quantitative real-time PCR was performed based on absolute quantitation using the Applied Biosystems 7900HT fast real-time PCR system (Applied Biosystems, Foster By, CA, USA). The copy numbers of *FRS2* (Hs02860563_cn), *ALB* (Hs05929625_cn) and *LSAMP* (Hs05902664_cn) were determined using TaqMan copy number assays from Applied Biosystems, *ALB* and *LSAMP* were used as endogenous controls, as these have low level of DNA copy number changes in a large panel of liposarcoma samples [16]. The copy numbers were determined using the CopyCaller Software v2.1 program (Applied Biosystems) as described by the manufacturer, and the FRS2 data were normalized to *LSAMP*. The copy numbers were validated using *ALB* as another endogenous reference gene (data not shown).

2.9. Quantitative Real-Time PCR Based Expression Assay

RNA was isolated from cells using the AllPrep DNA/RNA Mini Kit (QIAGEN) according to the manufacturers protocol. cDNA was prepared using 1 μg of RNA and the SuperScript VILO Master Mix (Invitrogen, Carlsbad, CA, USA). Quantitative real-time PCR was performed based on $\Delta\Delta$Ct relative quantitation using the Applied Biosystems 7900HT fast real-time PCR system (Applied Biosystems). The expression levels of FRS2 were determined using TaqMan gene expression assays (Hs00183614_m1) with human B2M (VIC®/MGB probe) (Applied Biosystems) as internal control for normalization. The relative expression levels were determined using the comparative $\Delta\Delta$Ct method as described by the manufacturer. Human Adipose Tissue Total RNA was used as reference (Clontech, Mountain View, CA, USA).

2.10. RNA Sequencing

RNA was isolated from the cell lines using the AllPrep DNA/RNA Mini Kit (QIAGEN). mRNA sequencing libraries were prepared using 100 ng of total RNA and the Illumina TruSeq Stranded mRNA Library Prep kit for NeoPrep following the supplier's instructions. The libraries were sequenced on a NextSeq 500 Illumina sequencer (Illumina, San Diego, CA, USA) using a High Output v2 kit chemistry, generating 2 × 75 bp paired-end sequence reads. RNA-Seq reads were aligned using STAR aligner (v.2.5.0b) against the human reference genome (UCSC hg19, RefSeq and Gencode gene annotations), and FPKM estimation was generated by Cufflinks 2 using the RNA-seq alignment app at Illumina BaseSpace.

3. Results

3.1. Improved Efficacy Using LY2874455

When treating NRH-LS1 cells at equivalent concentrations, LY2874455 inhibited growth of the cells more efficiently than did NVP-BGJ398 (Figure 1A). NRH-LS1 cells exposed to 100 nM of LY2874455 were completely growth inhibited after 72 h (Figure 1A), while treatment with 100 nM NVP-BGJ398 gave only partial growth inhibition at that time point. In contrast to NVP-BGJ398, we found that LY2874455 induced apoptosis in a subpopulation of NRH-LS1 cells (Figure 1B). Furthermore, 100 nM LY2874455 induced, on average, four times higher levels of apoptosis after 96 h, as compared to cells treated with 100 nM NVP-BGJ398 (Figure 1C). As shown in Figure 2, LY2874455 inhibited cell growth in a dose-dependent manner (Figure 2A), with the full effect at approximately 100 nM. The IC50 for

LY2874455 in NRH-LS1 cells was estimated to 7 nM compared to 47 nM for NVP-BGJ398 (Figure S2), based on viability after 120 h of treatment (Figure 2B).

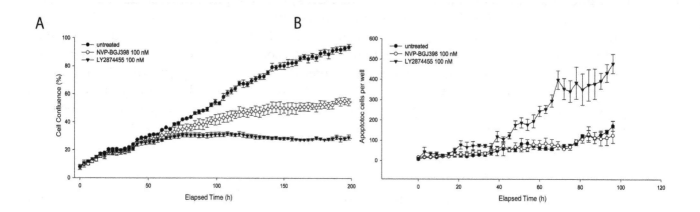

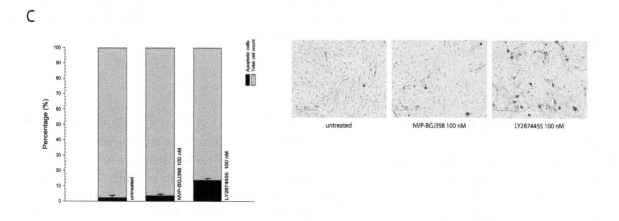

Figure 1. Comparison of the effect of NVP-BGJ398 and LY2874455 on proliferation and apoptosis of NRH-LS1 cells. (**A**) Proliferation of NRH-LS1 cells after inhibition of FGFR with either NVP-BGJ398 or LY2874455; one representative experiment is shown ($n = 3$), error bars represent the standard error of the mean (SEM) of technical replicates; (**B**) The number of cells with active caspase 3/7 during 96 h of treatment with either 100 nM of NVP-BGJ398 or 100 nM LY2874455, one representative experiment is shown ($n = 3$); (**C**) The percentage of apoptotic cells after 96 h of treatment with NVP-BGJ398 or LY2874455; the mean of experiments is shown ($n = 3$), error bars represent the standard deviation (SD) of the experiments. Representative images show apoptotic cells outlined in purple based on measured apoptotic signal. For all experiments untreated is with DMSO concentration corresponding to that of the highest drug concentration.

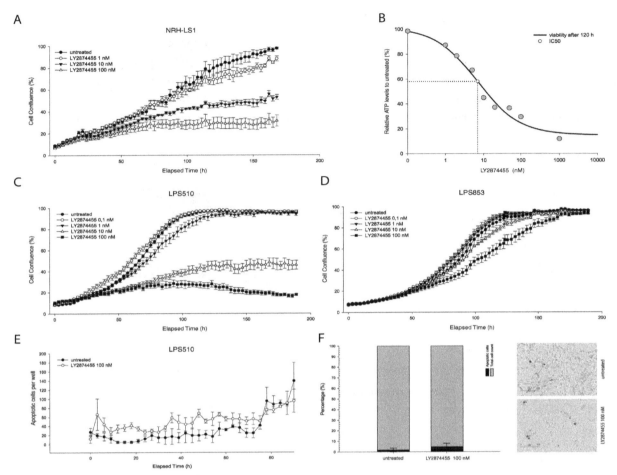

Figure 2. The effect of LY2874455 on the proliferation and viability of NRH-LS1, LPS510 and LPS853 cells. (**A**) Proliferation of NRH-LS1 cells at different concentrations of LY2874455; one representative experiment is shown ($n = 3$), error bars represent the standard error of the mean (SEM) of technical replicates; (**B**) The IC50 was estimated at 7 nM based on NRH-LS1 cell viability after 120 h of treatment with LY2874455; Proliferation of LPS510 (**C**) and LPS853 (**D**) cells in the presence of LY2874455. (**E**) The number of LPS510 cells with active caspase 3/7 during 96 h of treatment with 100 nM of LY2874455; (**C–E**) One representative experiment is shown ($n = 3$), error bars represent the standard error of the mean (SEM) of technical replicates; (**F**) The percentage of apoptotic LPS510 cells after 96 h of treatment with 100 nM of LY2874455. The mean of experiments is shown ($n = 4$), error bars represent the standard deviation (SD) of the experiments. Representative images show apoptotic cells outlined in purple based on measured apoptotic signal. For all experiments untreated is with DMSO concentration corresponding to that of the highest drug concentration.

3.2. The Response to FGFR Inhibition in FRS2-Amplified Cell Lines Is Variable

We next investigated the effect of LY2874455 in two additional *FRS2*-amplified DDLPS cell lines, LPS510 and LPS853, which have *FRS2* copy number and gene expression levels comparable to NRH-LS1 (Figure S1). The expression levels of FRS2 were 10–20 fold higher in all three cell lines compared to both human adipocyte tissue and an undifferentiated immortalized mesenchymal progenitor cell line (iMSC#3), [17] (Figure S1). LY2874455 inhibited the growth of LPS510 similar to NRH-LS1 (Figure 2C), with IC50 values of 5.5 and 6.9 respectively (Figure S2). In contrast, LPS853 was only modestly inhibited at 100 nM (Figure 2D). Similar levels of apoptosis were observed for both LPS510 and NRH-LS1 after treatment with 100 nM LY2874455, with 9% for LPS510 cells (Figure 2E,F) and 13% for NRH-LS1 (Figure 1C) at 96 h.

3.3. LY2874455 Induces Long-Lasting Growth Inhibition

In order to assess duration of the growth inhibitory effects of LY2874455 and NVP-BGJ398, we discontinued drug treatment after 96 h or 264 h. In contrast to NVP-BGJ398, the growth inhibition of NRH-LS1 and LPS510 was strong after withdrawal of LY2874455, although it was quite significant with LPS510 cells, especially at the highest doses (Figure 3A–D). Both LPS510 and NRH-LS1 maintained growth arrest within the time-frame of the experiment after 264 h of treatment with LY2874455.

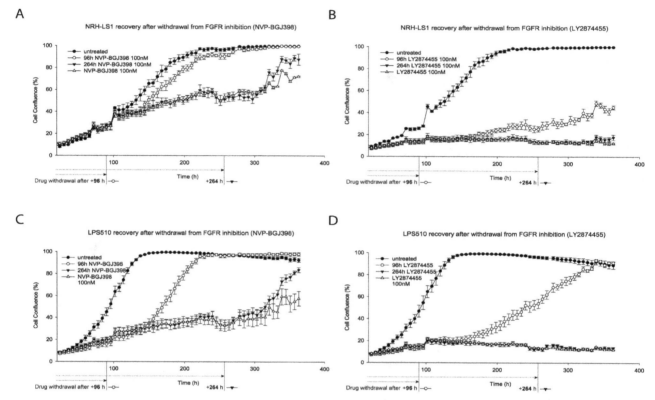

Figure 3. NRH-LS1 and LPS510 cells resume proliferation after withdrawal of treatment with NVP-BGJ398, but not LY2874455. Proliferation of NRH-LS1 (**A**,**B**) or LPS510 (**C**,**D**) cells treated with 100 nM of NVP-BGJ398 (**A**,**C**) or LY2874455 (**B**,**D**) continuously or upon withdrawal of the drug after 96 h or 264 h of treatment. One representative experiment is shown ($n = 3$), error bars represent the standard error of the mean (SEM) of technical replicates. For all experiments untreated with DMSO concentration corresponding to that of the highest drug concentration.

3.4. Both FGFR Inhibitors Inactivate MEK/ERK-Dependent Signaling in FRS2-Amplified Cells

To determine whether the different effects of these drugs could be attributed to the activation or inhibition of different signaling components in the FGFR pathway, we investigated the status of several signaling proteins downstream of FGFR and FRS2 in NRH-LS1 cells upon stimulation with FGF and treatment with NVP-BGJ398 or LY2874455 (Figure 4, quantified in Figure S4A). Although phosphorylation of FRS2 was expected to drive growth in these cells, we only detected phosphorylated FRS2 upon stimulation with exogenous FGF, probably because pFRS2 levels in unstimulated cells were below the detection limit of the pFRS2 antibody used in this western blot assay. An increased phosphorylation of the downstream signaling protein ERK was also found upon stimulation with FGF1. When NRH-LS1 cells were treated for 24 h with either 100 nM LY2874455 or 100 nM NVP-BGJ398, FGF1-induced phosphorylation of FRS2 was completely abolished (Figure 4A), supporting the expected drug action, and also endogenous pERK was reduced. Stimulation with FGF1 also induced phosphorylation of PLCγ1, an FRS2-independent component of the FGFR pathway, which was completely abolished by the treatment with either FGFR inhibitor. The expression and phosphorylation of AKT was unaffected by FGF1 stimulation or FGFR inhibition (Figure 4A). Thus, no clear differences in FGF signaling could be seen that explained the difference in growth inhibition of the two inhibitors in NRH-LS1.

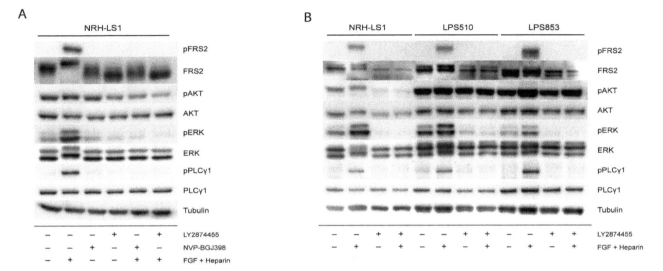

Figure 4. Signaling pathway analysis after FGFR stimulation and inhibition in NRH-LS1, LPS510 and LPS853. Western blots showing (**A**) The level of phosphorylated and total protein for the indicated proteins in NRH-LS1 cells treated for 24 h with 100 nM of LY2874455 or 100 nM NVP-BGJ398, with or without FGF stimulation as indicated; (**B**) The level of phosphorylated and total protein for the indicated proteins in NRH-LS1, LPS510 and LPS853 cells treated for 24 h with 100 nM of LY2874455, with or without FGF as indicated. In all Western blot experiments α-tubulin was used as a loading control.

We further compared the effect of LY2874455 on the same proteins on the three cell lines to understand their difference in sensitivity to FGFR inhibition (Figure 4B, quantified in Figure S4B). The western blots confirmed that FRS2 protein was expressed in all three cell lines, while pFRS2 was below detection in unstimulated cells. Phosphorylation of FRS2, PLCγ1, and ERK was induced by stimulation with FGF1 and was inhibited upon treatment with LY2874455 in all three cell lines, showing the ability of these cell lines to respond to FGF stimulation and FGFR inhibition. The phosphorylation levels of AKT were higher in both LPS510 and LPS853 compared to NRH-LS1 but remained unaffected by stimulation or inhibition of FGFR signaling (Figure 4B).

3.5. The Expression of FGF Receptors Varies in FRS2-Amplified LPS Lines

In order to investigate why only two out of three LPS cell lines responded to the treatment with FGFR inhibitors and whether this could be explained due to a differential expression of some components of the FGFR pathways, we analyzed transcriptome sequencing data of NRH-LS1, LPS510, and LPS853 for expression of FGFR signaling components upstream of FRS2. Similar relative expression (FPKM) values of 52.0, 51.8 and 44.3 for *FRS2* were found in NRH-LS1, LPS510 and LPS853, respectively. All three cell lines also had similar expression levels for *FGFR1* with FPKMs of 41.5, 51.5 and 36.1, while *FGFR4* was comparably higher expressed in LPS853 with an FPKM of 77.5, compared to 1.2 and 0 in NRH-LS1 and LPS510, respectively. All the cell lines had very low expression of *FGFR2* and *FGFR3*. *FGFRL1* is another member of the fibroblast growth factor receptor family, however it lacks the cytoplasmic tyrosine kinase domain and can act as a negative regulator of FGFR signaling [18]. Interestingly, *FGFRL1* was higher expressed in LPS853 than in LPS510 and NRH-LS1, with FPKM values of 50.0, 25.8 and 11.1, respectively.

We also found *FGF2* to be higher expressed in NRH-LS1 and LPS510 with FPKM values of 19.6 and 21.0 respectively, compared to LPS853 with FPKM of 0.5. In turn, *FGF11* expression was higher in LPS853 with FPKM of 24.5, compared to NRH-LS1 and LPS510 with values of 0.4 and 1.2 respectively.

A heat map that shows the expression of FGF receptors and ligands, as well as adapter proteins, is provided in Figure S3.

3.6. LY2874455 Inhibits Tumor Growth In Vivo

Having observed promising therapeutic potential *in vitro* for two out of three *FRS2*-amplified LPS cell lines treated with LY2874455, we investigated the effect *in vivo*. Patient-derived xenograft (PDX) models derived directly from patient material are more representative for drug responses, but the NRH-LS1 PDX model, from which the cell line is derived, grows slowly and was not suitable for preclinical testing. Rather than making a less representative cell line-derived PDX from LPS510, we used the DDLPS patient-derived LS70x PDX [14], which also provided an additional independent model. LS70x has amplification and increased expression of *FRS2* comparable to the three cell lines (Figure S1), and grows reasonably well in NSG mice. When tumors reached 150 mm^3, mice were treated with 3 mg/kg twice per day. Already from day 4 of treatment we observed significant inhibition of tumor growth compared to control-treated mice (Figure 5A). To confirm that LY2874455 reduced FGFR signaling *in vivo*, we performed a kinetic study of FGFR signaling proteins. Tumors were harvested at 3, 24 and 48 h after last treatment (end of study) and protein lysates were subjected to Western blotting to analyze the phosphorylation levels of FRS2 and ERK. The endogenous level of phosphorylated FRS2 in the LS70x tumors was, as in the cell lines, below detection (data not shown). However, ERK phosphorylation was clearly reduced 3 h after treatment and remained reduced for at least 24 h (Figure 5B, quantified in Figure S4C).

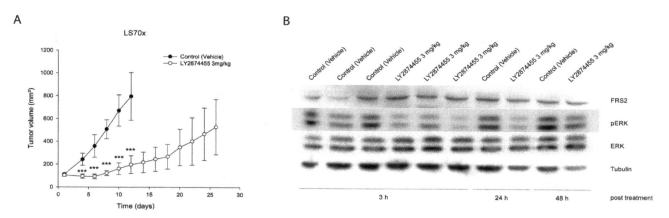

Figure 5. Growth inhibitory effect of LY2874455 *in vivo* on LS70x, a *FRS2*-amplified xenograft. (**A**) *in vivo* study with *FRS2*-amplified tumors of LS70x xenografts (*n* = 6) treated twice per day (BID) with LY2874455 3 mg/kg or control vehicle for up to 28 days until tumor size reaches a limit of 1 cm^3. Data shown as means ± SEM *** $p \leq 0.001$; unpaired two-tailed *t*-test treated versus vehicle treated; (**B**) Western blots showing the level of phosphorylated and total ERK in lysates extracted from LS70x tumors treated *in vivo* with vehicle or LY2874455 until endpoint. The tumors were harvested at the indicated time after last treatment. α-tubulin is used as loading control.

4. Discussion

WD/DDLPS tumors almost invariably have *FRS2* amplified [8] and are in great need of new therapies. We cannot expect many new drugs specific for rare cancers, therefore the possibility to repurpose existing drugs for these patients deserves thorough consideration. FGF receptor inhibitors have shown tolerable toxicities in rodent and patients [10,11], thus showing efficacy in preclinical models could pave the way to clinical trials on sarcoma patients.

Our previous *in vitro* study demonstrated limitations of NVP-BGJ398, since the drug was only transiently cytostatic and the cells quickly regained growth capacity when drug was removed [3]. In this study we further investigated the potential of FGFR inhibition as treatment for *FRS2*-amplified DDLPS.

Results were more promising using the FGFR inhibitor LY2874455, which gave improved efficacy *in vitro*. LY2874455 is reported to have a similar potency against all four FGFRs in biochemical assays and has shown potent activity against FGFR signaling in preclinical studies of several cancer types such as lung, gastric, bladder and multiple myeloma [12]. This drug had a stronger effect on the growth of NRH-LS1 cells, and induced apoptosis more efficiently than NVP-BGJ398 did. Interestingly, LY2874455 had a long-lasting effect on the responding cell lines, with cell growth inhibited several days after the drug treatment was discontinued. The observed effect and higher potency of LY2874455 compared to NVP-BGJ398 could potentially be due to off-target inhibition, since LY2874455 was shown to act as a multi-kinase inhibitor and consequently inhibits a wide range of different kinases [19]. However, the similarity of the response of FGFR signaling by the two inhibitors indicates a similar mode of action.

We hypothesized that amplified *FRS2* would potentiate FGFR signaling and drive growth of *FRS2*-amplified DDLPS. Thus, *FRS2* amplification could be a biomarker predicting sensitivity to FGFR inhibitors. Although the three *FRS2* amplified cell lines had similar levels of *FRS2*, LPS853 was resistant, suggesting that *FRS2* amplification alone is not sufficient as predictive biomarker. The lack of response of one cell line out of four independent models might not be representative of the patient population, but more refined biomarkers detecting such tumors would be valuable. To identify possible differences in FGFR signaling that could explain the different responses to FGFR inhibition, we analyzed the status of FGFR signaling pathway proteins. The basal level of phosphorylated FRS2 in unstimulated cells was undetectable but was drastically increased upon addition of FGF1 for all the three cell lines, confirming functional FGFR signaling. Both AKT and ERK are downstream of FRS2 and the strong inhibition of ERK phosphorylation when cells were treated with FGFR inhibitor indicated that the FGFR pathway is the predominant activator of ERK in these cells. This is consistent with a study in human bladder cancer, which showed dephosphorylation of FRS2 and ERK by NVP-BGJ398 [10]. We could not detect significant changes in levels of AKT or phosphorylated AKT, which is also consistent with previous studies [20].

We did not observe any FGFR mutations or translocations that could explain the different response to FGFR inhibition (data not shown). Although we observed equal expression of *FGFR1* among the cell lines, the non-responding LPS853 cells had considerably higher expression of *FGFR4*. Furthermore, LPS853 also had a high expression of *FGFRL1*, which lacks the kinase domain and has been suggested to be a negative regulator of FGFR1 signaling [21]. Although LPS853 cells do not respond to FGF inhibitors with reduced proliferation, FGFR inhibition still prevents endogenous and FGF1-induced ERK phosphorylation in these cells. This implies that FGFR signaling is maintained in LPS853 cells, but that these cells are not dependent on phosphorylated ERK for proliferation. This is known for other cell types as well. Only a subset of KRAS mutated colon, pancreatic and lung cancer depends on MEK/ERK signaling for proliferation, although the cells responds to MEK inhibition with reduced pERK [22]. Often this is due to other rescuing mechanisms. This could explain why LPS853 cells grow independent of this pathway, and also why they respond to higher doses of LY2874455, but not to high doses of NVP-BGJ398 (data not shown) [3], since LY2874455, unlike NVP-BGJ398, has a similar potency against all four FGF receptors in biochemical assays [12,23].

We expect the levels of FGFs in fetal calf serum to be low, but the responding cell lines may produce autocrine FGFs, as was indeed indicated by the RNA-seq data. NRH-LS1 and LPS510 had high expression of *FGF2*, a ligand for several of the FGF receptors, while LPS853 had high expression of the intracellular FGF11. While the function of FGF11 is not fully known, FGF2 is known to be secreted by adipocytes and to stimulate proliferation upon binding to FGFR1. We hypothesize that the lack of expression of any extracellular FGF ligands in LPS853 cells indicate that they grow independently of FGFR *in vitro*, but we cannot exclude that they depended on exogeneous FGF *in vivo* and have adapted to conditions without FGF *in vitro*. Further functional investigations including knock-down of the different ligands and receptors might solve these issues.

In summary, our results support LY2874455 as a better drug candidate than NVP-BGJ398 for treatment of *FRS2*-amplified liposarcoma. LY2874455 also showed significant efficacy *in vivo*, which is an important finding, since the FGF-regulatory landscape in tissues is different from cell cultures. Whether efficacy of LY2874455 can also be translated to other sarcomas with aberrations in the FGFR pathway, such as amplified, fused or mutated *FGFR* genes [24], needs to be investigated. We hope these studies could result in a clinical trial for DDLPS patients in great need of new treatments now that phase I clinical trials have shown tolerable toxicities [11].

Author Contributions: R.H.: conceptualization, drug studies, data curation and analysis, project administration, visualization, writing—original draft, review, and editing. I.G.: drug studies, data curation and analysis, writing—review and editing. E.M. (Else Munthe): protein studies, data curation and analysis, writing—review and editing. J.H.: animal studies, data curation and analysis, manuscript review. I.K.: animal studies, data curation and analysis, manuscript review. E.M. (Emmet McCormack): animal studies, data curation and analysis, manuscript review. L.A.M.-Z.: supervision, genomics, data curation and analysis, writing—review and editing. E.W.S.: supervision, design of animal studies, writing—review and editing. O.M.: conceptualization, funding acquisition, project administration, resources, supervision, writing—review and editing.

Acknowledgments: We are grateful to Mona Mari Lindeberg for DNA and RNA isolation and Susanne Lorenz from the Genomics Core Facility at Oslo University Hospital for the sequencing of the cell lines. We would also like to thank Mihaela Popa for contributing to the design and planning of the *in vivo* experiment. This work was supported by grants from the Norwegian Research Council, The Regional Health Authority of South-Eastern Norway, and the Norwegian Cancer Society.

References

1. Burningham, Z.; Hashibe, M.; Spector, L.; Schiffman, J.D. The epidemiology of sarcoma. *Clin. Sarcoma Res.* **2012**, *2*, 14. [CrossRef] [PubMed]

2. Barretina, J.; Taylor, B.S.; Banerji, S.; Ramos, A.H.; Lagos-Quintana, M.; Decarolis, P.L.; Shah, K.; Socci, N.D.; Weir, B.A.; Ho, A.; et al. Subtype-specific genomic alterations define new targets for soft-tissue sarcoma therapy. *Nat. Genet.* **2010**, *42*, 715–721. [CrossRef] [PubMed]

3. Hanes, R.; Grad, I.; Lorenz, S.; Stratford, E.W.; Munthe, E.; Reddy, C.C.S.; Meza-Zepeda, L.A.; Myklebost, O. Preclinical evaluation of potential therapeutic targets in dedifferentiated liposarcoma. *Oncotarget* **2016**, *7*, 54583–54595. [CrossRef] [PubMed]

4. Wang, X.; Asmann, Y.W.; Erickson-Johnson, M.R.; Oliveira, J.L.; Zhang, H.; Moura, R.D.; Lazar, A.J.; Lev, D.; Bill, K.; Lloyd, R.V.; et al. High-resolution genomic mapping reveals consistent amplification of the fibroblast growth factor receptor substrate 2 gene in well-differentiated and dedifferentiated liposarcoma. *Genes Chromosomes Cancer* **2011**, *50*, 849–858. [CrossRef] [PubMed]

5. Wesche, J.; Haglund, K.; Haugsten, E.M. Fibroblast growth factors and their receptors in cancer. *Biochem. J.* **2011**, *437*, 199–213. [CrossRef] [PubMed]

6. Korc, M.; Friesel, R.E. The role of fibroblast growth factors in tumor growth. *Curr. Cancer Drug Targets* **2009**, *9*, 639–651. [CrossRef] [PubMed]

7. Eswarakumar, V.P.; Lax, I.; Schlessinger, J. Cellular signaling by fibroblast growth factor receptors. *Cytokine Growth Factor Rev.* **2005**, *16*, 139–149. [CrossRef] [PubMed]

8. Zhang, K.; Chu, K.; Wu, X.; Gao, H.; Wang, J.; Yuan, Y.-C.; Loera, S.; Ho, K.; Wang, Y.; Chow, W.; et al. Amplification of FRS2 and activation of FGFR/FRS2 signaling pathway in high-grade liposarcoma. *Cancer Res.* **2013**, *73*, 1298–1307. [CrossRef] [PubMed]

9. Guagnano, V.; Kauffmann, A.; Wöhrle, S.; Stamm, C.; Ito, M.; Barys, L.; Pornon, A.; Yao, Y.; Li, F.; Zhang, Y.; et al. FGFR genetic alterations predict for sensitivity to NVP-BGJ398, a selective pan-FGFR inhibitor. *Cancer Discov.* **2012**, *2*, 1118–1133. [CrossRef] [PubMed]

10. Guagnano, V.; Furet, P.; Spanka, C.; Bordas, V.; Le Douget, M.; Stamm, C.; Brueggen, J.; Jensen, M.R.; Schnell, C.; Schmid, H.; et al. Discovery of 3-(2,6-dichloro-3,5-dimethoxy-phenyl)-1-{6-[4-(4-ethyl-piperazin-1-yl)-phenylamino]-pyrimidin-4-yl}-1-methyl-urea (NVP-BGJ398), a potent and selective inhibitor of the fibroblast growth factor receptor family of receptor tyrosine kinase. *J. Med. Chem.* **2011**, *54*, 7066–7083. [CrossRef]

11. Michael, M.; Bang, Y.-J.; Park, Y.S.; Kang, Y.-K.; Kim, T.M.; Hamid, O.; Thornton, D.; Tate, S.C.; Raddad, E.; Tie, J. A Phase 1 Study of LY2874455, an Oral Selective pan-FGFR Inhibitor, in Patients with Advanced Cancer. *Target Oncol.* **2017**, *12*, 463–474. [CrossRef] [PubMed]

12. Zhao, G.; Li, W.-Y.; Chen, D.; Henry, J.R.; Li, H.-Y.; Chen, Z.; Zia-Ebrahimi, M.; Bloem, L.; Zhai, Y.; Huss, K.; et al. A novel, selective inhibitor of fibroblast growth factor receptors that shows a potent broad spectrum of antitumor activity in several tumor xenograft models. *Mol. Cancer Ther.* **2011**, *10*, 2200–2210. [CrossRef] [PubMed]

13. Sebaugh, J.L. Guidelines for accurate EC50/IC50 estimation. *Pharm. Stat.* **2011**, *10*, 128–134. [CrossRef] [PubMed]

14. Kresse, S.H.; Meza-Zepeda, L.A.; Machado, I.; Llombart-Bosch, A.; Myklebost, O. Preclinical xenograft models of human sarcoma show nonrandom loss of aberrations. *Cancer* **2012**, *118*, 558–570. [CrossRef] [PubMed]

15. Wesche, J.; Wiedlocha, A.; Falnes, P.O.; Choe, S.; Olsnes, S. Externally added aFGF mutants do not require extensive unfolding for transport to the cytosol and the nucleus in NIH/3T3 cells. *Biochemistry* **2000**, *39*, 15091–15100. [CrossRef] [PubMed]

16. Kanojia, D.; Nagata, Y.; Garg, M.; Lee, D.H.; Sato, A.; Yoshida, K.; Sato, Y.; Sanada, M.; Mayakonda, A.; Bartenhagen, C.; et al. Genomic landscape of liposarcoma. *Oncotarget* **2015**, *6*, 42429–42444. [CrossRef] [PubMed]

17. Skårn, M.; Noordhuis, P.; Wang, M.-Y.; Veuger, M.; Kresse, S.H.; Egeland, E.V.; Micci, F.; Namløs, H.M.; Håkelien, A.-M.; Olafsrud, S.M.; et al. Generation and characterization of an immortalized human mesenchymal stromal cell line. *Stem Cells Dev.* **2014**, *23*, 2377–2389. [CrossRef] [PubMed]

18. Sleeman, M.; Fraser, J.; McDonald, M.; Yuan, S.; White, D.; Grandison, P.; Kumble, K.; Watson, J.D.; Murison, J.G. Identification of a new fibroblast growth factor receptor, FGFR5. *Gene* **2001**, *271*, 171–182. [CrossRef]

19. Hagel, M.; Miduturu, C.; Sheets, M.; Rubin, N.; Weng, W.; Stransky, N.; Bifulco, N.; Kim, J.L.; Hodous, B.; Brooijmans, N.; et al. First Selective Small Molecule Inhibitor of FGFR4 for the Treatment of Hepatocellular Carcinomas with an Activated FGFR4 Signaling Pathway. *Cancer Discov.* **2015**, *5*, 424–437. [CrossRef]

20. Luo, L.Y.; Kim, E.; Cheung, H.W.; Weir, B.A.; Dunn, G.P.; Shen, R.R.; Hahn, W.C. The tyrosine kinase adaptor protein FRS2 is oncogenic and amplified in high-grade serous ovarian cancer. *Mol. Cancer Res.* **2014**, *13*, 502–509. [CrossRef]

21. Trueb, B. Biology of FGFRL1, the fifth fibroblast growth factor receptor. *Cell. Mol. Life Sci.* **2011**, *68*, 951–964. [CrossRef] [PubMed]

22. Halilovic, E.; She, Q.-B.; Ye, Q.; Pagliarini, R.; Sellers, W.R.; Solit, D.B.; Rosen, N. PIK3CA mutation uncouples tumor growth and cyclin D1 regulation from MEK/ERK and mutant KRAS signaling. *Cancer Res.* **2010**, *70*, 6804–6814. [CrossRef] [PubMed]

23. Wu, D.; Guo, M.; Philips, M.A.; Qu, L.; Jiang, L.; Li, J.; Chen, X.; Chen, Z.; Chen, L. Crystal Structure of the FGFR4/LY2874455 Complex Reveals Insights into the Pan-FGFR Selectivity of LY2874455. *PLoS ONE* **2016**, *11*, e0162491. [CrossRef]

24. Asano, N.; Yoshida, A.; Mitani, S.; Kobayashi, E.; Shiotani, B.; Komiyama, M.; Fujimoto, H.; Chuman, H.; Morioka, H.; Matsumoto, M.; et al. Frequent amplification of receptor tyrosine kinase genes in welldifferentiated/dedifferentiated liposarcoma. *Oncotarget* **2017**, *8*, 12941. [CrossRef] [PubMed]

12

FGF19–*FGFR4* Signaling in Hepatocellular Carcinoma

Aroosha Raja [1], Inkeun Park [2], Farhan Haq [1,*] and Sung-Min Ahn [2,3,*]

[1]   Department of Biosciences, Comsats University, Islamabad 45550, Pakistan; aroosha.raja@gmail.com
[2]   Division of Medical Oncology, Department of Internal Medicine, Gachon University Gil Medical Center, Incheon 21565, Korea; ingni79@hanmail.net
[3]   Department of Genome Medicine and Science, College of Medicine, Gachon University, Incheon 21565, Korea
*   Correspondence: farhan.haq@comsats.edu.pk (F.H.); smahn@gachon.ac.kr or ahnsungmin@gmail.com (S.-M.A.)

Abstract: Hepatocellular carcinoma (HCC) is the sixth most common type of cancer, with an increasing mortality rate. Aberrant expression of fibroblast growth factor 19–fibroblast growth factor receptor 4 (*FGF19–FGFR4*) is reported to be an oncogenic-driver pathway for HCC patients. Thus, the *FGF19–FGFR4* signaling pathway is a promising target for the treatment of HCC. Several pan-*FGFR* (1–4) and *FGFR4*-specific inhibitors are in different phases of clinical trials. In this review, we summarize the information, recent developments, binding modes, selectivity, and clinical trial phases of different available *FGFR4*/pan-*FGF* inhibitors. We also discuss future perspectives and highlight the points that should be addressed to improve the efficacy of these inhibitors.

Keywords: prognosis; *FGF19*; *FGFR4*; HCC; inhibitors

1. Introduction

Hepatocellular carcinoma (HCC) is the sixth most common type of cancer, with the fourth highest mortality rate [1]. Despite advancements in therapeutic strategies, the response rate and overall survival rate are still low [2]. The most common cause of HCC is liver cirrhosis from any etiology including hepatitis B and hepatitis C infection, excessive alcohol consumption, diabetes mellitus, and non-alcoholic fatty liver disease [3]. Moreover, various molecular pathways are involved in the initiation and progression of HCC [4]. With respect to these pathways, there is evidence demonstrating the role of fibroblast growth factor pathway genes in HCC prognosis [5].

The fibroblast growth factors (*FGFs*) family comprises a large family of growth factors that are found in different multicellular organisms [6]. The FGFs signal through four transmembrane tyrosine kinase fibroblast growth factor receptors (*FGFRs*) namely *FGFR1*, *FGFR2*, *FGFR3*, and *FGFR4* [7]. *FGFs–FGFRs* are involved in regulation of many biological processes such as embryonic development, cell proliferation, differentiation, and tissue repair [8]. *FGF–FGFR* dysregulation is also widely reported in different types of diseases, disorders, and cancers [9]. Notably, aberrant expression of *FGF19/FGFR4* contributes to HCC progression [10].

Since sorafenib marked a new era in molecularly targeted therapy in advanced HCC [11], various drugs such as lenvatinib, regorafenib, cabozantinib, nivolumab, and ramucirumab have subsequently demonstrated overall survival benefits for patients [12–16]. However, the treatment outcome of metastatic HCC is still unsatisfactory, with a median overall survival below 15 months [12]. Thus, more effective treatment options for advanced HCC are needed. This can be achieved by a better understanding of the underlying genetic mechanisms involved in HCC. This review aims to provide comprehensive landscape of current information available on the *FGF19–FGFR4* pathway. It also discusses recent advancements on *FGF19–FGFR4* inhibitors in HCC. The data is obtained by systematic analysis of the literature and by using different text-mining approaches.

2. Overview of *FGFR4* and *FGF19*

2.1. Structure and Function of FGFR4

Fibroblast growth factor receptor 4 (*FGFR4*) is a protein coding gene and is a member of tyrosine kinase receptors family. The human *FGFR4* gene is located on chromosome 5 and measures 11.41 bp in length [17]. The *FGFR4* protein coded by two full transcripts of *FGFR4* gene consists of ~800 amino acids, with molecular weight of around 95–110 kDa [18]. The structure of *FGFR4* proteins contains three immunoglobin-like domains (D1–D3), a transmembrane domain, and the kinase domain [19]. (Figure 1) Among these immunoglobin-like domains, first two have role in receptor auto-inhibition, while the third domain is involved in specific binding of ligands [20]. The kinase domain (intracellular) is important in activation of downstream pathways [21]. Further, the kinase domain comprises the N-terminal (smaller) and C-terminal (larger) canonical domains [22]. FGF receptors differ from each other in tissue specificity and ligand-binding affinity. However, good identity scores are found between the kinase domains of *FGFR4* and other FGF receptors [22]. The expression of *FGFR4* is highly tissue-specific due to its unique ligand binding affinity [23]. At a functional level, *FGFR4* is predominantly involved in regeneration of muscles, regulation of lipid metabolism, bile acid biosynthesis, cell proliferation, differentiation, glucose uptake, and myogenesis [24]. Of note, it is reported that *FGFR4* is mostly expressed in liver tissue [25].

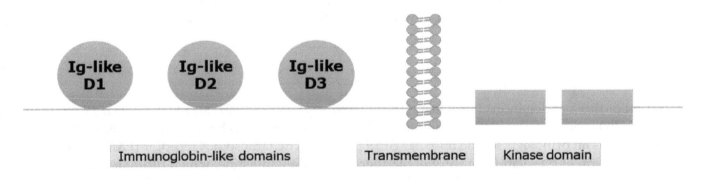

Figure 1. Structural overview of fibroblast growth factor receptor 4 (*FGFR4*) protein.

2.2. FGFR4 in Cancer

FGFR4 exerts a combination of biological effects that contribute to different hallmarks of cancer (Figure 2) [26]. Functional analysis demonstrated induction of both increased local growth and enhanced metastasis by mutated *FGFR4* [27]. Xu et al. described germline mutations in *FGFR4* i.e., glycine to arginine transition at position 388 in the transmembrane domain of *FGFR4* receptor, which results in the formation of *FGFR4* arg388 allele, leading to higher cancer risk [28]. Due to broad ligand binding spectrum of *FGFR4*, it is reportedly involved in multiple tumor types including HCC, breast cancer, colorectal cancer, rhabdomyosarcoma, and lung cancer [29–33].

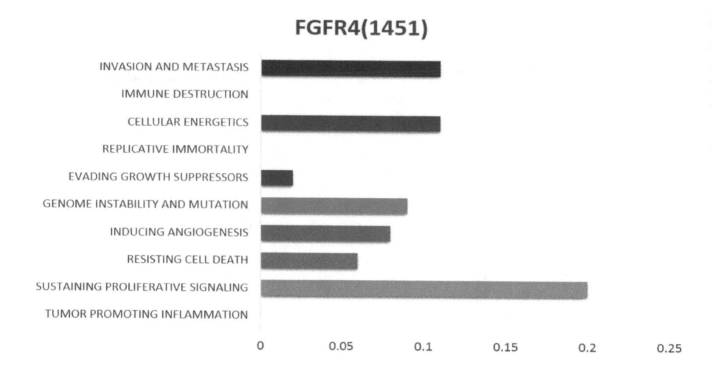

Figure 2. The association of *FGFR4* with different hallmarks of cancer, as reported in the literature. (Scales of bars from left to right represent the lowest to highest number of associations reported)

2.3. Structure and Function of FGF19

Out of three endogenous fibroblast growth factors (*FGF19*, *FGF21*, and *FGF23*), *FGF19* binds to *FGFR4* with highest affinity [34]. The human *FGF19* gene is located on chromosome 11q13. In mice, the *FGF15* gene is an orthologue of the human *FGF19* gene [6]. The farnesoid X receptor (FXR) is activated by the secretion of bile acid from the gall bladder to the small intestine, which ultimately stimulates *FGF19* secretion from the ileum [35,36]. The primary roles of *FGF19* are found in bile acid synthesis, gallbladder filling, glycogen synthesis, gluconeogenesis, and protein synthesis [37]. *FGF19* contributes to several hallmarks of cancer (Figure 3). Interestingly, *FGF19* and *FGF21* (endogenous fibroblast growth factors) are also most commonly involved in regulation of different functions occurring in liver [38]. Nicholes et al. demonstrated in transgenic mice that overexpression of *FGF19* is involved in liver dysplasia [39]. In our recent study, amplification of *FGF19* was found to be significantly associated with cirrhosis and also increased the risk of HCC [40]. Similarly, in our other study we used the fluorescence in situ hybridization technique and found the similar oncogenic patterns of *FGF19* in HCC [41]. Copy number amplification of *FGF19* is also highly reported in The Cancer Genome Atlas (TCGA) data [42]. Notably, the role of *FGF19* at expression level is also frequently reported in HCC prognosis [43,44].

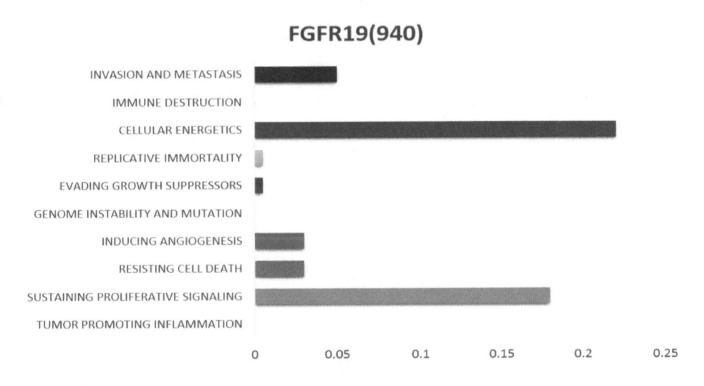

Figure 3. The association of *FGF19* with different hallmarks of cancer, as reported in the literature. (Scales of bars from left to right represent the lowest to highest number of associations reported)

2.4. Mechanism of FGFR4 Activation

Specific ligand receptor binding spectrum in FGFs lead to autophosphorylation and formation of multiple complex [45]. *FGFR4* is regulated using its co-receptor klotho-beta (*KLB*) (a transmembrane protein) [46]. The involvement of *KLB* co-receptor is reported in hepatocytes and adipose and pancreatic tissues [47]. *FGFR4* and *KLB* are found to be overexpressed in mature hepatocytes [48]. In addition, *KLB* is required for *FGF19–FGFR4* complex activation [49] (Figure 4).

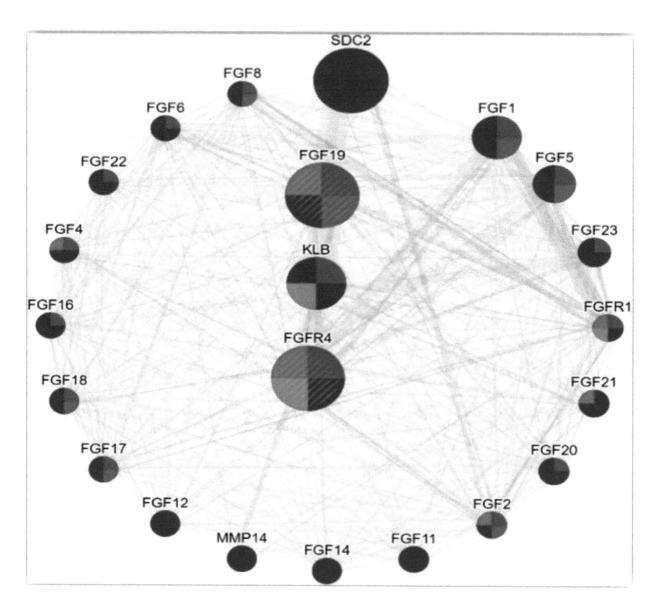

Figure 4. Interaction network of *FGFR4* with different genes with high potency and functional similarity. The interaction network is based on various parameters including co-expression, genetic interactivity, shared protein domains, co-localization and physical interactions.

FGFR4 related pathways have predominant involvement in proliferation, differentiation, survival, and migration of cells. (Figure 4) Multiple signaling cascades such as GSK3β/β-catenin, PI3K/AKT, PLCγ/DAG/PKC, and RAS/RAF/MAPK are modulated by *FGFR4* activation [10,50,51] (Figure 5).

FGFR4 selectively binds *FGF19* ligand [49,52]. *FGF19* is also reported as a functional partner of *FGFR4*, with the highest score in analysis through the STRING (https://string-db.org/) database.

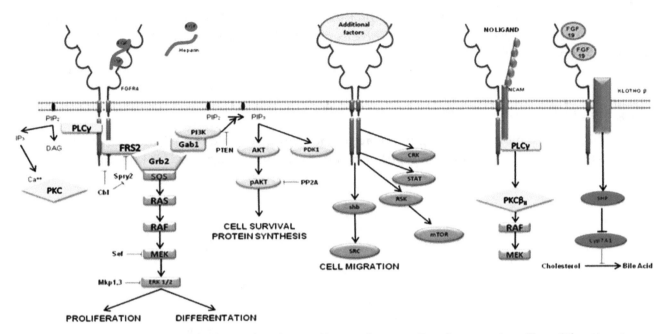

Figure 5. Involvement of *FGFR4*-related signaling pathways. Involvement in cell proliferation is depicted on the far left; next to it the cell survival signaling pathway is shown, and on the right side the cell migration pathway is explained (adapted from Atlas of Genetics and Cytogenetics in Oncology and Haematology).

2.5. FGF19–FGFR4 Pathway in HCC

FGF19/FGFR4 activation leads to the formation of FGF receptor substrate 2 (*FRS2*) and growth factor receptor-bound protein 2 (*GRB2*) complex, ultimately activating *Ras–Raf–ERK1/2MAPK* and *PI3K–Akt* pathways. (Figure 6) These pathways are predominantly involved in tumor proliferation and anti-apoptosis. (Figure 6).

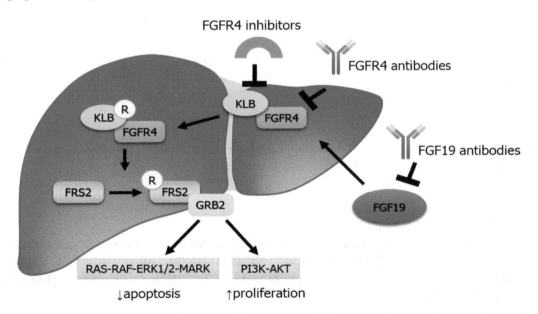

Figure 6. Binding mechanism of *FGF19* to *FGFR4* leads to *FRS2* along with recruitment of growth factor receptor-bound protein 2 (*GRB2*), ultimately leading to activation of the *Ras–Raf–ERK1/2 MAPK* and *PI3K–Akt* pathways.

As discussed, frequent studies reported the anomalous expression of *FGF19–FGFR4* complex enhances the progression of HCC [31,44]. In a study conducted on mice model, Cui et al. suggested

FGF19 as a potential therapeutic target for the treatment of HCC [53]. *FGFR4* dysregulation and its correlation with *TGF-β1* also suggested *FGFR4* as potential therapeutic target of HCC patients with invasiveness and metastasis [43,54].

3. Targeting *FGF19–FGFR4* in HCC

FGF19/FGFR4 inhibition is thought to lead to anti-tumor activities [55]. Thus, several FGFR (1–4) inhibitors are under trial for different types of malignancies including HCC [56] (Figure 7).

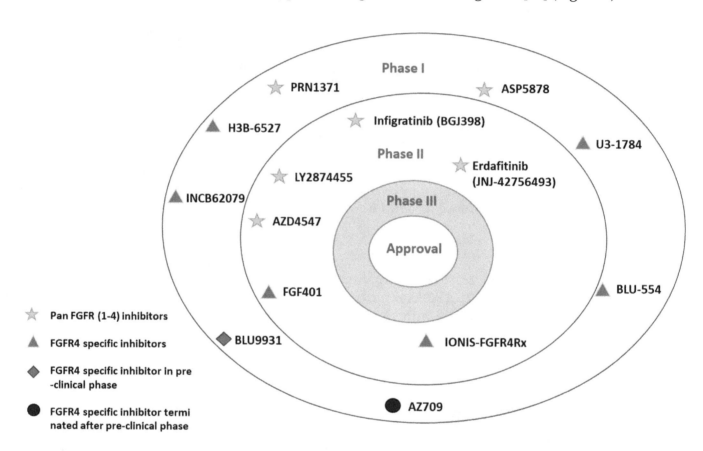

Figure 7. Selected overview of pan-*FGFRs* and *FGFR4*-specific inhibitors in different stages of clinical trials for hepatocellular carcinoma (HCC).

3.1. Pan-FGFR (1–4) Inhibitors

Multiple pan-*FGFR (1–4)* inhibitors are under-development in different phases of clinical trials (Figure 7). **LY2874455** (NCT01212107), **AZD4547** (NCT02038673), **infigratinib** (NCT02160041), and **erdafitinib** (**NCT02365597**) drugs are designed to target pan-FGFRs and are in phase II of development and clinical trials (Table 1).

LY2874455 is a small molecule inhibitor developed by Eli Lilly [53] (Figure 8). It has shown promising effects against advanced and metastatic cancers such as myelomas, lung, bladder, and gastric cancer [57]. Its highly effective inhibitory action suggests that it can be effective potential drug for HCC in the near future.

Table 1. Pan-FGFR inhibitors in different phases of clinical trials.

Drug	Company	Indication	Drug Target	Study Phase	Route of Administration	Clinical Trial ID
LY2874455	Eli Lilly	Advanced and metastatic cancers	Pan-FGFR (1–4) inhibitor	Phase II	Oral	NCT01212107
AZD4547	Astra Zeneca	Stage IV squamous cell lung cancer	Pan-FGFR (1–4) inhibitor	Phase II	Oral	NCT02965378
		ER+ breast cancer				NCT01791985
		Muscle-invasive bladder cancer (MIBC)		Phase I		NCT02546661
Infigratinib (BGJ398)	Novartis Pharmaceuticals	Tumors with FGFR genetic alterations	Pan-FGFR (1–4) inhibitor	Phase II	Oral	NCT02160041
		Advanced or metastatic cholangiocarcinoma		Phase II		NCT02150967
		Recurrent resectable or unresectable glioblastoma		Phase II		NCT01975701
		Solid tumor		Phase I		NCT01697605
		Advanced solid malignancies		Phase I		NCT01004224
Erdafitinib (JNJ-42756493)	Janssen Pharmaceuticals	Urothelial cancer Advanced hepatocellular carcinoma	Pan-FGFR (1–4) inhibitor	Phase II	Oral	NCT02365597
		Advanced non-small lung cancer Esophageal cancer				NCT02699606
		Lymphoma				NCT02952573
PRN1371	Prinicipia Biopharma Inc.	Solid tumor	Pan-FGFR (1–4) inhibitor	Phase I	Oral	NCT02608125
ASP5878	Astellas	Solid tumor	Pan-FGFR (1–4) inhibitor	Phase I	Oral	NCT02038673

ER+ breast cancer: estrogen-receptor-positive breast cancer.

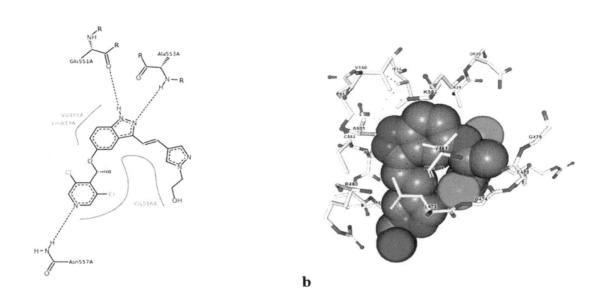

a   b

Figure 8. (**a**) Structure of LY2874455, and (**b**) binding mode of LY2874455 with the *FGFR4* kinase domain (PDB code 5JKG).

AZD4547 was developed to specifically target pan-*FGFR (1–4)* in solid tumors. However, **AZD4547** showed good efficacy against *FGFR (1–3)* but weaker activity against *FGFR4* [58], suggesting low efficacy when specifically targeting *FGFR4*.

Infigratinib (BGJ398), which targets *FGFR (1–3)* with high affinity and *FGFR4* with less affinity, was developed by Novartis Pharmaceuticals. It is currently in phase II for tumors with alteration of *FGFR* and for glioblastomas, solid tumors, hematologic malignancies, and advanced cholangiocarcinoma.

Infigratinib showed an effective response against *FGFR* signaling pathways in HCC [59]. However, FDA-approved clinical trials are yet to be conducted for infigratinib in HCC [59].

Janssen Pharmaceuticals reported **erdafitinib** (JNJ-42756493), a pan-*FGFR (1–4)* inhibitor (Figure 9), which is currently under phase II of clinical trials for advanced HCC. It significantly inhibited *FGFR*-overexpressing tumor cells in HCC [60].

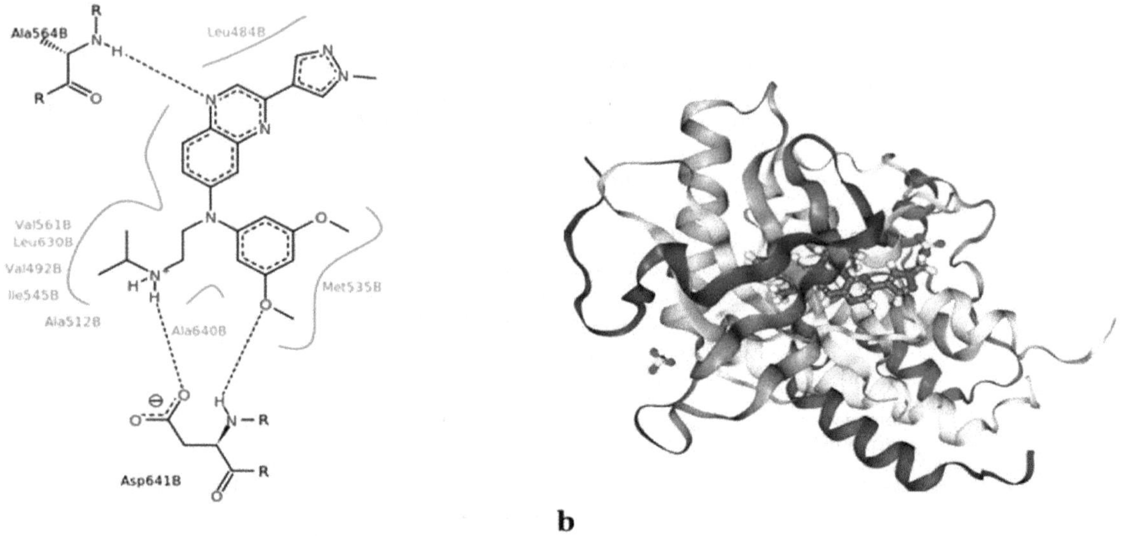

a   **b**

Figure 9. (**a**) Structure of JNJ-42756493 (**b**) Interaction of JNJ-42756493 with FGFR1 (PDB code 5EW8).

PRN1371 (NCT02608125) and **ASP5878** (NCT02038673) are drugs designed to target pan-*FGFRs* and are in phase I of development and clinical trials. PRN1371 was developed by Principia Biopharma Inc. for solid tumors. It is an irreversible inhibitor that specifically targets *FGFRs*. The inhibitory action of this drug has been reported in many tumor types like HCC, gastric, and lung cancer [61]. Astellas developed **ASP5878** to target pan-*FGFRs (1–4)* in solid tumors. Importantly, ASP5878 also inhibited HCC cell lines exhibiting overexpression of *FGF19* in the pre-clinical phase. In addition, this small inhibitor molecule improved the efficacy of sorafenib [62].

3.2. FGFR4-Specific Inhibitors

As discussed, the overexpression of *FGFR4* is most frequently reported receptor compared to FGFR (1–3) in HCC initiation and progression. However, selectivity of pan-*FGFR* inhibitors is comparatively lower for *FGFR4*. Thus, Prieto-Dominguez et al. outlined different targeted therapeutics available for the *FGF19–FGFR4* complex [29]. A number of drugs are under different phases of clinical trials which specifically target *FGF19/FGFR4*. Two potential drug candidates in the phase II stage of clinical trials, namely **IONIS-*FGFR4*Rx** (NCT02476019) and **FGF-401** (NCT02325739), are reported (Table 2).

Table 2. *FGFR4*-specific inhibitors under different phases of clinical trials.

Drug	Company	Indication	Drug Target	Study Phase	Route of Administration	Clinical Trial ID
IONIS-*FGFR4*Rx	Ionis Pharmaceuticals	Obesity and insulin sensitivity	*FGFR4*-specific	Phase II	Subcutaneous	NCT02476019
FGF401	Novartis AG	Hepatocellular carcinoma Solid malignancies	*FGFR4*-specific	Phase II (recruiting status)	Oral	NCT02325739
H3B-6527	H3 Biomedicine Inc.	Hepatocellular carcinoma	*FGFR4*-specific	Phase I	Oral	NCT02834780
U3-1784	Daiichi Sankyo Inc.	Advanced solid tumor Hepatocellular carcinoma	*FGFR4*-specific	Phase I (Terminated)	Intravenous	NCT02690350
BLU-554	Blueprint Medicines Corp.	Hepatocellular carcinoma (orphan drug designation for HCC by the U.S. FDA)	*FGFR4*-specific	Phase I	Oral	NCT02508467
AZ709	AstraZeneca	Hepatocellular carcinoma	*FGFR4*-specific	Inactive (Pre-clinical)	Unspecified	

U.S. FDA: U.S. Food and Drug Administration.

IONIS-*FGFR4*Rx, previously known as ISIS-*FGFR4*Rx, exhibited antisense inhibitor activity against *FGFR4* [59]. IONIS-*FGFR4*Rx has undergone a phase II clinical trial for obesity, specifically targeting *FGFR4* in liver and fat tissues. It is not only effective in reducing obesity but also improves insulin sensitivity [63]. Thus, we suggest that conducting trials with IONIS-*FGFR4*Rx in HCC patients may give significant results.

FGF401 was developed by Novartis and specifically targets *FGFR4* in HCC patients. According to the most recent update, FGF401 is in phase II of clinical trials for HCC, expected to be completed by the year 2020. FGF401, with an IC_{50}, exhibited at least 1000-fold potency for inhibiting *FGFR4* kinase activity compared to other *FGFRs (1–3)* [64].

H3B-6527 (NCT02834780), **U3-1784** (NCT02690350), and **BLU-554** (NCT02508467) are reported to be in phase I clinical trials to specifically target *FGFR4*.

H3B-6527 is a small inhibitor molecule developed by H3 Biomedicine Inc for targeting *FGFR4*-overexpression in advanced HCC and cholangiocarcinoma (IHCC) patients. In preclinical trials, **H3B-6527** proved to be effective in terms of repressing tumor growth in a xenograft model of HCC which exhibited activated aberrant *FGF19–FGFR4* signaling [65].

The human monoclonal drug **U3-1784** is under-development by Daiichi Sankyo Inc for HCC and other solid tumors. This antibody specifically binds to *FGFR4* and is most effective (approximately 90%) in *FGF19*-expressing models, suggesting it as a potential drug for HCC with an activated *FGF19–FGFR4* pathway. However, according to a recent update, the clinical trials for this drug have been terminated [66].

BLU-554, a *FGFR4*-specific inhibitor, is under recruiting phase by Blueprint Medicines Corp. for HCC and cholangiocarcinoma patients. In addition, it was also granted an orphan drug designation in 2015 by the U.S. FDA for HCC [67].

Lastly, **AZ709** showed good selective inhibition of *FGFR4* in HCC, as recently reported by AstraZeneca, and is in the preclinical stage of development. However, no progress has been reported on this drug to date (reported at the 2013 NCRI Cancer Conference, Liverpool, UK).

3.3. Irreversible FGFR4 Inhibitors

Two irreversible *FGFR4* inhibitors have also been recently reported, including **INCB62079** (ClinicalTrials.gov Identifier: NCT03144661) and **BLU9931** [68] (Figure 7, Table 3). **INCB62079**, developed by the Incyte Corporation, showed effective dose-dependent and compound-selective activity against cancer cells exhibiting active *FGF19–FGFR4*. Additionally, it showed good efficacy in

Hep3b hepatocellular cancer xenograft model in pre-clinical trial phase. **INCB62079** is currently in phase I clinical trials (ClinicalTrials.gov Identifier: NCT03144661) for HCC.

Table 3. *FGFR4*-specific irreversible inhibitors under different phases of clinical trials.

Drug	Company	Indication	Drug Target	Study Phase	Route of Administration	Clinical Trial ID
INCB62079	Incyte Corporation	Liver cancer	*FGFR4*-specific (irreversible)	Phase I	Unspecified	NCT03144661
BLU9931	Blueprint Medicines Corp.	Hepatocellular carcinoma	*FGFR4*-specific (irreversible)	Pre-clinical	Oral	

Blueprint Medicines Corp reported the remarkable drug **BLU9931**, a small irreversible inhibitor of *FGFR4*. It is currently in the pre-clinical stage of development for HCC and has not been approved by the U.S. FDA. In the preclinical trial phase, BLU9931 exhibited potent antitumor activity in mice with an HCC tumor xenograft with amplified *FGF19* and high expression of *FGF19* at the mRNA level. Recently, it has been reported that *FGF19* shows resistance to sorafenib, but BLU9931 is involved in improving sorafenib efficacy by inactivating *FGFR4* signaling [68].

Apart from the drugs reported in different clinical trials, different studies are underway to find new potent inhibitors against *FGF19*/FGFR. For instance, Cheuk et al. developed a chimeric antibody **3A11ScFvFc** (mice antibody Fv + Human IgG1Fc) to specifically target *FGFR4* in HCC [69]. Chen et al. found ***ABSK-011*** to be involved in suppressing high *FGFR4* expression, which ultimately results in HCC tumor suppression. *ABSK-011*, acting as irreversible inhibitor, selectively modifies *cys552*, which is the residue present within the active site of *FGFR4*. Of note, safety studies have also been conducted for this inhibitor [70]. Lee et al. examined the effect of the **HM81422** inhibitor on the *FGFR4–FGFR19* pathway. They successfully demonstrated that **HM81422** can potentially target *FGFR4* activated pathways. However, further elucidation is still required to understand the role of this inhibitor in HCC [71]. Furthermore, different pharmacological approaches suggested significant involvement of the drug sorafenib in inhibiting tyrosine kinase pathways. Initially, Gao et al. reported **sorafenib** as potential tyrosine kinase inhibitor which improves overall survival rate in HCC patients [68]. Later, Matsuki et al. revealed that sorafenib has no particular effect on the oncogenic FGF signaling pathway. However, the involvement of the drug **lenvatinib** was also recently reported [68]. Lenvatinib reportedly inhibits *FGF* pathways in HCC cell lines. Of note, studies suggested that it can be used as a pan-*FGFR (1–4)* inhibitor [68]. However, the specificity of lenvatinib against the *FGF19–FGFR4* signaling pathway still remains unclear [72].

4. Discussion and Conclusions

Compelling evidence supports the involvement of the *FGF19–FGFR4* signaling pathway in HCC [43]. Therefore, this pathway is considered to be a promising therapeutic target for the treatment of HCC. Interestingly, a number of different inhibitors and drugs have been reported to target FGF and *FGFR* signaling pathways. Despite promising advancements, it is still challenging to completely address all the underlying perspectives of this pathway. These perspectives, if clearly addressed, can improve the efficacy and potency of drugs available for HCC. The detailed analysis of available data revealed that *FGFR4* is structurally distinct from other *FGF* receptors (1–3) and also exhibits variable inhibition potency towards different available *FGFR* drugs [73]. Perhaps, this distinct characteristic of *FGFR4* should be exploited in depth to develop *FGFR4*-specific inhibitors to improve drug efficacy for HCC. Importantly, the evidence derived from primates suggests that anti-*FGF19* antibody treatment is mostly accompanied with dose-related liver toxicity [74]. Therefore, the likelihood of adverse effects of FGF/FGFR drugs should be properly envisaged to assure best possible and safe outcomes along with reduced dose-dependent side effects.

In addition, the correlation of *FGF19* gene amplification and HCC is reported to be highly significant, and it is consequently thought to act as potential biomarker for HCC [75]. Therefore, copy number gain of *FGF19* and *FGFR4* should be taken into consideration when designing potential inhibitors of these genes and their pathways.

Conceptually, it is shown that the patients having elevated bile acid concentrations and diabetes have a higher risk of developing HCC [44,53]. Therefore, these complications should be taken into account along with the inhibition of *FGF19–FGFR4* pathways to avoid potential adverse impacts and minimize safety risks in HCC patients.

Overall, the degree of *FGF–FGFR* inhibition in HCC is not satisfactory. This perhaps gives an indication towards elucidating other factors that are simultaneously involved in the *FGF–FGFR* signaling pathway. For instance, *KLB* (the co-receptor of *FGFR4*) is reportedly considered as a novel drug candidate as it is mostly found involved in inducing *FGFR4* overexpression and is also found in an elevated state in HCC [46,76]. Thus, in the future klotho-specific inhibitors can be considered to potentially maximize antitumor and therapeutic benefits in HCC by terminating *FGF19*-binding to *FGFR4*. Lastly, developing drugs that act on key SNPs of *FGFR4* i.e., Gly388 to Arg388, may also be clinically relevant.

In conclusion, most of the *FGFR4*-specific inhibitors are in pre-clinical phases. Progression of these potential inhibitors to advance clinical trial phases coupled with comprehensive research and improvements can revolutionize the available therapeutic options for HCC.

Author Contributions: S.-M.A. and F.H. contributed to the conception and design of the study. A.R. performed the literature review. A.R. and F.H. wrote the manuscript. A.R., I.P., and S.-M.A. contributed in revising the manuscript.

References

1. Bray, F.; Ferlay, J.; Soerjomataram, I.; Siegel, R.L.; Torre, L.A.; Jemal, A. Global cancer statistics 2018: GLOBOCAN estimates of incidence and mortality worldwide for 36 cancers in 185 countries. *CA Cancer J. Clin.* **2018**, *68*, 394–424. [CrossRef] [PubMed]

2. Huang, S.; He, X. The role of microRNAs in liver cancer progression. *Br. J. Cancer* **2011**, *104*, 235–240. [CrossRef] [PubMed]

3. Kulik, L.; El-Serag, H.B. Epidemiology and Management of Hepatocellular Carcinoma. *Gastroenterology* **2019**, *156*, 477–491. [CrossRef] [PubMed]

4. Moeini, A.; Cornellà, H.; Villanueva, A. Emerging Signaling Pathways in Hepatocellular Carcinoma. *LIC* **2012**, *1*, 83–93. [CrossRef] [PubMed]

5. Zheng, N.; Wei, W.; Wang, Z. Emerging roles of FGF signaling in hepatocellular carcinoma. *Transl. Cancer Res.* **2016**, *5*, 1–6. [PubMed]

6. Ornitz, D.M.; Itoh, N. Fibroblast growth factors. *Genome Biol.* **2001**, *2*, reviews3005.1–reviews3005.12. [CrossRef] [PubMed]

7. Itoh, N.; Ornitz, D.M. Fibroblast growth factors: From molecular evolution to roles in development, metabolism and disease. *J. Biochem.* **2011**, *149*, 121–130. [CrossRef] [PubMed]

8. Wilkie, A.; Morriss-Kay, G.M.; Yvonne Jones, E.; Heath, J.K. Functions of fibroblast growth factors and their receptors. *Curr. Biol.* **1995**, *5*, 500–507. [CrossRef]

9. Turner, N.; Grose, R. Fibroblast growth factor signalling: From development to cancer. *Nat. Rev. Cancer* **2010**, *10*, 116–129. [CrossRef]

10. Zhao, H.; Lv, F.; Liang, G.; Huang, X.; Wu, G.; Zhang, W.; Yu, L.; Shi, L.; Teng, Y. FGF19 promotes epithelial-mesenchymal transition in hepatocellular carcinoma cells by modulating the GSK3β/β- catenin signaling cascade via *FGFR4* activation. *Oncotarget* **2015**, *7*, 13575–13586.

11.   Llovet, J.M.; Hilgard, P.; de Oliveira, A.C.; Forner, A.; Zeuzem, S.; Galle, P.R.; Häussinger, D.; Moscovici, M. Sorafenib in Advanced Hepatocellular Carcinoma. *N. Engl. J. Med.* **2008**, *13*. [CrossRef] [PubMed]

12.   Kudo, M.; Finn, R.S.; Qin, S.; Han, K.-H.; Ikeda, K.; Piscaglia, F.; Baron, A.; Park, J.-W.; Han, G.; Jassem, J.; et al. Lenvatinib versus sorafenib in first-line treatment of patients with unresectable hepatocellular carcinoma: A randomised phase 3 non-inferiority trial. *Lancet* **2018**, *391*, 1163–1173. [CrossRef]

13.   Bruix, J.; Qin, S.; Merle, P.; Granito, A.; Huang, Y.-H.; Bodoky, G.; Pracht, M.; Yokosuka, O.; Rosmorduc, O.; Breder, V.; et al. Regorafenib for patients with hepatocellular carcinoma who progressed on sorafenib treatment (RESORCE): A randomised, double-blind, placebo-controlled, phase 3 trial. *Lancet* **2017**, *389*, 56–66. [CrossRef]

14.   Abou-Alfa, G.K.; Meyer, T.; Cheng, A.-L.; El-Khoueiry, A.B.; Rimassa, L.; Ryoo, B.-Y.; Cicin, I.; Merle, P.; Chen, Y.; Park, J.-W.; et al. Cabozantinib in Patients with Advanced and Progressing Hepatocellular Carcinoma. *N. Engl. J. Med.* **2018**, *379*, 54–63. [CrossRef] [PubMed]

15.   Zhu, A.X.; Kang, Y.-K.; Yen, C.-J.; Finn, R.S.; Galle, P.R.; Llovet, J.M.; Assenat, E.; Brandi, G.; Pracht, M.; Lim, H.Y.; et al. Ramucirumab after sorafenib in patients with advanced hepatocellular carcinoma and increased α-fetoprotein concentrations (REACH-2): A randomised, double-blind, placebo-controlled, phase 3 trial. *Lancet Oncol.* **2019**, *20*, 282–296. [CrossRef]

16.   El-Khoueiry, A.B.; Sangro, B.; Yau, T.; Crocenzi, T.S.; Kudo, M.; Hsu, C.; Kim, T.-Y.; Choo, S.-P.; Trojan, J.; Welling, T.H.; et al. Nivolumab in patients with advanced hepatocellular carcinoma (CheckMate 040): An open-label, non-comparative, phase 1/2 dose escalation and expansion trial. *Lancet* **2017**, *389*, 2492–2502. [CrossRef]

17.   Transcript: FGFR4-201 (ENST00000292408.8)—Protein summary—Homo sapiens—Ensembl genome browser 95. Available online: https://asia.ensembl.org/Homo_sapiens/Transcript/ProteinSummary?g=ENSG00000160867;r=5:177086886-177098144;t=ENST00000292408 (accessed on 19 January 2019).

18.   Partanen, J.; Mäkelä, T.P.; Eerola, E.; Korhonen, J.; Hirvonen, H.; Claesson-Welsh, L.; Alitalo, K. FGFR-4, a novel acidic fibroblast growth factor receptor with a distinct expression pattern. *EMBO J.* **1991**, *10*, 1347–1354. [CrossRef]

19.   Mohammadi, M.; Olsen, S.K.; Ibrahimi, O.A. Structural basis for fibroblast growth factor receptor activation. *Cytok. Growth Factor Rev.* **2005**, *16*, 107–137. [CrossRef]

20.   Wang, F.; Kan, M.; Yan, G.; Xu, J.; McKeehan, W.L. Alternately Spliced NH2-terminal Immunoglobulin-like Loop I in the Ectodomain of the Fibroblast Growth Factor (FGF) Receptor 1 Lowers Affinity for both Heparin and FGF-1. *J. Biol. Chem.* **1995**, *270*, 10231–10235. [CrossRef]

21.   Ornitz, D.M.; Itoh, N. The Fibroblast Growth Factor signaling pathway. *Wiley Interdiscip. Rev. Dev. Biol.* **2015**, *4*, 215–266. [CrossRef]

22.   Tucker, J.A.; Klein, T.; Breed, J.; Breeze, A.L.; Overman, R.; Phillips, C.; Norman, R.A. Structural Insights into FGFR Kinase Isoform Selectivity: Diverse Binding Modes of AZD4547 and Ponatinib in Complex with FGFR1 and *FGFR4*. *Structure* **2014**, *22*, 1764–1774. [CrossRef] [PubMed]

23.   Horlick, R.A.; Stack, S.L.; Cooke, G.M. Cloning, expression and tissue distribution of the gene encoding rat fibroblast growth factor receptor subtype 4. *Gene* **1992**, *120*, 291–295. [CrossRef]

24.   Reference, G.H. FGFR4 Gene. Available online: https://ghr.nlm.nih.gov/gene/FGFR4 (accessed on 19 January 2019).

25.   Hughes, S.E. Differential Expression of the Fibroblast Growth Factor Receptor (FGFR) Multigene Family in Normal Human Adult Tissues. *J. Histochem. Cytochem.* **1997**, *45*, 1005–1019. [CrossRef] [PubMed]

26.   Gross, S.; Rahal, R.; Stransky, N.; Lengauer, C.; Hoeflich, K.P. Targeting cancer with kinase inhibitors. *J. Clin. Investig.* **2015**, *125*, 1780–1789. [CrossRef]

27.   Greenman, C.; Stephens, P.; Smith, R.; Dalgliesh, G.L.; Hunter, C.; Bignell, G.; Davies, H.; Teague, J.; Butler, A.; Stevens, C.; et al. Patterns of somatic mutation in human cancer genomes. *Nature* **2007**, *446*, 153–158. [CrossRef] [PubMed]

28.   Xu, W.; Li, Y.; Wang, X.; Chen, B.; Wang, Y.; Liu, S.; Xu, J.; Zhao, W.; Wu, J. *FGFR4* transmembrane domain polymorphism and cancer risk: A meta-analysis including 8555 subjects. *Eur. J. Cancer* **2010**, *46*, 3332–3338. [CrossRef] [PubMed]

29.   Ye, Y.-W.; Zhang, X.; Zhou, Y.; Wu, J.; Zhao, C.; Yuan, L.; Wang, G.; Du, C.; Wang, C.; Shi, Y. The correlations between the expression of *FGFR4* protein and clinicopathological parameters as well as prognosis of gastric cancer patients. *J. Surg. Oncol.* **2012**, *106*, 872–879. [CrossRef]

30. Spinola, M.; Leoni, V.P.; Tanuma, J.; Pettinicchio, A.; Frattini, M.; Signoroni, S.; Agresti, R.; Giovanazzi, R.; Pilotti, S.; Bertario, L.; et al. *FGFR4* Gly388Arg polymorphism and prognosis of breast and colorectal cancer. *Oncol. Rep.* **2005**, *14*, 415–419. [CrossRef]

31. Matakidou, A.; el Galta, R.; Rudd, M.F.; Webb, E.L.; Bridle, H.; Eisen, T.; Houlston, R.S. Further observations on the relationship between the *FGFR4* Gly388Arg polymorphism and lung cancer prognosis. *Br. J. Cancer* **2007**, *96*, 1904–1907. [CrossRef]

32. Vi, J.G.T.; Cheuk, A.T.; Tsang, P.S.; Chung, J.-Y.; Song, Y.K.; Desai, K.; Yu, Y.; Chen, Q.-R.; Shah, K.; Youngblood, V.; et al. Identification of *FGFR4*-activating mutations in human rhabdomyosarcomas that promote metastasis in xenotransplanted models. *J. Clin. Investig.* **2009**, *119*, 3395–3407. [CrossRef]

33. Sheu, M.-J.; Hsieh, M.-J.; Chiang, W.-L.; Yang, S.-F.; Lee, H.-L.; Lee, L.-M.; Yeh, C.-B. Fibroblast Growth Factor Receptor 4 Polymorphism Is Associated with Liver Cirrhosis in Hepatocarcinoma. *PLoS ONE* **2015**, *10*, e0122961. [CrossRef] [PubMed]

34. Lee, K.J.; Jang, Y.O.; Cha, S.-K.; Kim, M.Y.; Park, K.-S.; Eom, Y.W.; Baik, S.K. Expression of Fibroblast Growth Factor 21 and β-Klotho Regulates Hepatic Fibrosis through the Nuclear Factor-κB and c-Jun N-Terminal Kinase Pathways. *Gut Liver* **2018**, *12*, 449–456. [CrossRef] [PubMed]

35. Liu, W.-Y.; Xie, D.-M.; Zhu, G.-Q.; Huang, G.-Q.; Lin, Y.-Q.; Wang, L.-R.; Shi, K.-Q.; Hu, B.; Braddock, M.; Chen, Y.-P.; et al. Targeting fibroblast growth factor 19 in liver disease: A potential biomarker and therapeutic target. *Expert Opin. Ther. Targets* **2015**, *19*, 675–685. [CrossRef] [PubMed]

36. Kurosu, H.; Choi, M.; Ogawa, Y.; Dickson, A.S.; Goetz, R.; Eliseenkova, A.V.; Mohammadi, M.; Rosenblatt, K.P.; Kliewer, S.A.; Kuro-o, M. Tissue-specific Expression of βKlotho and Fibroblast Growth Factor (FGF) Receptor Isoforms Determines Metabolic Activity of *FGF19* and FGF21. *J. Biol. Chem.* **2007**, *282*, 26687–26695. [CrossRef] [PubMed]

37. Kir, S.; Kliewer, S.A.; Mangelsdorf, D.J. Roles of *FGF19* in Liver Metabolism. *Cold Spring Harb. Symp. Quant. Biol.* **2011**, *76*, 139–144. [CrossRef] [PubMed]

38. Fukumoto, S. Actions and Mode of Actions of *FGF19* Subfamily Members. *Endocr. J.* **2008**, *55*, 23–31. [CrossRef] [PubMed]

39. Nicholes, K.; Guillet, S.; Tomlinson, E.; Hillan, K.; Wright, B.; Frantz, G.D.; Pham, T.A.; Dillard-Telm, L.; Tsai, S.P.; Stephan, J.-P.; et al. A Mouse Model of Hepatocellular Carcinoma: Ectopic Expression of Fibroblast Growth Factor 19 in Skeletal Muscle of Transgenic Mice. *Am. J. Pathol.* **2002**, *160*, 2295–2307. [CrossRef]

40. Ahn, S.-M.; Jang, S.J.; Shim, J.H.; Kim, D.; Hong, S.-M.; Sung, C.O.; Baek, D.; Haq, F.; Ansari, A.A.; Lee, S.Y.; et al. Genomic portrait of resectable hepatocellular carcinomas: Implications of RB1 and *FGF19* aberrations for patient stratification. *Hepatology* **2014**, *60*, 1972–1982. [CrossRef]

41. Kang, H.J.; Haq, F.; Sung, C.O.; Choi, J.; Hong, S.-M.; Eo, S.-H.; Jeong, H.J.; Shin, J.; Shim, J.H.; Lee, H.C.; et al. Characterization of Hepatocellular Carcinoma Patients with *FGF19* Amplification Assessed by Fluorescence in situ Hybridization: A Large Cohort Study. *LIC* **2019**, *8*, 12–23. [CrossRef]

42. Ally, A.; Balasundaram, M.; Carlsen, R.; Chuah, E.; Clarke, A.; Dhalla, N.; Holt, R.A.; Jones, S.J.M.; Lee, D.; Ma, Y.; et al. Comprehensive and Integrative Genomic Characterization of Hepatocellular Carcinoma. *Cell* **2017**, *169*, 1327–1341. [CrossRef]

43. Miura, S.; Mitsuhashi, N.; Shimizu, H.; Kimura, F.; Yoshidome, H.; Otsuka, M.; Kato, A.; Shida, T.; Okamura, D.; Miyazaki, M. Fibroblast growth factor 19 expression correlates with tumor progression and poorer prognosis of hepatocellular carcinoma. *BMC Cancer* **2012**, *12*, 56. [CrossRef] [PubMed]

44. Wu, X.; Ge, H.; Lemon, B.; Vonderfecht, S.; Weiszmann, J.; Hecht, R.; Gupte, J.; Hager, T.; Wang, Z.; Lindberg, R.; et al. *FGF19*-induced Hepatocyte Proliferation Is Mediated through *FGFR4* Activation. *J. Biol. Chem.* **2010**, *285*, 5165–5170. [CrossRef] [PubMed]

45. Powers, C.J.; McLeskey, S.W.; Wellstein, A. Fibroblast growth factors, their receptors and signaling. *Endocr. Relat. Cancer* **2000**, *7*, 165–197. [CrossRef] [PubMed]

46. Poh, W.; Wong, W.; Ong, H.; Aung, M.O.; Lim, S.G.; Chua, B.T.; Ho, H.K. Klotho-beta overexpression as a novel target for suppressing proliferation and fibroblast growth factor receptor-4 signaling in hepatocellular carcinoma. *Mol. Cancer* **2012**, *11*, 14. [CrossRef] [PubMed]

47. Ito, S.; Kinoshita, S.; Shiraishi, N.; Nakagawa, S.; Sekine, S.; Fujimori, T.; Nabeshima, Y. Molecular cloning and expression analyses of mouse βklotho, which encodes a novel Klotho family protein. *Mech. Dev.* **2000**, *98*, 115–119. [CrossRef]

48. Li, Y.; Zhang, W.; Doughtie, A.; Cui, G.; Li, X.; Pandit, H.; Yang, Y.; Li, S.; Martin, R. Up-regulation of fibroblast growth factor 19 and its receptor associates with progression from fatty liver to hepatocellular carcinoma. *Oncotarget* **2016**, *7*, 52329–52339. [CrossRef]

49. Lin, B.C.; Wang, M.; Blackmore, C.; Desnoyers, L.R. Liver-specific Activities of *FGF19* Require Klotho beta. *J. Biol. Chem.* **2007**, *282*, 27277–27284. [CrossRef]

50. Pratsinis, H.; Armatast, A.A.; Kletsas, D. Response of Fetal and Adult Cells to Growth Factors. In *Human Fetal Tissue Transplantation*; Bhattacharya, N., Stubblefield, P., Eds.; Springer: London, UK, 2013; pp. 65–77. ISBN 978-1-4471-4171-6.

51. Tiong, K.H.; Tan, B.S.; Choo, H.L.; Chung, F.F.-L.; Hii, L.-W.; Tan, S.H.; Khor, N.T.W.; Wong, S.F.; See, S.-J.; Tan, Y.-F.; et al. Fibroblast growth factor receptor 4 (*FGFR4*) and fibroblast growth factor 19 (*FGF19*) autocrine enhance breast cancer cells survival. *Oncotarget* **2016**, *7*, 57633–57650. [CrossRef] [PubMed]

52. Xie, M.-H.; Holcomb, I.; Deuel, B.; Dowd, P.; Huang, A.; Vagts, A.; Foster, J.; Liang, J.; Brush, J.; Gu, Q.; et al. FGF-19, a Novel Fibroblast Growth Factor with Unique Specificity for *FGFR4*. *Cytokine* **1999**, *11*, 729–735. [CrossRef]

53. Cui, G.; Martin, R.C.; Jin, H.; Liu, X.; Pandit, H.; Zhao, H.; Cai, L.; Zhang, P.; Li, W.; Li, Y. Up-regulation of FGF15/19 signaling promotes hepatocellular carcinoma in the background of fatty liver. *J. Exp. Clin. Cancer Res.* **2018**, *37*, 136. [CrossRef] [PubMed]

54. Ho, H.K.; Pok, S.; Streit, S.; Ruhe, J.E.; Hart, S.; Lim, K.S.; Loo, H.L.; Aung, M.O.; Lim, S.G.; Ullrich, A. Fibroblast growth factor receptor 4 regulates proliferation, anti-apoptosis and alpha-fetoprotein secretion during hepatocellular carcinoma progression and represents a potential target for therapeutic intervention. *J. Hepatol.* **2009**, *50*, 118–127. [CrossRef] [PubMed]

55. Lin, B.C.; Desnoyers, L.R. FGF19 and Cancer. In *Endocrine FGFs and Klothos*; Kuro-o, M., Ed.; Springer: New York, NY, USA, 2012; pp. 183–194. ISBN 978-1-4614-0887-1.

56. Touat, M.; Ileana, E.; Postel-Vinay, S.; André, F.; Soria, J.-C. Targeting FGFR Signaling in Cancer. *Clin. Cancer Res.* **2015**, *21*, 2684–2694. [CrossRef] [PubMed]

57. Michael, M.; Bang, Y.-J.; Park, Y.S.; Kang, Y.-K.; Kim, T.M.; Hamid, O.; Thornton, D.; Tate, S.C.; Raddad, E.; Tie, J. A Phase 1 Study of LY2874455, an Oral Selective pan-FGFR Inhibitor, in Patients with Advanced Cancer. *Target. Oncol.* **2017**, *12*, 463–474. [CrossRef] [PubMed]

58. Saka, H.; Kitagawa, C.; Kogure, Y.; Takahashi, Y.; Fujikawa, K.; Sagawa, T.; Iwasa, S.; Takahashi, N.; Fukao, T.; Tchinou, C.; et al. Safety, tolerability and pharmacokinetics of the fibroblast growth factor receptor inhibitor AZD4547 in Japanese patients with advanced solid tumours: A Phase I study. *Investig. New Drugs* **2017**, *35*, 451–462. [CrossRef] [PubMed]

59. Huynh, H.; Lee, L.Y.; Goh, K.Y.; Ong, R.; Hao, H.-X.; Huang, A.; Wang, Y.; Porta, D.G.; Chow, P.; Chung, A. Infigratinib mediates vascular normalization, impairs metastasis and improves chemotherapy in hepatocellular carcinoma. *Hepatology* **2018**. [CrossRef] [PubMed]

60. Nishina, T.; Takahashi, S.; Iwasawa, R.; Noguchi, H.; Aoki, M.; Doi, T. Safety, pharmacokinetic, and pharmacodynamics of erdafitinib, a pan-fibroblast growth factor receptor (FGFR) tyrosine kinase inhibitor, in patients with advanced or refractory solid tumors. *Investig. New Drugs* **2018**, *36*, 424–434. [CrossRef]

61. Brameld, K.A. Abstract SY30-01: Discovery of the highly selective covalent FGFR1-4 inhibitor PRN1371, currently in development for the treatment of solid tumors. *Cancer Res.* **2016**, *76*, SY30-01.

62. Futami, T.; Okada, H.; Kihara, R.; Kawase, T.; Nakayama, A.; Suzuki, T.; Kameda, M.; Shindoh, N.; Terasaka, T.; Hirano, M.; et al. ASP5878, a Novel Inhibitor of FGFR1, 2, 3, and 4, Inhibits the Growth of *FGF19*-Expressing Hepatocellular Carcinoma. *Mol. Cancer Ther.* **2017**, *16*, 68–75. [CrossRef]

63. Martinussen, C.; Bojsen-Moller, K.N.; Svane, M.S.; Dejgaard, T.F.; Madsbad, S. Emerging drugs for the treatment of obesity. *Expert Opin. Emerg. Drugs* **2017**, *22*, 87–99. [CrossRef]

64. Weiss, A.; Porta, D.G.; Reimann, F.; Buhles, A.; Stamm, C.; Fairhurst, R.A.; Kinyamu-Akunda, J.; Sterker, D.; Murakami, M.; Wartmann, M.; et al. Abstract 2103: NVP-FGF401: Cellular and in vivo profile of a novel highly potent and selective *FGFR4* inhibitor for the treatment of *FGF19/FGFR4/KLB*+ tumors. *Cancer Res.* **2017**, *77*, 2103.

65. Selvaraj, A.; Corcoran, E.; Coffey, H.; Prajapati, S.; Hao, M.-H.; Larsen, N.; Tsai, J.; Satoh, T.; Ichikawa, K.; Joshi, J.J.; et al. Abstract 3126: H3B6527, a selective and potent *FGFR4* inhibitor for *FGF19*-driven hepatocellular carcinoma. *Cancer Res.* **2017**, *77*, 3126.

66. Bartz, R.; Fukuchi, K.; Lange, T.; Gruner, K.; Ohtsuka, T.; Watanabe, I.; Hayashi, S.; Redondo-Müller, M.; Takahashi, M.; Agatsuma, T.; et al. Abstract 3852: U3-1784, a human anti-*FGFR4* antibody for the treatment of cancer. *Cancer Res.* **2016**, *76*, 3852.

67. Kim, R.; Sharma, S.; Meyer, T.; Sarker, D.; Macarulla, T.; Sung, M.; Choo, S.P.; Shi, H.; Schmidt-Kittler, O.; Clifford, C.; et al. First-in-human study of BLU-554, a potent, highly-selective *FGFR4* inhibitor designed for hepatocellular carcinoma (HCC) with *FGFR4* pathway activation. *Eur. J. Cancer* **2016**, *69*, S41. [CrossRef]

68. Gao, L.; Shay, C.; Lv, F.; Wang, X.; Teng, Y. Implications of *FGF19* on sorafenib-mediated nitric oxide production in hepatocellular carcinoma cells—A short report. *Cell Oncol.* **2018**, *41*, 85–91. [CrossRef] [PubMed]

69. Cheuk, A.; Shivaprasad, N.; Skarzynski, M.; Baskar, S.; Azorsa, P.; Khan, J. Abstract 5618: Anti-*FGFR4* antibody drug conjugate for immune therapy of rhabdomyosarcoma and hepatocellular carcinoma. *Cancer Res.* **2018**, *78*, 5618.

70. Chen, Z. Abstract LB-272: Discovery and characterization of a novel *FGFR4* Inhibitor for the treatment of hepatocellular carcinoma. *Cancer Res.* **2018**, *78*, LB–272.

71. Lee, J.; Kang, H.; Koo, K.; Ha, Y.; Lim, S.Y.; Byun, J.-Y.; Yu, H.; Song, T.; Lee, M.; Jung, S.H.; et al. Abstract 4780: A novel, potent and selective *FGFR4* inhibitor, HM81422 in hepatocellular carcinoma with *FGFR4*-driven pathway activation. *Cancer Res.* **2018**, *78*, 4780.

72. Matsuki, M.; Hoshi, T.; Yamamoto, Y.; Ikemori-Kawada, M.; Minoshima, Y.; Funahashi, Y.; Matsui, J. Lenvatinib inhibits angiogenesis and tumor fibroblast growth factor signaling pathways in human hepatocellular carcinoma models. *Cancer Med.* **2018**, *7*, 2641–2653. [CrossRef]

73. Ho, H.K.; Yeo, A.H.L.; Kang, T.S.; Chua, B.T. Current strategies for inhibiting FGFR activities in clinical applications: Opportunities, challenges and toxicological considerations. *Drug Discov. Today* **2014**, *19*, 51–62. [CrossRef]

74. Pai, R.; French, D.; Ma, N.; Hotzel, K.; Plise, E.; Salphati, L.; Setchell, K.D.R.; Ware, J.; Lauriault, V.; Schutt, L.; et al. Antibody-Mediated Inhibition of Fibroblast Growth Factor 19 Results in Increased Bile Acids Synthesis and Ileal Malabsorption of Bile Acids in Cynomolgus Monkeys. *Toxicol. Sci.* **2012**, *126*, 446–456. [CrossRef]

75. Kaibori, M.; Sakai, K.; Ishizaki, M.; Matsushima, H.; De Velasco, M.A.; Matsui, K.; Iida, H.; Kitade, H.; Kwon, A.-H.; Nagano, H.; et al. Increased *FGF19* copy number is frequently detected in hepatocellular carcinoma with a complete response after sorafenib treatment. *Oncotarget* **2016**, *7*, 49091–49098. [CrossRef] [PubMed]

76. Tang, X.; Wang, Y.; Fan, Z.; Ji, G.; Wang, M.; Lin, J.; Huang, S.; Meltzer, S.J. Klotho: A tumor suppressor and modulator of the Wnt/β-catenin pathway in human hepatocellular carcinoma. *Lab. Investig.* **2016**, *96*, 197–205. [CrossRef] [PubMed]

13

Cancer Mutations in FGFR2 Prevent a Negative Feedback Loop Mediated by the ERK1/2 Pathway

Patrycja Szybowska [1,2], Michal Kostas [2,3], Jørgen Wesche [2,3], Antoni Wiedlocha [1,2,4] and Ellen Margrethe Haugsten [2,3,*]

[1] Department of Molecular Cell Biology, Institute for Cancer Research, The Norwegian Radium Hospital, Oslo University Hospital, Montebello, 0379 Oslo, Norway; Patrycja.Szybowska@rr-research.no (P.S.); Antoni.Wiedlocha@rr-research.no (A.W.)

[2] Centre for Cancer Cell Reprogramming, Institute of Clinical Medicine, Faculty of Medicine, University of Oslo, Montebello, 0379 Oslo, Norway; Michal.Janusz.Kostas@rr-research.no (M.K.); Jorgen.Wesche@rr-research.no (J.W.)

[3] Department of Tumor Biology, Institute for Cancer Research, The Norwegian Radium Hospital, Oslo University Hospital, Montebello, 0379 Oslo, Norway

[4] Military Institute of Hygiene and Epidemiology, 01-163 Warsaw, Poland

* Correspondence: Ellen.M.Haugsten@rr-research.no

Abstract: Tight regulation of signaling from receptor tyrosine kinases is required for normal cellular functions and uncontrolled signaling can lead to cancer. Fibroblast growth factor receptor 2 (FGFR2) is a receptor tyrosine kinase that induces proliferation and migration. Deregulation of FGFR2 contributes to tumor progression and activating mutations in FGFR2 are found in several types of cancer. Here, we identified a negative feedback loop regulating FGFR2 signaling. FGFR2 stimulates the Ras/MAPK signaling pathway consisting of Ras-Raf-MEK1/2-ERK1/2. Inhibition of this pathway using a MEK1/2 inhibitor increased FGFR2 signaling. The putative ERK1/2 phosphorylation site at serine 780 (S780) in FGFR2 corresponds to serine 777 in FGFR1 which is directly phosphorylated by ERK1/2. Substitution of S780 in FGFR2 to an alanine also increased signaling. Truncated forms of FGFR2 lacking the C-terminal tail, including S780, have been identified in cancer and S780 has been found mutated to leucine in bladder cancer. Substituting S780 in FGFR2 with leucine increased FGFR2 signaling. Importantly, cells expressing these mutated versions of S780 migrated faster than cells expressing wild-type FGFR2. Thus, ERK1/2-mediated phosphorylation of S780 in FGFR2 constitutes a negative feedback loop and inactivation of this feedback loop in cancer cells causes hyperactivation of FGFR2 signaling, which may result in increased invasive properties.

Keywords: FGFR2; ERK1/2; phosphorylation; serine; negative feedback loop; cancer

1. Introduction

Tight regulation of receptor tyrosine kinase signaling is required for specific cellular responses, such as cell growth, differentiation, migration, and apoptosis. Inadequate regulation of signaling is a common event in cancer development and enhanced receptor signaling promotes tumor growth [1]. The receptor tyrosine kinase, FGFR2 (fibroblast growth factor receptor 2) is a transmembrane, cell-surface localized receptor that belongs to a family of four related receptors [2]. FGFR2 is activated by FGF ligands and induces various downstream signaling molecules. Deregulation of FGFR2 contributes to tumor progression and activating mutations in FGFR2 have been found in different types of cancer, like gastric cancer, breast cancer, and endometrial carcinoma [3,4]. In addition, activating mutations have been found in skeletal disorders, like Apert syndrome and Crouzon syndrome [5]. Clearly, precise regulation of FGFR2 signaling is important to prevent diseases.

Upon ligand binding, FGFRs dimerize. This, in turn, activates the tyrosine kinase domain of the receptor by trans-autophosphorylation [2]. FGFRs mediate signaling by recruiting specific molecules that bind to phosphorylated tyrosines, triggering a number of signaling pathways. The docking protein FRS2 (FGFR substrate 2) is phosphorylated by the activated receptor, creating phosphotyrosine docking sites for proteins containing SH2-domains. By binding to FRS2, the adaptor protein Grb2 (growth factor receptor-bound protein 2) activates the Ras/ mitogen-activated protein kinase (MAPK) pathway and the phosphoinositide 3-kinase (PI3K)/Akt pathway [2]. Ras activates the kinase activity of Raf, which phosphorylates MEK1/2. MEK then phosphorylates ERK1/2 (extracellular signal-regulated kinase) which activates 90 kDa Ribosomal S6 Kinase 2 (RSK2), among other downstream targets. Activated FGFRs also recruit and phosphorylate phospholipase Cγ (PLCγ), which culminates in the activation of protein kinase C (PKC) [2].

In comparison to the well-studied activation of FGFRs, the mechanisms leading to deactivation of the receptor are not fully understood. It is known that the signal from the activated receptor can be attenuated by internalization and degradation in lysosomes [6,7]. After internalization, FGFR ubiquitination marks the receptor for degradation [6,8]. Depending on the receptor type, the bound ligand, and possibly also the cell type and the context, FGFRs might also be recycled back to the cell surface instead of being transported to lysosomes, which may result in prolonged signaling [7].

FGFR signaling is also regulated by phosphatases. Recently, we have shown that a phosphatase, PTPRG, directly dephosphorylates activated FGFRs [9]. Proteins that regulate FGFR signaling, such as MAPK phosphatase 3 (MKP3) and Sprouty 1/2, are negative regulators that are induced or activated by FGF signaling and act on downstream signaling molecules [10]. In addition, FGFR signaling can be regulated by inhibitory phosphorylation, forming negative feedback loops that attenuate the signals. It has been shown that active ERK1/2 can phosphorylate FRS2 on threonine residues. This leads to reduced tyrosine phosphorylation of FRS2 and therefore reduced downstream signaling [11]. On the receptor level, two such negative feedback loops have been identified for FGFR1 [12,13]. It has been shown that upon FGFR1 activation/tyrosine phosphorylation, the receptor is also phosphorylated at serine 777 (S777) directly by activated ERK1/2. S777 phosphorylation reduces the tyrosine phosphorylation in the kinase domain of the receptor and thus also reduces signaling [12]. In addition, the serine/threonine kinase RSK2, which is activated through the Ras-MAPK pathway, can also bind to FGFR1 and phosphorylate FGFR1 at serine 789 [13]. This phosphorylation seems to be required for proper endocytosis and ubiquitination of FGFR1. Preventing RSK2 activation or mutation of S789 leads to increased signaling [13]. It is not clear if the other FGFRs are also regulated by such negative feedback loops.

Here, we have investigated whether a similar negative feedback loop mediated by ERK1/2 also exists for FGFR2. Inhibition of the ERK1/2 signaling pathway, using a MEK1/2 inhibitor (U0126), led to sustained FGFR2 phosphorylation. Moreover, substitution of serine 780 (S780) in FGFR2 for alanine also resulted in sustained FGFR2 activation. S780 in FGFR2 is equivalent to the ERK1/2 substrate S777 in FGFR1. Several truncated forms of FGFR2 lacking the C-terminal tail, including S780, have been identified in cancer. In addition, S780 has been found mutated to leucine in a patient with bladder cancer. Substituting S780 in FGFR2 with leucine also increased FGFR2 signaling. More importantly, cells expressing the mutated versions of S780 were migrating faster than cells expressing wild-type FGFR2. Possibly, the lack of MAPK-dependent negative feedback gives FGFR2-expressing cancer cells an advantage. These results also indicate that care should be taken when the MAPK-pathway is inhibited in cancer.

2. Materials and Methods

2.1. Materials, Antibodies, and Compounds

The following antibodies were used: Mouse anti-phospho-ERK1/2 (Thr202/Tyr204) (#9106), rabbit anti-ERK1/2 (#9102), mouse anti-phospho-FGFR (Tyr653/654) (#3476), rabbit anti-FGFR2 (#11835), rabbit anti-FGFR2 (N-terminal) (#23328), rabbit anti-FGFR1 (#9749), rabbit anti-FGFR3 (#4574), rabbit anti-FGFR4 (#8562), rabbit anti-phospho-PLCγ (Tyr783) (#14008), and rabbit anti-phospho-RSK2

(Ser 227) (#3556) from Cell Signaling Technology (Leiden, The Netherlands) and mouse anti-γ-tubulin (T6557) from Sigma-Aldrich (St. Louis, MO, USA). Fluorescently labelled secondary antibodies were from Jackson ImmunoResearch Laboratories (Cambridgeshire, UK). HRP-conjugated secondary antibodies were from Jackson ImmunoResearch Laboratories and Agilent (Santa Clara, CA, USA).

U0126 (1144) was from Tocris Bioscience (Bristol, UK). PD173074 was from Calbiochem (San Diego, CA, USA). Cycloheximide, recombinant EGF, mowiol, heparin, and protein-G-sepharose were from Sigma Aldrich. Restriction enzymes were from New England Biolabs (Ipswich, MA, USA). Adenosine triphosphate [γ-^{32}P] 3000 Ci/mmol EasyTides was purchased from PerkinElmer (Norwalk, CT, USA). PhosSTOP phosphatase inhibitor cocktail and cOmplete EDTA-free protease inhibitor cocktail were from Roche (Basel, Switzerland). Hoechst 33342, DyLight 550 NHS Ester, and recombinant active ERK1 with glutathione S-transferase (GST) tag (#PV3311) were from Thermo Fisher Scientific (Waltham, MA, USA). Recombinant FGF1 was prepared as previously described [14]. FGF1 was labelled with DyLight 550 (DL550-FGF1) following the manufacturer's procedures.

2.2. Plamids and siRNAs

cDNA encoding full-length human FGFR2 (IIIc) (NCBI: NM_000141) was cut out from the pCMV6-XL4 cDNA clone (Origene Technologies, Rockville, MD, USA) as an *EcoRI-XbaI* fragment and ligated into pcDNA3 (Thermo Fisher Scientific, Waltham, MA, USA). The resulting plasmid was further cut with *KpnI* to remove the upstream untranslated region. To remove the untranslated region downstream of the gene, the plasmid was partially cut with *Tth111I*, followed by cutting with *XbaI*. The plasmid was furthermore treated with T4 DNA polymerase (New England Biolabs, Ipswich, MA, USA) to make blunt ends and then ligated. Note that the *XbaI* and the *Tth111I* sites were destroyed. After sequencing, a point mutation in the N-terminal region was discovered (G183V). This point mutation was mutated back (generating a glycine at the 138 position) using site-directed mutagenesis with the following primer: 5-CGCTGCCCAGCCGGGGGGAACCCAATGCCAACC-3. pcDNA3 hFGFR2 was used as a template to generate pcDNA3 hFGFR2 S780A, S780D, and S780L. The following primers were used: S780A; 5-CCTCTCGAACAGTATGCACCTAGTTACCCTGAC-3, S780D; 5-CCTCTCGAACAGTATGACCCTAGTTACCCTGAC-3, S780L; and 5-CCTCTCGAACAGTATCTACC TAGTTACCCTGAC-3. All constructs were verified by sequencing (Eurofins Genomics, Ebersberg, Germany). pcDNA3 hFGFR1 and pcDNA3 hFGFR4 have been described previously [7,15] and pcDNA3 hFGFR3 was a generous gift from Dr. A. Yayon (ProChon Biotech, Ness Ziona, Israel).

2.3. Cell Lines and Transfection

To generate U2OS cells stably expressing FGFR2, FGFR2 S780A, FGFR2 S780D, and FGFR2 S780L, Fugene 6 transfection reagent (Promega, Madison, WI, USA) was used according to the manufacturer's protocol. Clones were selected with 1 mg/mL geneticin and then the clones were chosen based on their receptor expression levels analyzed by immunofluorescence and Western blotting. Throughout the paper, clone #1 of the particular stable cell line is used if nothing else is stated.

The cells were propagated in Dulbecco's Modified Eagle Medium (DMEM) supplemented with 10% fetal bovine serum, 100 U/mL penicillin, and 100 µg/mL streptomycin in a 5% CO_2 atmosphere at 37 °C.

Transient transfection was performed using Fugene 6 transfection reagent according to the manufacturer's protocol. Cells were analyzed 16–24 h after transfection.

2.4. Western Blotting

Cells were treated as indicated and then lysed in Laemmli sample buffer (Bio-Rad, Oxford, UK). Proteins in the cell lysates were separated on a gradient (4–20%) sodium dodecyl sulfate-polyacrylamide gel electrophoresis (SDS-PAGE) and then blotted onto a membrane using the TransBlot® Turbo Transfer system (Bio-Rad). Membranes were then incubated with indicated primary antibodies followed by corresponding secondary antibody coupled to HRP. Bands were visualized by chemiluminscence using SuperSignal™ West Dura Extended Duration Substrate (Thermo Fisher Scientific, Waltham,

MA, USA) or SuperSignal™ West Femto Maximum Sensitivity Substrate (Thermo Fisher Scientific). In some cases, antibodies were stripped from the membranes using Pierce Stripping buffer and the membranes were reprobed. The images were prepared using ImageLab Software (Bio-Rad) and Adobe Illustrator CS4 14.0.0 (San Jose, CA, USA). Quantification of bands of interest was performed in Fiji ImageJ software [16]. Lane normalization factor (LNF) was determined by dividing the intensity of the γ-tubulin bands on its highest signal in each blot.

2.5. Microscopy

Cells, seeded onto coverslips, were treated as indicated and fixed in 4% formaldehyde. The cells were then permeabilized with 0.1% triton X-100, stained with indicated antibodies and Hoechst 33342 and mounted in mowiol. Confocal images were acquired with a 63X objective on a Zeiss confocal Laser Scanning Miscroscope (LSM) 780 (Jena, Germany). Images were prepared in Fiji Image J software and Adobe Illustrator CS4 14.0.0. Images for quantification of p-FGFR and DL550-FGF1 signal intensities were taken with identical settings and the quantification was performed with Fiji Image J software. The same threshold was used for all images in the same experiment. Due to background staining in the nuclei, p-FGFR intensities in the nuclei were subtracted from the total intensities in the corresponding cell.

2.6. In Vitro Phosphorylation Assay

The cells were starved for 2 h in serum-free media and lysed in lysis buffer (20 mM phosphate-Na pH 7.4, 150 mM NaCl, 1 mM Ethylenediaminetetraacetic acid (EDTA), 1% Triton X-100, protease inhibitors). The receptors were immunoprecipitated for 1 h using anti-N-terminal-FGFR2 antibodies pre-bound to protein-G-sepharose, washed 3 times with 1 M NaCl and treated with 1 μM PD173074 for 30 min. The kinase reaction was performed on beads using 50 ng recombinant active ERK1 and 50 μCi ATP-γ-^{32}P (per 100 μL reaction) in 50 mM HEPES-Na pH 7.5, 20 mM $MgCl_2$, 5 mM Ethylene Glycol Tetraacetic Acid (EGTA), and phosphatase inhibitors for 30 min at 30 °C. The reaction was quenched with 20 mM EDTA. Then, the immunoprecipitated receptors were washed 3 times (25 mM HEPES-Na pH 7.5, 1 mM EDTA) and released from the beads in SDS-loading buffer by 15 min at 95 °C and subjected to SDS-PAGE before analysis with autoradiography and immunoblotting.

2.7. Cell Migration

Cells sparsely seeded in IncuCyte Image Lock 96-well plates (Essen BioSciences, Hertfordshire, UK) were imaged every 10 min for 21 h by IncuCyte® S3 Live Cell Analysis System with IncuCyte® S3 Software (V2018B) (Essen BioSciences). In all experiments, cells were either left untreated or treated with FGF1 (100 ng/mL) and heparin (20 U/mL). Images were analyzed with IncuCyte® S3 Software (V2018B) and Fiji ImageJ software with Manual Tracking and Chemotaxis and Migration Tool (ibidi GmbH, Planegg, Germany).

3. Results

3.1. Inhibition of MEK1/2 Increases FGFR2 Signaling

Signaling from FGFRs is regulated by mechanisms such as endocytic trafficking [6,7] and dephosphorylation by phosphatases (PTPRG) [9]. Recently, we identified a negative feedback loop that involves direct phosphorylation of serine 777 (S777) in the C-terminal tail of FGFR1 by active ERK1/2 [12]. Phosphorylation of S777 in FGFR1 is necessary for proper attenuation of FGFR1 signaling and treatment of cells with U0126, a MEK1/2 inhibitor, leads to increased activation of FGFR1. To investigate if a similar ERK1/2-mediated negative feedback loop also exists for FGFR2, we treated cells with U0126 MEK1/2 inhibitor and investigated tyrosine phosphorylation status of FGFR2 at different time-points after addition of FGF1. Since FGFR levels are low in many cells and endogenous FGFRs can be difficult to detect, we generated U2OS cells stably expressing FGFR2 IIIc (U2OS-R2). In contrast to parental U2OS cells, our U2OS-R2 cells endocytose detectable amounts of DL550-FGF1 (FGF1 labelled with DyLight550) and are strongly stained with anti-FGFR2 antibodies (Figure 1a). U2OS cells do not express detectable levels of any of the four FGFRs (Figure S1a and [17]) and, although the antibody that we use against phosphorylated FGFR (p-FGFR antibody) recognizes all four receptors, only the ectopic FGFR in the stably transfected U2OS cells is detected.

First, we investigated which doses of the MEK1/2 inhibitor (U0126) efficiently inhibit ERK1/2 activation upon FGF1 stimulation in U2OS-R2 cells (Figure 1b). MEK1/2 is upstream of ERK1/2 in the Ras/MAPK signaling pathway. Incubating the cells with increasing concentrations of U0126 demonstrated that 20 μM U0126 efficiently blocked ERK1/2 activation. Next, we treated the cells with 20 μM U0126 and compared the receptor activation in treated cells versus untreated cells. The levels of tyrosine-phosphorylated FGFR2 were increased in U0126 treated cells compared to untreated cells (Figure 1c). Similar effects were observed in two additional clones of U2OS-R2 (Figure S1b). To investigate this effect further, we also stained cells with antibodies against tyrosine phosphorylated FGFR (p-FGFR) and compared the intensity of p-FGFR staining between indicated treatments (Figure 1d). When resting cells or cells treated with FGF1 together with FGFR inhibitor (PD173074) were stained with p-FGFR antibodies, we could detect a bright signal in the nucleus. We considered this as unspecific staining by the antibody. Thus, upon quantification, the intensity of the nuclear p-FGFR antibody staining was subtracted from that of the total cell. Interestingly, in cells treated with FGF1, we observed a clear increase in p-FGFR antibody intensity in the cytosol compared to resting cells or PD173074 treated cells. As expected, we could also observe a high degree of co-localization between DL550-FGF1 and p-FGFR antibody staining (Figure 1d, second panel). When cells were treated with FGF1 and U0126 (to prevent ERK1/2 signaling), we detected an increase in p-FGFR antibody staining compared to FGF1 treatment alone. Taken together, our data indicates that a similar feedback mechanism as to that found for FGFR1 might also exist in the case of FGFR2. We conclude that ERK1/2 signaling is required for attenuation of FGFR2 signaling.

Since ERK1/2 signaling can be activated by other receptor tyrosine kinases as well, we investigated if activation of ERK1/2 prior to FGFR2 activation would influence the response to FGF1. To test this, we treated cells with EGF 30 min prior to stimulation with FGF1 and compared the levels of FGFR2 tyrosine phosphorylation to that in cells not pretreated with EGF. First, we investigated whether EGF activates ERK1/2 signaling in U2OS-R2 cells. We observed a peak of ERK1/2 phosphorylation 10–20 min after addition of EGF. Indeed, ERK1/2 is active in U2OS-R2 cells during this 30 min period of stimulation with EGF (Figure 2a).

Next, we stimulated cells for 30 min with EGF before activation of FGFR by addition of FGF1. Interestingly, reduced levels of tyrosine-phosphorylated FGFR2 was observed in cells pretreated with EGF (Figure 2b). These data indicate a dual role for the ERK1/2 signaling-mediated feedback loop in FGFR2 signaling. Not only does it function to ensure proper attenuation of FGFR2 signaling, it also ensures accurate responses to FGF1 stimulation. In an environment where the ERK1/2 pathway is activated by other receptor tyrosine kinases, the response to FGF1 is less pronounced than in resting cells. In this way, different receptors may cross-talk to prevent excess signaling.

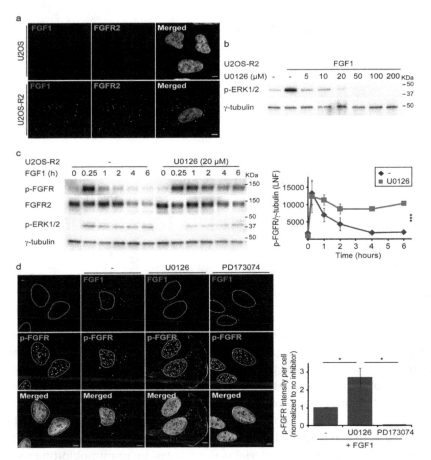

Figure 1. Inhibition of the ERK1/2 pathway prolongs FGFR2 signaling. (**a**) U2OS cells or U2OS cells stably transfected with FGFR2IIIc (U2OS-R2) were treated with 200 ng/mL DL550-FGF1 in the presence of heparin (50 U/mL) for 30 min. The cells were then fixed, stained with anti-FGFR2 antibodies and Hoechst, and analyzed by confocal microscopy. The images were taken at fixed intensity settings, and brightness/contrast was adjusted in the same way for all images. Representative images are shown. Scale bar: 5 μM. (**b**) U2OS-R2 cells were kept in serum-free media for 2 h prior to stimulation for 30 min with 200 ng/mL FGF1 in the presence of heparin (20 U/mL) and increasing concentrations of U0126. Cells were then lysed and the lysates were analyzed by immunoblotting using the indicated antibodies. A p in front of the name of the antibody indicates that it recognizes the phosphorylated form of the protein. One representative experiment is shown. (**c**) U2OS-R2 cells were kept in serum-free media for 2 h before addition of 100 ng/mL FGF1 and heparin (20 U/mL) in the presence or absence of U0126 (20 μM) for indicated periods of time. Cycloheximide (10 μg/mL) was added at the beginning of the starvation period and kept throughout the experiment. After lysis, the cellular material was analyzed with immunoblotting using the indicated antibodies. A p in front of the name of the antibody indicates that it recognizes the phosphorylated form of the protein. Quantifications of three independent experiments are presented in the graph. The bands corresponding to phosphorylated receptor were normalized to Lane normalization factor (LNF) (γ-tubulin). Error bars denote the standard deviation. The difference between U0126 treated cells versus untreated cells was significant ($p \leq 0.001$, 3-way ANOVA, Holm-Sidak test, n = 3). (**d**) U2OS-R2 cells were kept in serum-free media for 2 h before addition of 200 ng/mL DL550-FGF1 and heparin (50 U/mL) for 30 min. The cells were pretreated with U0126 (20 μM) or PD173074 (50 nM) 30 min before addition of FGF1 as indicated. The cells were then fixed, stained with anti-p-FGFR antibodies and Hoechst, and analyzed by confocal microscopy. Scale bar: 5 μM. Quantifications of two independent experiments were performed as described in materials and methods and are presented in the graph. In total, 60 cells treated with FGF1 alone, 60 cells treated with FGF1 and U0126, and 38 cells treated with FGF1 and PD173074 were quantified. Outliers were removed according to the 1.5*IQR outlier rule. Error bars denote the standard error of the mean (SEM), n = 2. Due to a general variation in the intensity between the two experiments, the means in each experiment were normalized to the mean of cells treated with FGF1 alone (no inhibitor) in the corresponding experiment. (* $p \leq 0.05$, two-sided t test on normalized data, n = 2).

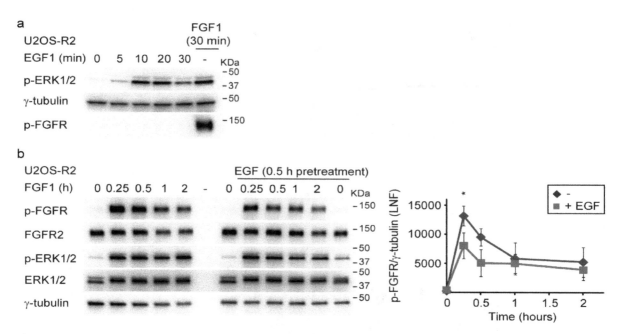

Figure 2. Pretreatment with EGF reduces the response to FGF1. (**a**) U2OS-R2 cells were kept in serum-free media for 2 h and then 100 ng/mL EGF or 100 ng/mL FGF1 was added to the cells. The cells were lysed after the indicated periods of time. A p in front of the name of the antibody indicates that it recognizes the phosphorylated form of the protein. (**b**) U2OS-R2 cells were kept in serum-free media for 2 h. Then, EGF (20–100 ng/mL) was added to the samples as indicated. After 30 min, the cells were stimulated with 20 ng/mL FGF1 and heparin (10 U/mL) and lysed after the indicated periods of time. Cycloheximide (10 μg/mL) was added at the beginning of the starvation period and kept throughout the experiment. After lysis, the cellular material was analyzed by immunoblotting using the indicated antibodies. A p in front of the name of the antibody indicates that it recognizes the phosphorylated form of the protein. Quantifications of three independent experiments are presented in the graph. The time-point of 30 min is only from two experiments. The bands corresponding to phosphorylated receptor were normalized to LNF (γ-tubulin). Error bars denote the standard deviation. The difference between EGF-pretreated cells versus cells not treated with EGF was significant at the time point of 15 min (* $p \leq 0.05$, 1-way ANOVA, Tukey test, n = 3).

3.2. Mutation of Serine 780 in FGFR2 Leads to Increased FGFR2 Activity

Since the phosphorylation site of ERK1/2 in FGFR1 (S777) is already identified, we wanted to investigate if the corresponding serine in FGFR2 is important for proper downregulation of FGFR2 signaling. By sequence alignment, we identified S780 in FGFR2 to correspond to FGFR1 S777 (Figure 3a). Interestingly, in both receptors, the particular serine is followed by a proline and thus forms an ERK1/2 phosphorylation motif (pS/T-P) [18]. We therefore decided to substitute serine 780 in FGFR2 with alanine. Alanine represents a site that cannot be phosphorylated. Next, we prepared U2OS cells stably expressing FGFR2 S780A (U2OS-R2 S780A).

We then investigated whether FGFR2 S780A is expressed to similar levels as the wild-type receptor and if it maintained normal FGFR2 properties. We therefore stimulated cells with FGF1 and analyzed the lysates using Western blotting. First of all, the levels of FGFR2 wild-type and FGFR2 S780A seem comparable in the two clones (Figure 3b). Secondly, we noticed that FGFR2 S780A is able to activate the main downstream signaling pathways similarly to wild-type FGFR2 (Figure 3b). In addition, the mutated receptor was able to bind FGF1 at the cell surface and internalize FGF1 into early endosomes similarly to wild-type FGFR2 (Figure 3c). Comparable results were confirmed in two additional clones of U2OS-R2 wild-type and U2OS-R2 S780A (Figure S2a,b). Moreover, FGFR2 S780A co-localizes with DL550-FGF1, similarly to FGFR2 wild-type (Figure 3d).

We then analyzed the level of FGFR tyrosine phosphorylation over time in FGFR2 S780A-expressing cells. Compared to wild-type expressing cells, FGFR2 activation was sustained in U2OS-R2 S780A

(Figure 4a and Figure S3). This effect was similar to the effect observed upon U0126 treatment. It is therefore likely that this serine, also in the case of FGFR2, is phosphorylated by ERK1/2.

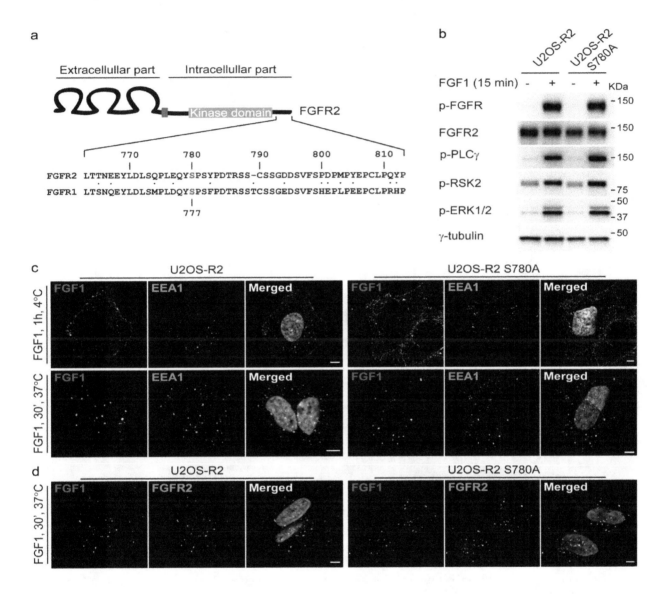

Figure 3. Characterization of cell lines stably expressing FGFR2 S780A mutant. (**a**) A pairwise sequence alignment tool from EMBL-EBI was used to align the C-terminal tails of FGFR2 and FGFR1. S780 in FGFR2 corresponds to S777 in FGFR1 (labelled in red in the figure). Numbers refer to the amino acid numbering used for human FGFR2 (NCBI: NM_000141). (**b**) U2OS-R2 cells or U2OS-R2 S780A cells were kept in serum-free media for two hours and then treated or not with 100 ng/mL FGF1 for 15 min in the presence of heparin (20 U/mL). After lysis, the cellular material was analyzed by immunoblotting using the indicated antibodies. A p in front of the name of the antibody indicates that it recognizes the phosphorylated form of the protein. One representative experiment is shown. (**c**) U2OS-R2 or U2OS-R2 S780A cells were kept at 4 °C with DL550-FGF1 for one hour in the presence of heparin (50 U/mL). Next, the cells were either fixed directly (upper panel) or incubated for 30 min at 37 °C before fixation (lower panel). The cells were then stained with anti-EEA1 antibodies and Hoechst and analyzed by confocal microscopy. Representative images are shown. Scale bar: 5 μM. (**d**) U2OS-R2 or U2OS-R2 S780A cells were treated with 200 ng/mL DL550-FGF1 in the presence of heparin (50 U/mL) for 30 min. The cells were then fixed, stained with anti-FGFR2 antibodies and Hoechst, and analyzed by confocal microscopy. Representative images are shown. Scale bar: 5 μM.

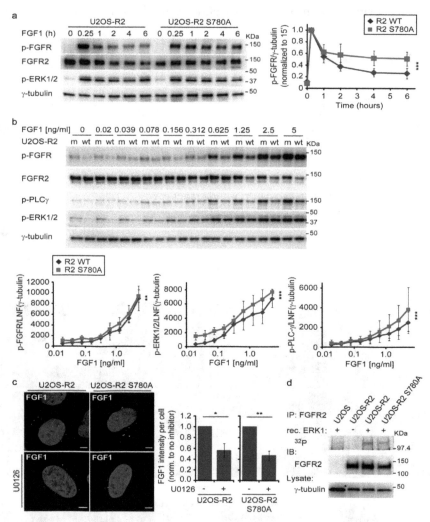

Figure 4. Signaling from FGFR2 S780A is prolonged compared to wild-type FGFR2. **(a)** U2OS-R2 wild-type or U2OS-R2 S780A cells were kept for two hours in serum-free media before addition of 100 ng/mL FGF1 and heparin (20 U/mL) for the indicated periods of time. Cycloheximide (10 μg/mL) was added at the beginning of the starvation period and kept throughout the experiment. After lysis, the cellular material was analyzed by immunoblotting using the indicated antibodies. A p in front of the name of the antibody indicates that it recognizes the phosphorylated form of the protein. Quantifications of five independent experiments are presented in the graph. The bands corresponding to phosphorylated receptor were normalized to γ-tubulin and within each experiment to the time point of 15 min. Error bars denote the standard deviation. The difference between FGFR2 wild-type and the S780A mutant was significant ($p \leq 0.001$, 3-way ANOVA, Holm-Sidak test, n = 5). **(b)** U2OS-R2 wild-type (wt) or U2OS-R2 S780A (m) cells were kept for two hours in serum-free media before addition of indicated concentrations of FGF1 in the presence of heparin (20 U/mL) for 15 min. After lysis, the cellular material was analyzed by immunoblotting using the indicated antibodies. A p in front of the name of the antibody indicates that it recognizes the phosphorylated form of the protein. Quantifications of three independent experiments are presented in the graph. The bands corresponding to the phosphorylated receptor/ERK1/2/PLCγ were normalized to LNF (γ-tubulin). Error bars denote the standard deviation. The difference between FGFR2 wild-type and S780A mutant was significant (** $p \leq 0.01$, *** $p \leq 0.001$, 3-way ANOVA, Holm-Sidak test, n = 3). **(c)** Internalization of DL550-FGF1 in FGFR2 and FGFR2 S780A cells is reduced upon U0126 treatment. U2OS-R2 and U2OS-R2 S780A cells were incubated with 200 ng/mL DL550-FGF1 and heparin (50 U/mL) for 30 min. The cells were pretreated as indicated with U0126 (20 μM) for 30 min before addition of FGF1.

The cells were then fixed, stained with Hoechst, and analyzed by confocal microscopy. The images were taken at fixed intensity settings, and brightness/contrast was adjusted in the same way for all images. Scale bar: 5 μM. Quantifications of four independent experiments were performed as described in materials and methods and are presented in the graph. In total, 269 U2OS-R2 cells, 262 U2OS-R2 cells treated with U0126, 247 U2OS-R2 S780A cells, and 213 U2OS-R2 S780A cells treated with U0126 were quantified. Outliers were removed according to the 1.5*IQR outlier rule. Error bars denote the SEM (n = 4). Due to a general variation in the intensity between experiments, the means of U0126 treated cells for each cell line in each experiment were normalized to the mean of the corresponding cell line in the same experiment (** $p \leq 0.05$, * $p \leq 0.05$, two-sided t test, n = 4). (**d**) In vitro phosphorylation of FGFR2 by active recombinant ERK1. Lysates from U2OS, U2OS-R2, and U2OS-R2 S780A cells were subjected to FGFR2 immunoprecipitation (IP). The immunoprecipitated materials were next incubated with [γ-^{32}P]-labelled adenosine triphosphate and recombinant active ERK1 (rec. ERK1) in the presence of PD173074. After washing, the samples were subjected to SDS-PAGE and analyzed by autoradiography and immunoblotting (IB). One representative experiment is shown.

In the previous experiments, higher concentrations of FGF1 were used to activate the receptor. We wanted to test if increased FGFR activity also occurred at lower concentrations of FGF1. We treated U2OS-R2 and U2OS-R2 S780A with different concentrations of FGF1 starting at 0.02 ng/mL (Figure 4b). Tyrosine phosphorylation of the receptor and its main downstream signaling pathways were then analyzed with Western blotting. We observed a slight increase in the levels of tyrosine-phosphorylated FGFR2 as well as in the levels of phosphorylated ERK1/2 and PLC-γ in S780A-expressing cells compared to wild-type cells with low concentrations of FGF1. This experiment was performed after 15 min of FGF1 treatment where the effect is not at its highest. However, although the increase in signaling in FGFR2 S780A cells was modest, it was consistent at all concentrations tested. We therefore conclude that the negative feedback loop is operational at both lower and higher concentrations of ligand and at early time points.

Increased signaling can be a result of reduced receptor endocytosis. In Figure 1d, we detected more surface staining and less uptake of DL550-FGF1 in cells treated with U0126 (MEK1/2 inhibitor). This indicates a decrease in endocytosis when MEK1/2-ERK1/2 signaling is inhibited. However, from our previous work on FGFR1, despite a decrease in endocytosis upon MEK1/2-ERK1/2 inhibition, this effect was not due to lack of ERK1/2 phosphorylation of the receptor but rather a lack of a second serine phosphorylation event in FGFR1 mediated by RSK2 [13]. To investigate this, we compared the uptake of DL550-FGF1 in U2OS-R2 wild-type and U2OS-R2 S780A mutant cells in the presence of U0126 (Figure 4c). Upon U0126 treatment, the uptake of DL550-FGF1 was reduced similarly in both cell lines. Thus, the lack of phosphorylation on S780 is probably not the reason for the reduced endocytosis upon MEK1/2 inhibition. Other phosphorylation events mediated by components of the MAPK signaling pathways might be important for proper FGFR2 endocytosis.

Next, we wanted to test if ERK1/2 directly phosphorylates FGFR2. We therefore immunoprecipitated FGFR2 from cell lysates and incubated the immunoprecipitated receptor with recombinant active ERK1 and radioactive [γ-^{32}P]-labelled adenosine triphosphate (ATP). The experiment was performed in the presence of PD173074 (FGFR inhibitor) to prevent autophosphorylation of the receptor. Using autoradiography, we could observe a band representing phosphorylated FGFR2 in the presence of active ERK1 (Figure 4d). This band was somewhat reduced in the sample from FGFR2 S780A cells. Thus, it seems that ERK1 directly phosphorylates FGFR2 on S780A. Since the phosphorylation of FGFR2 S780A is only partially reduced, we cannot exclude that other sites in FGFR2 might be phosphorylated by ERK1.

In order to study the role of S780 in FGFR2 further, we also prepared cell lines stably expressing FGFR2 S780D. The negatively charged aspartic acid might mimic constitutive phosphorylation of the residue. First, we verified that U2OS-R2 S780D cells were able to activate the main signaling pathways as wild-type FGFR2 (Figure S4a). We then investigated the tyrosine phosphorylation levels of FGFR2 upon FGF1 stimulation in U2OS-R2 S780D cells. Unfortunately, we observed the same effect of serine 780 mutated to an aspartic acid as we observed for FGFR2 S780A (Figure S4b). This is not surprising, as the mimicry of a phosphorylated serine by an aspartic acid often fails to reproduce the function of the phosphorylated serine [19]. We think that, instead of mimicking a constitutively phosphorylated serine, FGFR2 S780D rather displays a site that has lost its ability to become phosphorylated. Thus, FGFR2 S780D acts similarly to FGFR2 S780A and shows increased FGFR2 tyrosine phosphorylation.

3.3. Possible Role of Serine 780 in FGFR2 in Cancer Progression

We next investigated if the ERK1/2-mediated negative feedback loop possibly could play a role in cancer progression. By exploring databases reporting known alterations in cancer (cBioPortal; http://www.cbioportal.org and COSMIC (Catalogue of Somatic Mutations in Cancer); http://cancer. sanger.ac.uk) [20–22], we found several alterations that might influence the negative feedback loop in FGFR2 (Figure 5a). First, several truncated versions of FGFR2 lacking the C-terminal tail, including S780, have been identified in thyroid, skin, endometrial, and gastric cancers. In these cases, the negative feedback loop will not be operational and the receptor signaling may not be properly attenuated. This could potentially contribute to cancer progression. Secondly, several mutations in the close proximity of S780 have also been identified, including the glutamic acid at position 777 to a lysine and tyrosine 779 to a cysteine (Figure 5a). It is possible that these mutations influence S780 phosphorylation and receptor activity. Especially, the exchange of a negatively charged glutamic acid to a positively charged lysine might affect the properties of this region. Interestingly, serine 780 in FGFR2 has been found mutated to leucine in a patient with bladder cancer (Figure 5a). We decided to investigate the effect of this mutation further.

First of all, we generated U2OS cells stably expressing FGFR2 S780L (U2OS-R2 S780L) and confirmed that FGFR2 S780L cells were able to activate signaling pathways similarly to wild-type FGFR2 (Figure 5b and Figure S5a). Moreover, the levels of FGFR2 wild-type and FGFR2 S780L seem comparable and the mutated receptor is able to bind FGF1 at the cell surface and internalize FGF1 into early endosomes similarly to wild-type FGFR2 (Figure 5c and Figure S5b). In addition, the internalized DL550-FGF1 co-localizes well with anti-FGFR2 staining (Figure 5d).

We then analyzed the levels of FGFR2 tyrosine phosphorylation over time in FGFR2 S780L-expressing cells stimulated with FGF1. Compared to wild-type expressing cells, FGFR2 activation was prolonged in U2OS-R2 S780L (Figure 6a). Similar results were observed in two additional clones of U2OS-R2 S780L (Figure S6).

Clearly, the mutation of serine 780 to leucine leads to increased receptor signaling, which may be an advantage for cancer cells. Most cancer deaths (~90%) are caused by metastasis [23]. In order to metastasize and spread to distant organs, cancer cells need to be mobile and able to migrate. We therefore tested the mobility of U2OS cells stably expressing wild-type or S780 mutants. Since clonal variations might occur, we tested three different clones of each. Cells were seeded sparsely to allow for random migration and then imaged every 10 min for 21 h. We observed that stimulation of cells with FGF1 increased the migration velocities of all cell lines (Figure 6b and Videos S1 and S2). Moreover, U2OS cells expressing either of the mutant forms of S780 (A/L) migrated significantly faster than wild-type expressing cells in the presence of FGF1 (Figure 6b and Videos S2–S4). Preventing the negative feedback loop in FGFR2 by mutation of S780 causes increased signaling and, as a consequence, increased cell migration. In a cancer setting, this might contribute to disease progression.

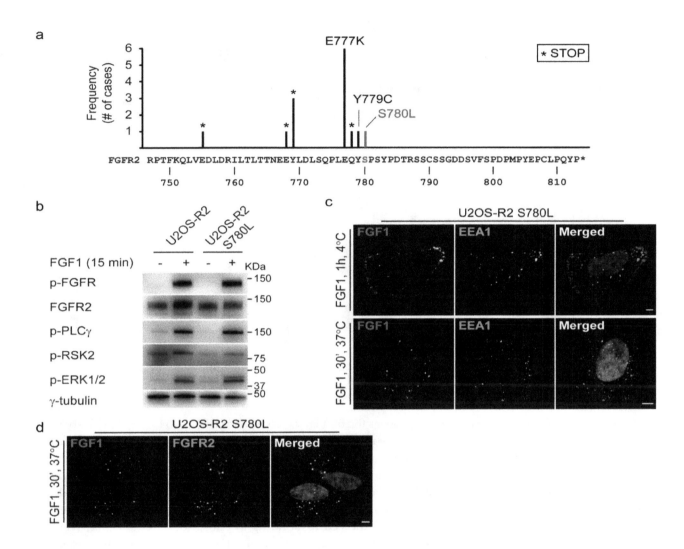

Figure 5. FGFR2 S780 is mutated in cancer. (**a**) The sequence of the C-terminal tail of FGFR2. Variations that might influence S780 phosphorylation and have been identified in cancer patients are indicated. An asterisk indicates a stop codon. Numbers refer to the amino acid numbering used for human FGFR2 (NCBI: NM_000141). The variations are reported in cBioPortal and COSMIC. (**b**) U2OS-R2 cells or U2OS-R2 S780L cells were kept in serum-free media for two hours and then treated or not with 100 ng/mL FGF1 for 15 min in the presence of heparin (20 U/mL). After lysis, the cellular material was analyzed by immunoblotting using the indicated antibodies. A p in front of the name of the antibody indicates that it recognizes the phosphorylated form of the protein. One representative experiment is shown. (**c**) U2OS-R2 S780L cells were kept at 4 °C with DL550-FGF1 for one hour. Next, the cells were either fixed directly (upper panel) or incubated for 30 min at 37 °C before fixation (lower panel). The cells were then stained with anti-EEA1 antibodies and Hoechst and analyzed by confocal microscopy. Representative images are shown. Scale bar: 5 μM. (**d**) U2OS-R2 S780L cells were treated with 200 ng/mL DL550-FGF1 in the presence of heparin (50 U/mL) for 30 min. The cells were then fixed, stained with anti-FGFR2 antibodies and Hoechst, and analyzed by confocal microscopy. Representative images are shown. Scale bar: 5 μM.

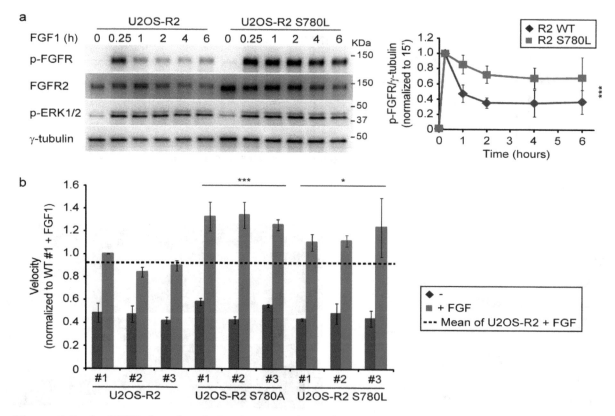

Figure 6. Lack of S780 phosphorylation increases the biological response to FGF1. (**a**) U2OS-R2 wild-type or U2OS-R2 S780L cells were kept in serum-free media for two hours before addition of 100 ng/mL FGF1 and heparin (20 U/mL) for indicated periods of time. Cycloheximide (10 μg/mL) was added at the beginning of the starvation period and kept throughout the experiment. After lysis, the cellular material was analyzed by immunoblotting using the indicated antibodies. A p in front of the name of the antibody indicates that it recognizes the phosphorylated form of the protein. Quantifications of three independent experiments are presented in the graph. The bands corresponding to phosphorylated receptor were normalized to γ-tubulin and within each experiment to the time point of 15 min. Error bars denote the standard deviation. The difference between FGFR2 wild-type and S780L mutant was significant ($p \leq 0.001$, 3-way ANOVA, Holm-Sidak test, n = 3). (**b**) Three different clones of U2OS-R2 wild-type, U2OS-R2 S780A, and U2OS-S780L cells were seeded into Image Lock 96-well plates. The cells were left untreated or stimulated with FGF1 (100 ng/mL) in the presence of heparin (20 U/mL) and imaged every 10 min over a period of 21 h by IncuCyte® S3 Live Cell Analysis System. The graph represents the mean velocities normalized to U2OS-R2 clone #1 with FGF1 of three independent experiments. The total number of cells tracked: U2OS-R2 #1 (-/+): 138/140, #2 (-/+): 111/138, #3 (-/+): 106/141, U2OS-R2 S780A #1 (-/+): 139/167, #2 (-/+): 135/142, #3 (-/+): 128/142, U2OS-R2 S780L #1 (-/+): 97/174, #2 (-/+): 110/153, #3 (-/+): 121/139. Error bars denote the SEM (n = 3). The difference between the wild-type and S780 mutants were significant (*** $p \leq 0.001$, * $p \leq 0.05$, student t test, n = 3).

4. Discussion

We have identified a negative feedback loop, mediated by the ERK1/2 pathway that regulates FGFR2 signaling. First, we found that inhibition of the ERK1/2-pathway leads to sustained FGFR2 signaling. Next, we found that substituting serine 780 in FGFR2 with alanine or leucine results in increased signaling. Serine 780 in FGFR2 is followed by a proline and thus forms an ERK1/2 phosphorylation motif (pS/T-P). In addition, S780 in FGFR2 corresponds to S777 in FGFR1. S777 in FGFR1 has previously been shown to be phosphorylated directly by ERK1/2. Taken together, we propose that ERK1/2-mediated phosphorylation of S780 in FGFR2 acts as a negative feedback loop to prevent excess signaling. This was evident when cells were pretreated with EGF to activate ERK1/2 prior to FGF1 stimulation. In activated cells, the response to FGF1 was lower than in resting cells. The feedback loop may function to fine-tune FGFR2 signaling in an environment where signaling is already on, preventing a further increase in signaling. We observed that cells lacking S780 (mutated to alanine or leucine) migrate faster than wild-type expressing cells. Since migration is important for spreading of cancer cells and metastasis, clearly the lack of this negative feedback loop gives cancer cells an advantage. Indeed, S780L has been identified in a patient with bladder cancer. In addition, several truncated forms of FGFR2 lacking S780 have also been identified in cancer. Maintaining the negative feedback loop ensures accurate signaling and preventing the feedback loop (either by mutation of S780 in FGFR2 or by inhibition of ERK1/2 signaling) could cause cancer progression.

Aberrant signaling through the Ras-Raf-MEK1/2-ERK1/2 pathway has been implicated in many types of cancer and is a promising therapeutic target. Although BRaf- and MEK-inhibitor mono- or combination-therapy have shown promising effects in cancer patients, many patients develop resistance and experience disease progression [24]. These resistance mechanisms include reactivation of the MAPK and/or the PI3K/Akt pathway. Examples of common resistance mechanisms include NRas mutation, BRaf v600 amplification, loss of PTEN, PI3KCA mutation, and RTK activation [24]. Since ERK1/2 is the only activator in the pathway with the ability to stimulate a wide variety of downstream substrates, it has emerged as an attractive therapeutic target. Despite the discovery of ERK1/2 many decades ago, ERK1/2 inhibitors have so far not been successfully implemented in the clinic [25]. One possible reason is that activated ERK1/2 stimulates inhibitory phosphorylation of many upstream factors and kinases, such as MEK, Raf, and different RTKs (including FGFR2), which prevent extensive signaling [25,26]. It is therefore worth considering that sole inhibition of the ERK1/2 signaling pathway in cancer could give rise to increased FGFR signaling through other signaling pathways (for example PI3K/Akt). Indeed, a recent study showed an increase in FGFR signaling upon MEK inhibition in KRas-driven lung cancer [27]. Therefore, caution should be taken when considering the use of MEK/ERK pathway inhibitors in cancer patients with FGFR2 expression.

Interestingly, a similar negative feedback loop involving ERK1/2-mediated phosphorylation has been identified for EGFR. In this case, ERK1/2-mediated signaling phosphorylates EGFR at threonine 669 (T669) [28,29]. Phosphorylation of this residue, which is localized in the juxtamembrane region of the receptor, was shown to reduce the tyrosine phosphorylation levels of EGFR. It seems that the T669-phosphorylated juxtamembrane region in EGFR has a reduced ability to cross-activate the other receptor of the dimer [30]. Although S780 is located in the C-terminal tail of FGFR2, it is possible that local conformational changes introduced by the phosphorylation at S780 reduce its cross-activation. It is also possible that a local conformational change in the receptor, caused by phosphorylation of S780, makes FGFR2 a better substrate for tyrosine dephosphorylation. Another attractive possibility is that phosphorylated S780 directly recruits a negative regulator such as a phosphatase or a scaffolding protein. Interestingly, serine 779 (S782 according to our numbering) in FGFR2 is phosphorylated by active PKCε and provides a docking site for the adaptor protein 14-3-3 [31]. However, in this case, phosphorylation of S779 (S782) seems to be required for sustained ERK1/2 activation and thus does not function as a negative feedback loop. It will be interesting to understand how these two phosphorylation events at S779 (S782) and S780 in FGFR2 work in partnership. It is also possible that S780 phosphorylation plays a role in receptor endocytosis and degradation. Previously, we found that S789 in FGFR1 is phosphorylated by RSK2 and seems to be required for proper internalization [13]. Interestingly, when we treated FGF1 stimulated cells with U0126 and stained for p-FGFR, we observed increased p-FGFR staining close to the cell surface (Figure 1d). We also observed less FGF1 internalized. However, when U2OS-R2 S780A cells were treated with U0126, the uptake of DL550-FGF1 was reduced to a similar extent as in U2OS-R2 wild-type cells (Figure 4c). It is possible that an RSK2-mediated feedback loop similar to that observed for FGFR1 exists also for FGFR2. U0126 inhibits both ERK1/2 and its downstream target, RSK2. There are also other examples of receptors that are serine-phosphorylated similarly to FGFR2, FGFR1, and EGFR. The Met receptor is phosphorylated by active PKCδ/ε at serine 985 in the juxtamembrane region [32]. Substitution of serine 985 by alanine resulted in increased tyrosine phosphorylation of Met. Similarly to FGFR2 and FGFR1, it is not clear what causes the reduced tyrosine phosphorylation in this case. Taken together, serine and threonine phosphorylation of receptor tyrosine kinases might be a common event that regulates receptor activity. A better understanding of these events will provide useful information when targeting receptor tyrosine kinases in cancer.

Although the mutation of serine 780 in FGFR2 to leucine clearly increases FGFR2 tyrosine phosphorylation levels and FGF1-stimulated cell migration, the role of this mutation in cancer is not clear. The mutation was found in a patient with bladder cancer. Although increased signaling and increased migration are traits that normally would benefit cancer cells, the role of FGFR2 signaling in bladder cancer is not fully understood and the FGFR2b isoform has been suggested to act as a tumor suppressor in the urothelium. It has been reported that reduction of FGFR2b levels in urothelial cancer samples correlate with decreased survival [33] and the chromosomal arm 10q, where the *FGFR2* gene is located, is often lost in advanced bladder cancer [34]. It should be noted that 10q also contains the tumor suppressor *PTEN* [35]. Moreover, expression of FGFR2b in urothelial cells lacking endogenous FGFR2 led to reduced proliferation and reduced tumorigenicity in nude mice [36]. On the other hand, increased FGFR2c expression has been reported in a model of epithelial-to-mesechymal transition (EMT) in bladder cancer cells [37] and recently, the U.S. Food and Drug Administration (FDA) approved Balversa (Erdafitinib), a pan FGFR-inhibitor for clinical use in patients with locally advanced or metastatic urothelial carcinoma with FGFR2 and FGFR3 aberrations [38]. This is the first targeted-FGFR therapy approved for clinical use and the first targeted therapy in advanced urothelial carcinoma. Alongside the approval of the drug, the FDA approved an RT-PCR-based diagnostic test to identify patients with FGFR3 mutations or FGFR2 fusions. It is possible that FGFR2 plays a tumor-suppressing role in earlier stages of bladder cancer, but could have tumor-promoting effects in certain patients with advanced bladder cancer.

The S780L mutation is only reported once in the COSMIC database and at such low frequency that the significance is questionable. On the other hand, truncated versions of FGFR2 lacking S780 have been identified in cancer patients (Figure 5a). In addition, an alternatively spliced form of FGFR2, FGFR2IIIb-C3, is also lacking S780 [39]. In contrast to full-length FGFR2IIIb, FGFR2IIIb-C3 is only identified in human cancer samples. Aberrant expression of FGFR2IIIb-C3 in SUM-52 breast cancer cells resulted in sustained signaling leading to transformation [40]. A tyrosine phosphorylation site (corresponding to Y766 in FGFR1) is also lacking in FGFR2IIIb-C3 and could explain the increased signaling and transformation capabilities of FGFR2IIIb-C3. However, mutation of only this tyrosine in full-length FGFR2 did not lead to increased signaling [41]. Thus, loss of other mechanisms maintained by the C-terminal tail of FGFR2 might cause the increased signaling and transforming potential of FGFR2IIIb-C3. We propose that lack of S780 in FGFR2IIIb-C3 could promote its transforming capabilities. Interestingly, a patient with endometrial cancer was identified with the activating mutation N549H in FGFR2 and a truncated C-terminal tail (cBioPortal; http://www.cbioportal.org) [20,21]. This combination of alterations in FGFR2 will clearly impact signaling output and could be even more cancer-promoting than versions with either mutation alone.

In summary, we have identified an ERK1/2-mediated negative feedback-loop in FGFR2. We propose that lack of this feedback loop could give cancer cells an advantage and, indeed, variants of FGFR2 lacking the feedback loop have been identified in several human cancers. We conclude therefore that, in addition to the previously reported activating mutations in the kinase domain of FGFRs [4], mutations in the C-terminal tail of FGFR2 may also cause hyperactivation of the receptors.

Supplementary Materials:
Figure S1: Inhibition of ERK1/2 prolongs FGFR2 signaling in several U2OS-R2 clones. Figure S2: Characterization of U2OS-R2 S780A clones. Figure S3: Signaling from FGFR2 S780A is prolonged compared to wild-type FGFR2. Figure S4: FGFR2 S780D mutant is not phosphomimetic. Figure S5: Characterization of U2OS-R2 S780L clones. Figure S6: Signaling from FGFR2 S780L is prolonged compared to wild-type FGFR2. Video S1: Migration of U2OS-R2 cells in the absence of FGF1. Video S2: Migration of U2OS-R2 cells stimulated with FGF1. Video S3: Migration of U2OS-R2 S780A cells stimulated with FGF1. Video S4: Migration of U2OS-R2 S780L cells stimulated with FGF1.

Author Contributions: Conceptualization, P.S., E.M.H., A.W., and J.W.; methodology, P.S. and E.M.H.; validation, P.S., and E.M.H.; formal analysis, E.M.H., P.S., and M.K.; investigation, P.S., J.W., A.W., and M.K.; writing—original draft preparation, P.S. and E.M.H.; writing—review and editing, P.S., E.M.H., M.K., A.W., J.W.; visualization, P.S., and E.M.H.; supervision, E.M.H., and A.W.; project administration, E.M.H.; funding acquisition, E.M.H., A.W., and J.W. All authors read and approved the manuscript.

Acknowledgments: The results shown here are in part based upon data generated by the TCGA Research Network: https://www.cancer.gov/tcga.

References

1. Sangwan, V.; Park, M. Receptor tyrosine kinases: Role in cancer progression. *Curr. Oncol.* **2006**, *13*, 191–193.
2. Ornitz, D.M.; Itoh, N. The Fibroblast Growth Factor signaling pathway. *Wiley Interdiscip. Rev. Dev. Biol.* **2015**, *4*, 215–266. [CrossRef] [PubMed]
3. Wesche, J.; Haglund, K.; Haugsten, E.M. Fibroblast growth factors and their receptors in cancer. *Biochem. J.* **2011**, *437*, 199–213. [CrossRef] [PubMed]
4. Babina, I.S.; Turner, N.C. Advances and challenges in targeting FGFR signalling in cancer. *Nat. Rev. Cancer* **2017**, *17*, 318–332. [CrossRef]
5. Ornitz, D.M.; Marie, P.J. Fibroblast growth factor signaling in skeletal development and disease. *Genes Dev.* **2015**, *29*, 1463–1486. [CrossRef]
6. Haugsten, E.M.; Malecki, J.; Bjorklund, S.M.; Olsnes, S.; Wesche, J. Ubiquitination of fibroblast growth factor receptor 1 is required for its intracellular sorting but not for its endocytosis. *Mol. Biol. Cell* **2008**, *19*, 3390–3403. [CrossRef] [PubMed]
7. Haugsten, E.M.; Sorensen, V.; Brech, A.; Olsnes, S.; Wesche, J. Different intracellular trafficking of FGF1 endocytosed by the four homologous FGF receptors. *J. Cell Sci.* **2005**, *118*, 3869–3881. [CrossRef] [PubMed]
8. Belleudi, F.; Leone, L.; Maggio, M.; Torrisi, M.R. Hrs regulates the endocytic sorting of the fibroblast growth factor receptor 2b. *Exp. Cell Res.* **2009**, *315*, 2181–2191. [CrossRef] [PubMed]
9. Kostas, M.; Haugsten, E.M.; Zhen, Y.; Sorensen, V.; Szybowska, P.; Fiorito, E.; Lorenz, S.; Jones, N.; de Souza, G.A.; Wiedlocha, A.; et al. Protein Tyrosine Phosphatase Receptor Type G (PTPRG) Controls Fibroblast Growth Factor Receptor (FGFR) 1 Activity and Influences Sensitivity to FGFR Kinase Inhibitors. *Mol. Cell Proteom.* **2018**, *17*, 850–870. [CrossRef]
10. Goetz, R.; Mohammadi, M. Exploring mechanisms of FGF signalling through the lens of structural biology. *Nat. Rev. Mol. Cell Biol.* **2013**, *14*, 166–180. [CrossRef]
11. Lax, I.; Wong, A.; Lamothe, B.; Lee, A.; Frost, A.; Hawes, J.; Schlessinger, J. The docking protein FRS2alpha controls a MAP kinase-mediated negative feedback mechanism for signaling by FGF receptors. *Mol. Cell* **2002**, *10*, 709–719. [CrossRef]
12. Zakrzewska, M.; Haugsten, E.M.; Nadratowska-Wesolowska, B.; Oppelt, A.; Hausott, B.; Jin, Y.; Otlewski, J.; Wesche, J.; Wiedlocha, A. ERK-mediated phosphorylation of fibroblast growth factor receptor 1 on Ser777 inhibits signaling. *Sci. Signal.* **2013**, *6*, ra11. [CrossRef]
13. Nadratowska-Wesolowska, B.; Haugsten, E.M.; Zakrzewska, M.; Jakimowicz, P.; Zhen, Y.; Pajdzik, D.; Wesche, J.; Wiedlocha, A. RSK2 regulates endocytosis of FGF receptor 1 by phosphorylation on serine 789. *Oncogene* **2014**, *33*, 4823–4836. [CrossRef] [PubMed]
14. Wesche, J.; Malecki, J.; Wiedlocha, A.; Ehsani, M.; Marcinkowska, E.; Nilsen, T.; Olsnes, S. Two nuclear localization signals required for transport from the cytosol to the nucleus of externally added FGF-1 translocated into cells. *Biochemistry* **2005**, *44*, 6071–6080. [CrossRef]
15. Klingenberg, O.; Wiedlocha, A.; Rapak, A.; Khnykin, D.; Citores, L.; Olsnes, S. Requirement for C-terminal end of fibroblast growth factor receptor 4 in translocation of acidic fibroblast growth factor to cytosol and nucleus. *J. Cell Sci.* **2000**, *113*, 1827–1838. [PubMed]
16. Schindelin, J.; Arganda-Carreras, I.; Frise, E.; Kaynig, V.; Longair, M.; Pietzsch, T.; Preibisch, S.; Rueden, C.; Saalfeld, S.; Schmid, B.; et al. Fiji: An open-source platform for biological-image analysis. *Nat. Methods* **2012**, *9*, 676–682. [CrossRef]

17. Haugsten, E.M.; Zakrzewska, M.; Brech, A.; Pust, S.; Olsnes, S.; Sandvig, K.; Wesche, J. Clathrin- and dynamin-independent endocytosis of FGFR3–implications for signalling. *PLoS ONE* **2011**, *6*, e21708. [CrossRef]

18. Unal, E.B.; Uhlitz, F.; Bluthgen, N. A compendium of ERK targets. *FEBS Lett.* **2017**, *591*, 2607–2615. [CrossRef] [PubMed]

19. Chen, Z.; Cole, P.A. Synthetic approaches to protein phosphorylation. *Curr. Opin. Chem. Biol.* **2015**, *28*, 115–122. [CrossRef]

20. Gao, J.; Aksoy, B.A.; Dogrusoz, U.; Dresdner, G.; Gross, B.; Sumer, S.O.; Sun, Y.; Jacobsen, A.; Sinha, R.; Larsson, E.; et al. Integrative analysis of complex cancer genomics and clinical profiles using the cBioPortal. *Sci. Signal.* **2013**, *6*, pl1. [CrossRef] [PubMed]

21. Cerami, E.; Gao, J.; Dogrusoz, U.; Gross, B.E.; Sumer, S.O.; Aksoy, B.A.; Jacobsen, A.; Byrne, C.J.; Heuer, M.L.; Larsson, E.; et al. The cBio cancer genomics portal: An open platform for exploring multidimensional cancer genomics data. *Cancer Discov.* **2012**, *2*, 401–404. [CrossRef] [PubMed]

22. Forbes, S.A.; Beare, D.; Boutselakis, H.; Bamford, S.; Bindal, N.; Tate, J.; Cole, C.G.; Ward, S.; Dawson, E.; Ponting, L.; et al. COSMIC: Somatic cancer genetics at high-resolution. *Nucleic Acids Res.* **2017**, *45*, D777–D783. [CrossRef] [PubMed]

23. Chaffer, C.L.; Weinberg, R.A. A perspective on cancer cell metastasis. *Science* **2011**, *331*, 1559–1564. [CrossRef] [PubMed]

24. Kakadia, S.; Yarlagadda, N.; Awad, R.; Kundranda, M.; Niu, J.; Naraev, B.; Mina, L.; Dragovich, T.; Gimbel, M.; Mahmoud, F. Mechanisms of resistance to BRAF and MEK inhibitors and clinical update of US Food and Drug Administration-approved targeted therapy in advanced melanoma. *Oncotargets Ther.* **2018**, *11*, 7095–7107. [CrossRef]

25. Liu, F.; Yang, X.; Geng, M.; Huang, M. Targeting ERK, an Achilles' Heel of the MAPK pathway, in cancer therapy. *Acta Pharm. Sin. B* **2018**, *8*, 552–562. [CrossRef] [PubMed]

26. Lake, D.; Correa, S.A.; Muller, J. Negative feedback regulation of the ERK1/2 MAPK pathway. *Cell. Mol. Life Sci.* **2016**, *73*, 4397–4413. [CrossRef] [PubMed]

27. Manchado, E.; Weissmueller, S.; Morris, J.P.; Chen, C.C.; Wullenkord, R.; Lujambio, A.; de Stanchina, E.; Poirier, J.T.; Gainor, J.F.; Corcoran, R.B.; et al. A combinatorial strategy for treating KRAS-mutant lung cancer. *Nature* **2016**, *534*, 647–651. [CrossRef]

28. Northwood, I.C.; Gonzalez, F.A.; Wartmann, M.; Raden, D.L.; Davis, R.J. Isolation and characterization of two growth factor-stimulated protein kinases that phosphorylate the epidermal growth factor receptor at threonine 669. *J. Biol. Chem.* **1991**, *266*, 15266–15276.

29. Takishima, K.; Griswold-Prenner, I.; Ingebritsen, T.; Rosner, M.R. Epidermal growth factor (EGF) receptor T669 peptide kinase from 3T3-L1 cells is an EGF-stimulated "MAP" kinase. *Proc. Natl. Acad. Sci. USA* **1991**, *88*, 2520–2524. [CrossRef]

30. Sato, K.; Shin, M.S.; Sakimura, A.; Zhou, Y.; Tanaka, T.; Kawanishi, M.; Kawasaki, Y.; Yokoyama, S.; Koizumi, K.; Saiki, I.; et al. Inverse correlation between Thr-669 and constitutive tyrosine phosphorylation in the asymmetric epidermal growth factor receptor dimer conformation. *Cancer Sci.* **2013**, *104*, 1315–1322. [CrossRef] [PubMed]

31. Lonic, A.; Powell, J.A.; Kong, Y.; Thomas, D.; Holien, J.K.; Truong, N.; Parker, M.W.; Guthridge, M.A. Phosphorylation of serine 779 in fibroblast growth factor receptor 1 and 2 by protein kinase C(epsilon) regulates Ras/mitogen-activated protein kinase signaling and neuronal differentiation. *J. Biol. Chem.* **2013**, *288*, 14874–14885. [CrossRef]

32. Gandino, L.; Longati, P.; Medico, E.; Prat, M.; Comoglio, P.M. Phosphorylation of serine 985 negatively regulates the hepatocyte growth factor receptor kinase. *J. Biol. Chem.* **1994**, *269*, 1815–1820.

33. Diez de Medina, S.G.; Chopin, D.; El, M.A.; Delouvee, A.; LaRochelle, W.J.; Hoznek, A.; Abbou, C.; Aaronson, S.A.; Thiery, J.P.; Radvanyi, F. Decreased expression of keratinocyte growth factor receptor in a subset of human transitional cell bladder carcinomas. *Oncogene* **1997**, *14*, 323–330. [CrossRef] [PubMed]

34. Cappellen, D.; Gil Diez de Medina, S.; Chopin, D.; Thiery, J.P.; Radvanyi, F. Frequent loss of heterozygosity on chromosome 10q in muscle-invasive transitional cell carcinomas of the bladder. *Oncogene* **1997**, *14*, 3059–3066. [CrossRef] [PubMed]

35. Hurst, C.D.; Platt, F.M.; Taylor, C.F.; Knowles, M.A. Novel tumor subgroups of urothelial carcinoma of the bladder defined by integrated genomic analysis. *Clin. Cancer Res. Off. J. Am. Assoc. Cancer Res.* **2012**, *18*, 5865–5877. [CrossRef]

36. Ricol, D.; Cappellen, D.; El, M.A.; Gil-Diez-de-Medina, S.; Girault, J.M.; Yoshida, T.; Ferry, G.; Tucker, G.; Poupon, M.F.; Chopin, D.; et al. Tumour suppressive properties of fibroblast growth factor receptor 2-IIIb in human bladder cancer. *Oncogene* **1999**, *18*, 7234–7243. [CrossRef] [PubMed]

37. Chaffer, C.L.; Brennan, J.P.; Slavin, J.L.; Blick, T.; Thompson, E.W.; Williams, E.D. Mesenchymal-to-epithelial transition facilitates bladder cancer metastasis: Role of fibroblast growth factor receptor-2. *Cancer Res.* **2006**, *66*, 11271–11278. [CrossRef]

38. Nadal, R.; Bellmunt, J. Management of metastatic bladder cancer. *Cancer Treat. Rev.* **2019**, *76*, 10–21. [CrossRef]

39. Cha, J.Y.; Maddileti, S.; Mitin, N.; Harden, T.K.; Der, C.J. Aberrant receptor internalization and enhanced FRS2-dependent signaling contribute to the transforming activity of the fibroblast growth factor receptor 2 IIIb C3 isoform. *J. Biol. Chem.* **2009**, *284*, 6227–6240. [CrossRef]

40. Moffa, A.B.; Tannheimer, S.L.; Ethier, S.P. Transforming potential of alternatively spliced variants of fibroblast growth factor receptor 2 in human mammary epithelial cells. *Mol. Cancer Res.* **2004**, *2*, 643–652.

41. Moffa, A.B.; Ethier, S.P. Differential signal transduction of alternatively spliced FGFR2 variants expressed in human mammary epithelial cells. *J. Cell Physiol.* **2007**, *210*, 720–731. [CrossRef] [PubMed]

14

Expression of FGF8, FGF18 and FGFR4 in Gastroesophageal Adenocarcinomas

Gerd Jomrich [1,2], **Xenia Hudec** [2], **Felix Harpain** [1,2], **Daniel Winkler** [3], **Gerald Timelthaler** [2], **Thomas Mohr** [2], **Brigitte Marian** [2,*] **and Sebastian F. Schoppmann** [1]

[1] Department of Surgery, Medical University of Vienna and Gastroesophageal Tumor Unit, Comprehensive Cancer Center (CCC), Spitalgasse 23, 1090 Vienna, Austria; gerd.jomrich@meduniwien.ac.at (G.J.); felix.harpain@meduniwien.ac.at (F.H.); sebastian.schoppmann@meduniwien.ac.at (S.F.S.)

[2] Department of Medicine I, Institute of Cancer Research, Medical University of Vienna, Borschkegasse 8a, 1090 Vienna, Austria; xenia.hudec@meduniwien.ac.at (X.H.); gerald.timelthaler@meduniwien.ac.at (G.T.); thomas.mohr@meduniwien.ac.at (T.M.)

[3] Department of Statistics and Operations Research, University of Vienna, Oskar Morgenstern Platz 1, 1090 Vienna, Austria; Daniel.Winkler@wu.ac.at

[*] Correspondence: brigitte.marian@meduniwien.ac.at

Abstract: Even though distinctive advances in the field of esophageal cancer therapy have occurred over the last few years, patients' survival rates remain poor. FGF8, FGF18, and FGFR4 have been identified as promising biomarkers in a number of cancers; however no data exist on expression of FGF8, FGF18, and FGFR4 in adenocarcinomas of the esophago-gastric junction (AEG). A preliminary analysis of the Cancer Genome Atlas (TCGA) database on FGF8, FGF18, and FGFR4 mRNA expression data of patients with AEG was performed. Furthermore, protein levels of FGF8, FGF18, and FGFR4 in diagnostic biopsies and post-operative specimens in neoadjuvantly treated and primarily resected patients using immunohistochemistry were investigated. A total of 242 patients was analyzed in this study: 87 patients were investigated in the TCGA data set analysis and 155 patients in the analysis of protein expression using immunohistochemistry. High protein levels of FGF8, FGF18, and FGFR4 were detected in 94 (60.7%), 49 (31.6%) and 84 (54.2%) patients, respectively. Multivariable Cox proportional hazard regression models revealed that high expression of FGF8 was an independent prognostic factor for diminished overall survival for all patients and for neoadjuvantly treated patients. By contrast, FGF18 overexpression was significantly associated with longer survival rates in neoadjuvantly treated patients. In addition, FGF8 protein level correlated with Mandard regression due to neoadjuvant therapy, indicating potential as a predictive marker. In summary, FGF8 and FGF18 are promising candidates for prognostic factors in adenocarcinomas of the esophago-gastric junction and new potential targets for new anti-cancer therapies.

Keywords: FGF8; FGF18; FGFR4; adenocarcinoma of the esophagogastric junction; neoadjuvant therapy

1. Introduction

Esophageal Cancer (EC) is the eighth most common cancer worldwide. Whereas the number of esophageal squamous cell carcinomas (ESCC) is decreasing, the number of adenocarcinomas of the esophago-gastric junction (AEG) is increasing dramatically [1]. Despite improvements in diagnostics and the use of multimodal approaches, combining surgical resection with perioperative chemo-(radio) therapy, overall prognosis of AEG remains poor [2,3]. Survival rates vary considerably among patients with AEG, and an appreciable proportion of patients with advanced stages develop recurrence, even after initially curative resection [4,5]. Therapy response is often limited due to a number of inherent mechanisms of resistance [6]. This problem is aggravated by the heterogeneity in malignant tumors, containing a small subpopulation of cancer stem-like cells (CSC), characterized by a long lifespan and enhanced survival capacity that supports drug resistance [7,8]. Stem cell characteristics of CSCs are governed by the activity of distinct stem cell specific regulatory pathways leading to cancer relapse as well as chemo- and radio-resistance [9]. The role of CD133- and CD44-positive subpopulation in EC has been described recently [10–12] and the wnt-, notch-, hedgehog-, and hippo-pathways have been identified as stem cell specific targets driving therapy resistance and relapse [10–13]. Both CSC-specific signaling pathways and the survival capacity of a larger tumor cell pool might influence the therapy response.

Specifically, FGFs have found their way into anti-cancer therapy as targets to overcome resistance to chemotherapy in a number of different malignancies [14]. FGFs play a major physiological role in embryonic development and tissue repair by mediating strong survival signals via activation of the direct receptor substrate FRS2α, and the RAS- and PI3K-pathways [15–17]. In cancers the pathway might be deregulated by manifold-mechanisms causing either hyperactivation or even constitutively active FGFR-dependent survival signaling [14]. Both expression of specific receptors and up-regulation of autocrine FGF ligands have been found to be associated with resistance to chemo-(radiation) as well as to targeted therapy [18–22]. Previously, our group has studied a CD44-positive stem-like population in colorectal cancer (CRC) and identified a wnt-driven FGF18-dependent autocrine-signaling loop as a strong driver of tumor cell survival [8,23]. Furthermore, we demonstrated a progressive up regulation of FGF18 in CRC [24]. The growth factor induces autocrine survival signaling via the FGF receptor FGFR3-IIIc and blocking of this receptor inhibits tumor growth by inducing apoptosis [25]. Alternatively, FGF18 effects may be mediated by FGFR4, a receptor for which a polymorphic variant exists that causes substitution of an arginine for a glycine at position 388 in the transmembrane domain [26–28]. FGF8 is known to play an important role in embryonic development [29,30]. In tumors, overexpression of FGF8 is associated with diminished survival based on stimulating anti-apoptotic pathways mediated by the IIIc splice variants of FGFR1, 2, 3 as well as FGFR4 [28,31,32]. Recently, we could show that the expression of FGF8 was strongly associated with the regression grade in neoadjuvantly treated colorectal cancer patients [33].

Until now, little has been known about the role of FGFs and their receptors in AEG, in particular to the best of our knowledge no data was published describing the expression of FGF 8, 18, and FGFR4 in AEG. Therefore, the aim of this study is to investigate the role of FGF 8, 18, and FGFR4 in AEG in order to define predictive markers and possibly identify suitable new targets for multimodal therapies.

2. Materials and Methods

2.1. Preliminary TCGA (The Cancer Genome Atlas) Analysis

Data (HTSeq counts) for AEG were downloaded from the TCGA-ESCA project, preprocessed, and normalized using the TCGABiolinks package of R [34]. Optimal cutoff values for gene expression were determined by maximizing the log-rank statistics using the survminer package of R [35]. Differentially expressed genes where determined using TCGABiolinks, employing the edgeR algorithm with exact testing [36]. Gene expression of relevant KEGG pathways was visualized using pathview [37,38].

2.2. Patient Selection

Patients who underwent a resection of gastroesophageal adenocarcinomas between January 1992 and April 2012 at the Department of Surgery at the Medical University Vienna were identified from a prospectively maintained database. Patients with distant metastasis at time of diagnosis were excluded. The study was approved by the Ethics Committee of the Medical University of Vienna, Austria, according to the declaration of Helsinki (EK 1652/2016). Patients with locally advanced AEG received neoadjuvant chemotherapy according to the recommendation of the interdisciplinary tumor board meeting. Regression grade to neoadjuvant chemotherapy was classified as defined by Mandard

A.M. et al. [39]. The tumor stage was conducted according to the pathological tumor-node-metastasis (TNM) classification of the Union for International Cancer Control (UICC), 7th edition.

2.3. Immunohistochemistry

Immunohistochemistry (IHC) was performed on paraffin-embedded specimens fixed in 4% buffered formalin, using 3-μm-thick histological sections. Furthermore, per case two tissue cylinders with a 2.0 millimeter diameter were punched from representative tissue areas to build a tissue micro array (TMA), as described previously [40]. Expression of FGF8, FGF18, and FGFR4 was detected by using polyclonal rabbit antibodies as follows: FGF8 antibody (Abcam, Cambridge, UK, ab203030) in a dilution of 1:600, FGF18 antibody (Assay Biotech, Fremont, CA, USA, C12364) in a dilution of 1:500, and FGFR4 antibody (Santa Cruz Biotechnology, Dallas, TX, USA, sc-124) in a dilution of 1:400, respectively. Secondary antibody was biotinylated and coupled to an avidin-biotin-HRP complex (Thermo Scientific™ Lab Vision™ UltraVision™ LP, Waltham, MA, USA). 3,30-diaminobenzidine (DAB; Chromogen) was used to visualize the staining and counterstaining was achieved with hematoxylin. Antibodies used in this study were optimized for gastroesophageal adenocarcinomas on colorectal cancer tissue with known expression from previously published studies [33,41]. Two observers (J.G. and H.F.) independently reviewed all slides. For the quantitative evaluation of expression, only epithelial cells were investigated. Immunostaining scores (0–12) of FGF8, FGF18, and FGFR4 were calculated as the products of the staining intensity (0 = negative, 1 = weak, 2 = moderate or 3 = strong expression) and points (0–4) were given for the percentages of tumor cells showing positive staining 0 (<1%), 1 (1–10%), 2 (10–50%), 3 (51–80%), and 4 (>80%). Tumors were considered to have high expression with final scores exceeding the median score. Tumors showing expression equal or below the median were considered as being low or absent.

2.4. Statistical Analysis

Statistical analysis was performed using the R Statistical Software, Vienna, Austria (Version 3.6) with the "survival" package [42,43]. Univariable and multivariable analyses were conducted using the Cox proportional hazard model. The graphical analysis was performed using the Kaplan-Meier estimator. Plotting was performed using the "survminer" package [35]. The significance of differences in survival times were determined with a log-rank test. Correlations between clinicopathological parameters and FGF8, FGF18, and FGFR4 expression levels were analyzed with the x^2 test. In order to measure statistical dependence between FGF 8 and FGF 18 the non-parametric Kendall's rank correlation was used.

Overall survival (OS) was defined as the time between surgery and the patients' death. Death from causes other than AEG or survival until the end of the observation was considered as censored observations.

3. Results

3.1. Preliminary TCGA (The Cancer Genome Atlas) Analysis

While investigating mRNA expression data of patients with AEG (n = 87) available from the TCGA data base, overexpression of FGF8, FGF18, and FGFR4 was found in 64, 43, 12 cases, respectively. No significant correlation of overexpression of FGF8, FGF18, and FGFR4 and clinicopathological parameters (tumor stage, lymph node status and age) was found. Survival analysis using Kaplan-Meier curves for visualization, found significantly better OS rates for patients with FGF18 overexpressing tumors ($p = 0.017$). No significance could be found for FGF8 and FGFR4 (Figure 1a–c).

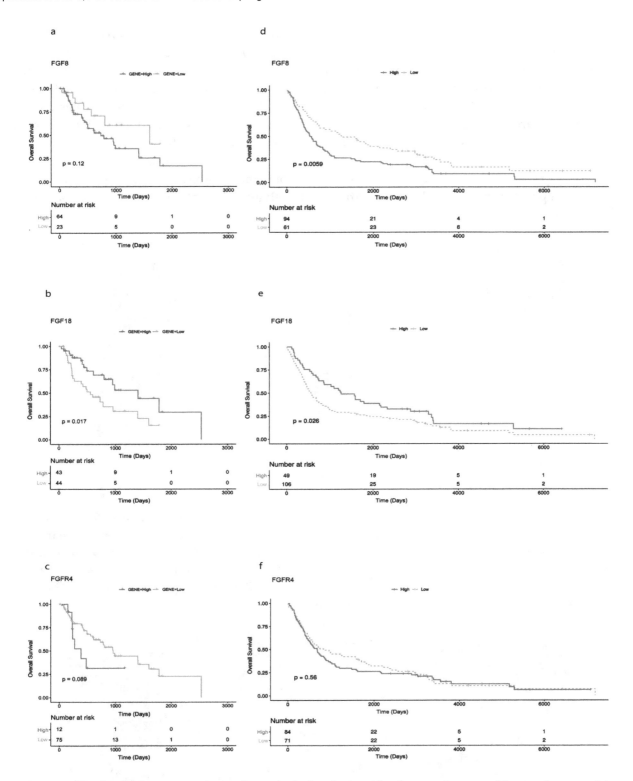

Figure 1. Kaplan-Meier curves of overall survival of patients with adenocarcinomas of the esophago-gastric junction. (**a–c**) Patients from TCGA data set analysis: high FGF8, FGF18, and FGFR4 expression compared with those with low/absent FGF8, FGF18, and FGFR4 expression. (**d–f**) Patients from the immunohistological analysis: high FGF8, FGF18, and FGFR4 expression compared with those with low/absent FGF8, FGF18, and FGFR4 expression.

3.2. Immunohistochemical Analysis of Tumor Tissue Samples

A total of 155 patients (124 males, 80%) with histologically verified AEG were investigated for this study. From 10 patients full section slides were investigated to confirm staining quality for all antibodies used in this study. Tissue specimens of the tumors were stained for FGF8, FGF18, FGFR4, cytokeratin 7 (CK7) and the proliferation marker Ki67 (Figure 2). For all 3 markers staining was predominantly seen in the cytoplasm of tumor cells. Weaker staining was also observed in the tumor stroma. For quantification, only tumor cell staining was assessed. High expression of FGF8, FGF18, and FGFR4 was found in 94 (60.7%), 49 (31.6%) and 84 (54.2%), respectively (Figure 3a–c) as compared to low expressing areas (Figure 3d–f). Each marker had a distinct expression pattern with no correlation between individual markers. Correlation of clinicopathological parameters and expression of FGF8, FGF18, and FGFR4 in the tumor tissue revealed significant correlations of the FGF8 protein level with tumor size ((y)pT), UICC stage, and Mandard regression grade (Table 1). FGFR4 protein level only correlates with gender and for FGF18 no relationship with any clinical parameter could be observed (compiled in Table 1).

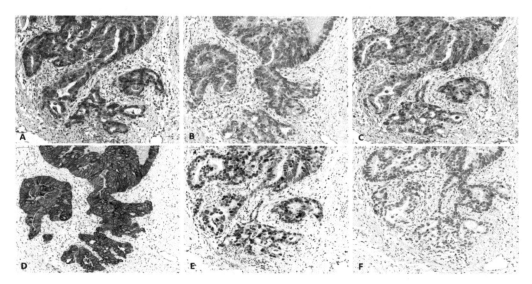

Figure 2. Specimen of adenocarcinomas of the esophago-gastric junction stained for (**a**) FGF8, (**b**) FGF18 and (**c**) FGFR4. Positive staining was found in the tumor cells and to a lesser degree in the microenvironment. FGF8 and FGFR4 expression were primarily found in the nucleus, while FGF18 expression was mainly found in the cytoplasm. For quantitative evaluation, only epithelial cells were investigated. Corresponding sections stained by CK7 (**d**), Ki67 (**e**), and negative control (**f**). (The bar corresponds to 50 μm.) Original magnification ×400 all).

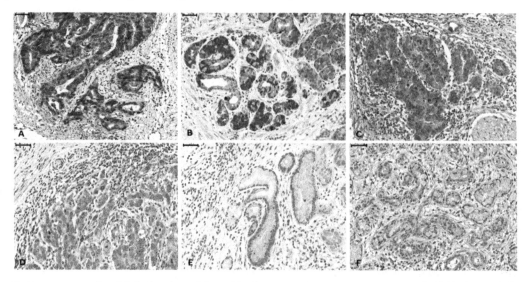

Figure 3. Representative high (**a–c**) and low (**d–f**) expressing tumor section of FGF8 (**a** and **d**), FGF18 (**b** and **e**), and FGFR4 (**c** and **f**).

Table 1. FGF8, FGF18, and FGFR4 expression and their correlation with clinicopathologic parameters in patients with adenocarcinoma of the esophago-gastric junction.

Factors	FGF8 high 65 (11)		FGF8 low/absent 65 (10)		FGF8 p-value	FGF18 High 66 (12)		FGF18 low/absent 62 (10)		FGF18 p-value	FGFR4 high 66 (11)		FGFR4 low/absent 64 (11)		FGFR4 p-value
Age (SD)					>0.05 >0.05					>0.05 >0.05					>0.05 0.008
Sex					>0.05					>0.05					>0.05
Male	75	(48.4%)	49	(31.6%)		38	(24.5%)	86	(55.5%)		74	(47.7%)	50	(32.3%)	
Female	19	(12.3%)	12	(7.7%)		11	(7.1%)	20	(12.9%)		10	(6.5%)	21	(13.5%)	
Neoadjuvant treatment					>0.05					>0.05					>0.05
Yes	37	(23.9%)	32	(20.6%)		26	(16.8%)	60	(38.7%)		31	(20.0%)	38	(24.5%)	
No	57	(36.8%)	29	(18.7%)		23	(14.8%)	46	(29.7%)		53	(34.2%)	33	(21.3%)	
(y)pT					0.003					>0.05					>0.05
0	0	(0.0%)	3	(1.9%)		0	(0.0%)	3	(1.9%)		0	(0.0%)	3	(1.9%)	
1	1	(0.6%)	2	(1.3%)		2	(1.3%)	1	(0.6%)		1	(0.6%)	2	(1.3%)	
2	17	(11.0%)	23	(14.8%)		11	(7.1%)	29	(18.7%)		22	(14.2%)	18	(11.6%)	
3	68	(43.9%)	32	(20.6%)		34	(21.9%)	66	(42.6%)		56	(36.1%)	44	(28.4%)	
4	8	(5.2%)	1	(0.6%)		2	(1.3%)	7	(4.5%)		5	(3.2%)	4	(2.6%)	
(y)pN					>0.05					>0.05					>0.05
0	17	(11.0%)	22	(14.2%)		16	(10.3%)	23	(14.8%)		18	(11.6%)	21	(13.5%)	
1	25	(16.1%)	13	(8.4%)		13	(8.4%)	25	(16.1%)		19	(12.3%)	19	(12.3%)	
2	24	(15.5%)	13	(8.4%)		7	(4.5%)	30	(19.4%)		25	(16.1%)	12	(7.7%)	
3	28	(18.1%)	13	(8.4%)		13	(8.4%)	28	(18.1%)		22	(14.2%)	19	(12.3%)	
Tumor differentiation					>0.05					>0.05					>0.05
0	0	(0.0%)	0	(0.0%)		0	(0.0%)	0	(0.0%)		0	(0.0%)	0	(0.0%)	
1	3	(1.9%)	4	(2.6%)		1	(0.6%)	6	(3.9%)		3	(1.9%)	4	(2.6%)	
2	29	(18.7%)	25	(16.1%)		20	(12.9%)	34	(21.9%)		26	(16.8%)	28	(18.1%)	
3	62	(40.0%)	32	(20.6%)		28	(18.1%)	66	(42.6%)		55	(35.5%)	39	(25.2%)	
Lymph node ratio					>0.05					>0.05					>0.05
<0.3	61	(39.4%)	42	(27.1%)		35	(22.6%)	68	(43.9%)		55	(35.5%)	48	(31.0%)	
≥0.3	33	(21.3%)	19	(12.3%)		15	(9.7%)	38	(24.5%)		29	(18.7%)	23	(14.8%)	
R					>0.05					>0.05					>0.05
0	72	(46.5%)	51	(32.9%)		39	(25.2%)	84	(54.2%)		64	(41.3%)	59	(38.1%)	
1	22	(14.2%)	10	(6.5%)		10	(6.5%)	22	(14.2%)		20	(12.9%)	12	(7.7%)	
UICC Staging					0.01					>0.05					>0.05
0	0	(0.0%)	3	(1.9%)		0	(0.0%)	3	(1.9%)		0	(0.0%)	3	(1.9%)	
I	6	(3.9%)	9	(5.8%)		7	(4.5%)	8	(5.2%)		7	(4.5%)	8	(5.2%)	
II	10	(6.5%)	11	(7.1%)		10	(6.5%)	11	(7.1%)		12	(7.7%)	9	(5.8%)	
III	50	(32.3%)	24	(15.5%)		19	(12.3%)	55	(35.5%)		42	(27.1%)	32	(20.6%)	
IV	28	(18.1%)	14	(9.0%)		13	(8.4%)	29	(18.7%)		23	(14.8%)	19	(12.3%)	
Mandard regression grade*					0.039					>0.05					>0.05
1	0	(0.0%)	3	(1.9%)		0	(0.0%)	3	(1.9%)		0	(0.0%)	3	(1.9%)	
2	2	(1.3%)	1	(0.6%)		2	(1.3%)	1	(0.6%)		1	(0.6%)	2	(1.3%)	
3	7	(4.5%)	9	(5.8%)		7	(4.5%)	9	(5.8%)		7	(4.5%)	9	(5.8%)	
4	12	(7.7%)	14	(9.0%)		9	(5.8%)	17	(11.0%)		14	(9.0%)	12	(7.7%)	
5	16	(10.3%)	5	(3.2%)		5	(3.2%)	16	(10.3%)		9	(5.8%)	12	(7.7%)	
Adjuvant Treatment					>0.05					>0.05					>0.05
yes	45	(29.0%)	39	(25.2%)		24	(15.5%)	47	(30.3%)		42	(27.1%)	60	(38.7%)	
no	49	(31.6%)	22	(14.2%)		25	(16.1%)	59	(38.1%)		42	(27.1%)	11	(7.1%)	

FGF = fibroblast growth factor; FGFR = fibroblast growth factor receptor; SD = standard deviation; R = resection margin; UICC = Union for International Cancer Control; NT = neoadjuvant therapy; AT = adjuvant therapy * 1 = Complete regression; 2 = Presence of rare residual cancer cells; 3 = increase of number of residual cancer cells, but fibrosis still predominant 4 = residual cancer outgrowing fibrosis; 5 = absence of regressive changes.

69 (44.5%) patients received neoadjuvant treatment. Median time of OS was 23 months (range 0.3–236.0 months) and 134 patients died during the time of observation. The rate of 3- and 5-year OS was 38.7% and 29.7%, respectively. However, Kaplan-Meier analysis shows a significant correlation between high FGF8 expression ($p = 0.006$) and reduced patients' OS, high FGF18 expression ($p = 0.026$) was significantly associated with longer patients' OS. No significance was found for FGFR4 expression and patients' survival (Figure 1d–f).

Univariable Cox proportional hazard regression revealed that high expression of FGF8 (HR 0.61, 95% CI 0.43-0.87, $p = 0.006$), advanced tumor stage (HR 2.74, 95% CI 1.37–5.50, $p = 0.005$), poor tumor differentiation (HR 0.57, 95% CI 0.39–0.83, $p = 0.003$), high lymph node ratio (HR 1.93, 95% CI 1.35–2.77, $p < 0.001$), positive resection margin (HR 2.06, 95% CI 1.36–3.10, $p < 0.001$), and receiving adjuvant treatment (HR 1.55, 95% CI 1.10–2.18, $p = 0.013$) were significantly associated with impaired patients OS, whereas high expression of FGF18 (HR 0.6, 95% CI 0.45-0.95, $p = 0.027$), negative lymph node status (HR 0.23, 95% CI 0.14–0.39, $p < 0.001$), and low UICC staging (HR 0.34, 95% CI 0.19–0.63, $p < 0.001$), were significantly associated with improved OS (Table 2). Univariable subgroup analysis revealed significant correlation for tumor size, lymph node status and UICC stage, and OS in both neoadjuvantly treated and primarily resected patients. Tumor differentiation and resection margin were found to be significantly associated with OS only in neoadjuvantly treated patients and lymph node ratio only in primarily resected patients (Table 2). Further results of univariable subgroup analysis of neoadjuvantly treated and primarily resected patients can be found in Table 2 as well.

For multivariable Cox proportional hazard regression analysis separate models for FGF8, FGF18, and FGFR4 were used. Besides FGF8, FGF18, and FGFR4 the factors age, gender, tumor differentiation, UICC stage, lymph node ratio, adjuvant treatment, and Mandard regression grade (in neoadjuvantly treated patients only) were included. In multivariable analysis, high FGF8 (HR 0.68, 95% CI 0.46–0.99, $p = 0.04$) was identified as the only independent predictor for shorter OS. Subgroup analysis of neoadjuvantly treated and primarily resected patients revealed that high FGF8 (HR 0.43, 95% CI 0.22–0.82, $p = 0.011$) and FGF18 (HR 0.44, 95% CI 0.22–0.86, $p = 0.017$) in neoadjuvantly treated but not in primarily resected patients remained as independent predictors for OS (Table 3).

4. Discussion

Despite significant improvements in diagnosis, surgical techniques and multimodal perioperative therapies over the last few years, survival rates of patients suffering from adenocarcinoma of the esophago-gastric junction remain poor. To investigate and understand the pathophysiological mechanisms of tumorigenesis in these cancers might be the key to better therapies and therefore to improved survival rates.

In this study, we show the utility of FGF8 and FGF18 as independent prognostic markers in AEG for the first time.

FGF and FGFR as targets have recently found their way into anti-cancer therapy, especially to overcome chemo- and radio-resistance [14]. Recently, our group investigated the role of FGF8, FGF18, and FGFR4 in colo-(rectal) and hepatocellular cancer [23,24,31,41,44]. However, data of FGF and FGFR expression in adenocarcinomas of the esophago-gastric junction were limited until now. Therefore, a TCGA analysis and immunohistochemistry, including tumor tissue from patients before and after neoadjuvant treatment, was performed to investigate the prognostic role of expression of FGF8, FGF18, and FGFR4 in adenocarcinomas of the esophago-gastric junction.

Table 2. Univariable Cox regression analysis estimating the influence of FGF 8, FGF 18 and FGFR 4 expression and clinicopathological parameters on overall survival for patients with adenocarcinoma of the esophago-gastric junction.

Factors	All Patients			Neoadjuvantly Treated Patients			Primarily Resected Patients		
	Hazard Ratio	95% CI	p Value	Hazard Ratio	95% CI	p Value	Hazard Ratio	95% CI	p Value
FGF 8 (ref.: high)									
low/absent	0.61	(0.43–0.87)	0.006	0.43	(0.27–0.83)	0.008	0.77	(0.48–1.24)	0.287
FGF 18 (ref.: low/absent)									
high	0.66	(0.45–0.95)	0.027	0.54	(0.30–0.97)	0.039	0.80	(0.49–1.29)	0.363
FGFR 4 (ref.: high)									
low/absent	0.9	(0.64–1.27)	0.562	1.05	(0.61–1.79)	0.871	0.89	(0.56–1.40)	0.615
Age (years)	1.00	(0.99–1.02)	0.887	0.98	(0.96–1.01)	0.217	1.01	(0.98–1.03)	0.341
Sex (ref.: Male)									
female	0.87	(0.56–1.35)	0.529	1.27	(0.62–2.63)	0.511	0.89	(0.51–1.54)	0.672
Neoadjuvant treatment (ref.: no)									
yes	0.81	(0.57–1.14)	0.224	/	/	/	/	/	/
(y)pT (ref.: T3)									
0	0.22	(0.03–1.60)	0.135	0.22	(0.03–1.64)	0.139	/	/	/
1	0.17	(0.02–1.22)	0.080	0.16	(0.02–1.20)	0.075	/	/	/
2	0.58	(0.39–0.86)	0.008	0.63	(0.33–1.20)	0.160	0.54	(0.31–0.91)	0.020
4	2.74	(1.37–5.50)	0.005	29.63	(5.72–153.40)	<0.001	1.69	(0.72–3.99)	0.228
(y)pN (ref.: N3)									
0	0.23	(0.14–0.39)	<0.001	0.32	(0.15–0.69)	0.004	0.20	(0.10–0.39)	<0.001
1	0.40	(0.25–0.64)	<0.001	0.51	(0.24–1.07)	0.075	0.36	(0.19–0.68)	0.002
2	0.63	(0.40–1.00)	0.049	0.81	(0.38–1.73)	0.585	0.54	(0.31–0.97)	0.040
Tumor differentiation (ref.: 3)									
0 + 1	0.44	(0.18–1.08)	0.074	0.44	(0.13–1.47)	0.185	0.45	(0.11–1.87)	0.273
2	0.57	(0.39–0.83)	0.003	0.53	(0.30–0.93)	0.027	0.64	(0.39–1.06)	0.080
Mandard regression grade * (ref.: 3 + 4)									
1 + 2	/	/	/	1.40	(0.49–4.02)	0.529	/	/	/
5	/	/	/	2.28	(0.76–6.86)	0.143	/	/	/
Lymph node ratio (ref.: <0.3)									
≥0.3	1.93	(1.35–2.77)	<0.001	1.62	(0.91–2.89)	0.100	2.05	(1.29–3.25)	0.002
R (ref.: 0)									
1	2.06	(1.36–3.10)	<0.001	3.43	(1.79–6.56)	<0.001	1.51	(0.88–2.58)	0.134
UICC Staging (ref.: II + III + IV)									
0 + I	0.34	(0.19–0.63)	<0.001	2.91	(1.14–7.43)	0.026	0.34	(0.15–0.75)	0.007°
Adjuvant treatment (ref.: no)									
yes	1.55	(1.10–2.18)	0.013	1.33	(0.75–2.38)	0.330	1.58	(0.99–2.50)	0.052

CI = confidence interval; FGF = fibroblast growth factor; FGFR = fibroblast growth factor receptor; R = resection margin; UICC = Union for International Cancer Control; * 1 = Complete regression; 2 = Presence of rare residual cancer cells; 3 = increase of number of residual cancer cells; 4 = residual cancer outgrowing fibrosis; 5 = absence of regressive changes; ° UICC I + II vs. III + IV in primarily resected patients.

Table 3. Multivariable Cox regression analysis estimating the influence of FGF 8, FGF 18 and FGFR 4 expression on overall survival for patients with adenocarcinoma of the esophago-gastric junction.

Factors	All Patients			Neoadjuvantly Treated Patients			Primarily Resected Patients		
	Hazard Ratio	95% CI	p Value	Hazard Ratio	95% CI	p Value	Hazard Ratio	95% CI	p Value
FGF 8 (ref.: high)									
low/absent	0.68	(0.46–0.99)	0.042	0.43	(0.22–0.82)	0.011	1.04	(0.63–1.72)	0.882
FGF 18 (ref.: low/absent)									
high	0.71	(0.48–1.04)	0.08	0.44	(0.22–0.86)	0.017	0.81	(0.49–1.33)	0.408
FGFR 4 (ref.: high)									
low/absent	1.04	(0.72–1.50)	0.834	1.02	(0.58–1.81)	0.945	1.03	(0.63–1.67)	0.908

CI = confidence interval; FGF = fibroblast growth factor; FGFR = fibroblast growth factor receptor.

The analysis revealed significantly shorter OS when tumors were highly abundant in FGF8 in the cohort of all patients ($p = 0.04$) and the subgroup of neoadjuvantly treated patients ($p = 0.011$). This goes in good accordance with published data: overexpression of androgen related FGF8 is known to play a crucial role in prostate and colorectal cancer. Furthermore, Harpain et al. recently found that FGF8 induced therapy resistance in neoadjuvantly treated rectal cancer patients [33,45]. In this study FGF8 expression was found to correlate with Mandard regression grade, suggesting a role of the growth factor in therapy response. Investigating this correlation in a larger cohort would be of high interest.

With regard to FGF18, our observations in AEG contradict older reports: our previously published findings on FGF18 expression demonstrated increased tumor cell survival and migration caused by FGF18 expression in colorectal cancer [25]. In gastric cancer cell lines, Zhang et al. published data of poor survival when tumors were high in FGF18 and related it this to ERK-MAPK signaling [23,24,46]. In contrast to these results, our data show a significantly improved OS in neoadjuvantly treated patients with adenocarcinomas of the esophago-gastric junction when the tumors are high in FGF18. Analysis of the TCGA data set of AEG patients showed similar results, supporting the conclusion that FGF18 is a positive prognostic factor. Previously, we published data on ETV1 and MK2 expression in adenocarcinomas of the esophago-gastric junction, showing that nuclear overexpression of ETV1 was associated with significantly better patients OS [40]. A potential explanation for the protective role of ETV1 and FGF18 overexpression might be found in the ERK-MAPK signaling pathway as mentioned above. However, our findings remain controversial and need further investigation.

Among all the samples tested, overexpression of FGF8 and FGF18 was found in 31 (20%) cases. However, analysis of a potential correlation between FGF8 and FGF18 overexpression in tumor tissue found no significant correlation between these two markers (Kendall's rank correlation).

Interestingly, no significant correlation between the overexpression of FGFR4 and OS was found in our analysis. Based on our previously published data on the prognostic role of FGFR4 in colorectal cancer and data on FGFR4 expression on esophageal squamous cell carcinoma, one would anticipate finding alike results in patients with adenocarcinomas of the esophago-gastric junction [41,44,47]. However, our findings are supported by our TCGA analysis of FGFR4 expression in patients with adenocarcinomas of the esophago-gastric junction, showing no significant correlation between expression and OS.

Even though the results of this study demonstrate that FGF8 and FGF18 are independent prognostic factors in patients with adenocarcinomas of the esophago-gastric junction, our study has certain limitations: one is the potential selection bias which was inevitably associated with only partial availability of tumor tissue, especially diagnostic biopsies before neoadjuvantly treated patients. Another limitation might be the retrospective nature of this single center research study. This is balanced by the fact that patient recruitment is ongoing and the patient database is maintained prospectively. Regarding the scoring method used in this study, one potential weak point has to be mentioned. Until now, no data exists on the expression of FGF8, FGF18, and FGFR4 on adenocarcinomas of the esophago-gastric junction investigated by immunohistochemistry and no signifier scoring method is available. Therefore, based on recently published recommendations for IHC scoring, conventional visual scoring based on our previously gained experiences using these antibodies on CRC tissue was used as appropriate [48–50]. However, further investigations on the expression of FGF8, FGF18, and FGFR4, using other methods including digital image analysis are urgently needed.

5. Conclusions

In conclusion, this is the first study that investigated the prognostic role of FGF8 and FGF18 including a subgroup analysis of neoadjuvantly treated and primarily resected patients using a preliminary TCGA analysis and immunohistochemistry demonstrating FGF8 and FGF18 as independent prognostic factors in resectable AEG. Furthermore, this is the first study comparing the expression of FGF8, FGF18, and FGFR4 in tumor tissue available before and after neoadjuvant treatment. However, due to the unexpected results of FGF18 overexpression and its protective nature, further investigations evaluating FGF8 and FGF18 as potential therapeutic targets are urgently needed.

Author Contributions: Conceptualization, G.J., B.M. and S.F.S.; methodology, G.J., X.H. and F.H.; software, D.W., G.T., T.M.; validation, B.M., S.F.S.; formal analysis, W.D. and T.M.; investigation, J.G., B.M.; resources, B.M., S.F.S.; data curation, G.J., F.H.; writing—original draft preparation, G.J.; writing—review and editing, G.J. and B.M.; visualization D.W., G.T. and T.M.; supervision, B.M., S.F.S.; project administration, B.M., S.F.S.

Acknowledgments: The authors would like to thank Andrea Beer for her help to acquire diagnostic biopsies.

References

1.   Ferlay, J.; Soerjomataram, I.; Dikshit, R.; Eser, S.; Mathers, C.; Rebelo, M.; Parkin, D.M.; Forman, D.; Bray, F. Cancer incidence and mortality worldwide: Sources, methods and major patterns in globocan 2012. *Int. J. Cancer* **2015**, *136*, E359–E386. [CrossRef] [PubMed]

2.   Burmeister, B.H.; Smithers, B.M.; Gebski, V.; Fitzgerald, L.; Simes, R.J.; Devitt, P.; Ackland, S.; Gotley, D.C.; Joseph, D.; Millar, J.; et al. Surgery alone versus chemoradiotherapy followed by surgery for resectable cancer of the oesophagus: A randomised controlled phase iii trial. *Lancet Oncol.* **2005**, *6*, 659–668. [CrossRef]

3.   Reynolds, J.V.; Muldoon, C.; Hollywood, D.; Ravi, N.; Rowley, S.; O'Byrne, K.; Kennedy, J.; Murphy, T.J. Long-term outcomes following neoadjuvant chemoradiotherapy for esophageal cancer. *Ann. Surg.* **2007**, *245*, 707–716. [CrossRef] [PubMed]

4.   Christein, J.D.; Hollinger, E.F.; Millikan, K.W. Prognostic factors associated with resectable carcinoma of the esophagus. *Am. Surg.* **2002**, *68*, 258–262; discussion 262–263. [PubMed]

5.   Gertler, R.; Stein, H.J.; Langer, R.; Nettelmann, M.; Schuster, T.; Hoefler, H.; Siewert, J.R.; Feith, M. Long-term outcome of 2920 patients with cancers of the esophagus and esophagogastric junction: Evaluation of the new union internationale contre le cancer/american joint cancer committee staging system. *Ann. Surg.* **2011**, *253*, 689–698. [CrossRef] [PubMed]

6.   Ku, G.Y.; Ilson, D.H. Esophagogastric cancer: Targeted agents. *Cancer Treat. Rev.* **2010**, *36*, 235–248. [CrossRef] [PubMed]

7.   Rassouli, F.B.; Matin, M.M.; Saeinasab, M. Cancer stem cells in human digestive tract malignancies. *Tumour Biol.* **2016**, *37*, 7–21. [CrossRef]

8.   Schulenburg, A.; Ulrich-Pur, H.; Thurnher, D.; Erovic, B.; Florian, S.; Sperr, W.R.; Kalhs, P.; Marian, B.; Wrba, F.; Zielinski, C.C.; et al. Neoplastic stem cells: A novel therapeutic target in clinical oncology. *Cancer* **2006**, *107*, 2512–2520. [CrossRef]

9.   Borah, A.; Raveendran, S.; Rochani, A.; Maekawa, T.; Kumar, D.S. Targeting self-renewal pathways in cancer stem cells: Clinical implications for cancer therapy. *Oncogenesis* **2015**, *4*, e177. [CrossRef]

10. Honing, J.; Pavlov, K.V.; Mul, V.E.; Karrenbeld, A.; Meijer, C.; Faiz, Z.; Smit, J.K.; Hospers, G.A.; Burgerhof, J.G.; Kruyt, F.A.; et al. Cd44, shh and sox2 as novel biomarkers in esophageal cancer patients treated with neoadjuvant chemoradiotherapy. *Radiother Oncol.* **2015**, *117*, 152–158. [CrossRef]

11. Sui, Y.P.; Jian, X.P.; Ma, L.I.; Xu, G.Z.; Liao, H.W.; Liu, Y.P.; Wen, H.C. Prognostic value of cancer stem cell marker cd133 expression in esophageal carcinoma: A meta-analysis. *Mol. Clin. Oncol.* **2016**, *4*, 77–82. [CrossRef] [PubMed]

12. Hang, D.; Dong, H.C.; Ning, T.; Dong, B.; Hou, D.L.; Xu, W.G. Prognostic value of the stem cell markers cd133 and abcg2 expression in esophageal squamous cell carcinoma. *Dis. Esophagus* **2012**, *25*, 638–644. [CrossRef] [PubMed]

13. Qian, X.; Tan, C.; Wang, F.; Yang, B.; Ge, Y.; Guan, Z.; Cai, J. Esophageal cancer stem cells and implications for future therapeutics. *Onco. Targets Ther.* **2016**, *9*, 2247–2254. [PubMed]

14. Heinzle, C.; Sutterluty, H.; Grusch, M.; Grasl-Kraupp, B.; Berger, W.; Marian, B. Targeting fibroblast-growth-factor-receptor-dependent signaling for cancer therapy. *Expert Opin. Ther. Targets* **2011**, *15*, 829–846. [CrossRef] [PubMed]

15. Beenken, A.; Mohammadi, M. The fgf family: Biology, pathophysiology and therapy. *Nat. Rev. Drug Discov.* **2009**, *8*, 235–253. [CrossRef] [PubMed]

16. Knights, V.; Cook, S.J. De-regulated fgf receptors as therapeutic targets in cancer. *Pharmacol. Ther.* **2010**, *125*, 105–117. [CrossRef]
17. Powers, C.J.; McLeskey, S.W.; Wellstein, A. Fibroblast growth factors, their receptors and signaling. *Endocr. Relat. Cancer* **2000**, *7*, 165–197. [CrossRef] [PubMed]
18. Kono, S.A.; Marshall, M.E.; Ware, K.E.; Heasley, L.E. The fibroblast growth factor receptor signaling pathway as a mediator of intrinsic resistance to egfr-specific tyrosine kinase inhibitors in non-small cell lung cancer. *Drug Resist. Updat.* **2009**, *12*, 95–102. [CrossRef] [PubMed]
19. Motomura, K.; Hagiwara, A.; Komi-Kuramochi, A.; Hanyu, Y.; Honda, E.; Suzuki, M.; Kimura, M.; Oki, J.; Asada, M.; Sakaguchi, N.; et al. An fgf1:Fgf2 chimeric growth factor exhibits universal fgf receptor specificity, enhanced stability and augmented activity useful for epithelial proliferation and radioprotection. *Biochim. Biophys. Acta* **2008**, *1780*, 1432–1440. [CrossRef]
20. Pardo, O.E.; Arcaro, A.; Salerno, G.; Raguz, S.; Downward, J.; Seckl, M.J. Fibroblast growth factor-2 induces translational regulation of bcl-xl and bcl-2 via a mek-dependent pathway: Correlation with resistance to etoposide-induced apoptosis. *J. Biol. Chem.* **2002**, *277*, 12040–12046. [CrossRef]
21. Pardo, O.E.; Lesay, A.; Arcaro, A.; Lopes, R.; Ng, B.L.; Warne, P.H.; McNeish, I.A.; Tetley, T.D.; Lemoine, N.R.; Mehmet, H.; et al. Fibroblast growth factor 2-mediated translational control of iaps blocks mitochondrial release of smac/diablo and apoptosis in small cell lung cancer cells. *Mol. Cell Biol.* **2003**, *23*, 7600–7610. [CrossRef] [PubMed]
22. Roidl, A.; Berger, H.J.; Kumar, S.; Bange, J.; Knyazev, P.; Ullrich, A. Resistance to chemotherapy is associated with fibroblast growth factor receptor 4 up-regulation. *Clin. Cancer Res.* **2009**, *15*, 2058–2066. [CrossRef] [PubMed]
23. Koneczny, I.; Schulenburg, A.; Hudec, X.; Knofler, M.; Holzmann, K.; Piazza, G.; Reynolds, R.; Valent, P.; Marian, B. Autocrine fibroblast growth factor 18 signaling mediates wnt-dependent stimulation of cd44-positive human colorectal adenoma cells. *Mol. Carcinog.* **2015**, *54*, 789–799. [CrossRef] [PubMed]
24. Sonvilla, G.; Allerstorfer, S.; Stattner, S.; Karner, J.; Klimpfinger, M.; Fischer, H.; Grasl-Kraupp, B.; Holzmann, K.; Berger, W.; Wrba, F.; et al. Fgf18 in colorectal tumour cells: Autocrine and paracrine effects. *Carcinogenesis* **2008**, *29*, 15–24. [CrossRef] [PubMed]
25. Sonvilla, G.; Allerstorfer, S.; Heinzle, C.; Stattner, S.; Karner, J.; Klimpfinger, M.; Wrba, F.; Fischer, H.; Gauglhofer, C.; Spiegl-Kreinecker, S.; et al. Fibroblast growth factor receptor 3-iiic mediates colorectal cancer growth and migration. *Br. J. Cancer* **2010**, *102*, 1145–1156. [CrossRef] [PubMed]
26. Bange, J.; Prechtl, D.; Cheburkin, Y.; Specht, K.; Harbeck, N.; Schmitt, M.; Knyazeva, T.; Muller, S.; Gartner, S.; Sures, I.; et al. Cancer progression and tumor cell motility are associated with the fgfr4 arg(388) allele. *Cancer Res.* **2002**, *62*, 840–847.
27. Heinzle, C.; Erdem, Z.; Paur, J.; Grasl-Kraupp, B.; Holzmann, K.; Grusch, M.; Berger, W.; Marian, B. Is fibroblast growth factor receptor 4 a suitable target of cancer therapy? *Curr. Pharm. Des.* **2014**, *20*, 2881–2898. [CrossRef]
28. Zhang, X.; Ibrahimi, O.A.; Olsen, S.K.; Umemori, H.; Mohammadi, M.; Ornitz, D.M. Receptor specificity of the fibroblast growth factor family. The complete mammalian fgf family. *J. Biol. Chem.* **2006**, *281*, 15694–15700. [CrossRef]
29. Brewer, J.R.; Mazot, P.; Soriano, P. Genetic insights into the mechanisms of fgf signaling. *Genes Dev.* **2016**, *30*, 751–771. [CrossRef]
30. Tickle, C.; Munsterberg, A. Vertebrate limb development–the early stages in chick and mouse. *Curr. Opin. Genet. Dev.* **2001**, *11*, 476–481. [CrossRef]
31. Gauglhofer, C.; Sagmeister, S.; Schrottmaier, W.; Fischer, C.; Rodgarkia-Dara, C.; Mohr, T.; Stattner, S.; Bichler, C.; Kandioler, D.; Wrba, F.; et al. Up-regulation of the fibroblast growth factor 8 subfamily in human hepatocellular carcinoma for cell survival and neoangiogenesis. *Hepatology* **2011**, *53*, 854–864. [CrossRef] [PubMed]
32. Mattila, M.M.; Harkonen, P.L. Role of fibroblast growth factor 8 in growth and progression of hormonal cancer. *Cytokine Growth Factor Rev.* **2007**, *18*, 257–266. [CrossRef] [PubMed]
33. Harpain, F.; Ahmed, M.A.; Hudec, X.; Timelthaler, G.; Jomrich, G.; Mullauer, L.; Selzer, E.; Dorr, W.; Bergmann, M.; Holzmann, K.; et al. Fgf8 induces therapy resistance in neoadjuvantly radiated rectal cancer. *J. Cancer Res. Clin. Oncol.* **2019**, *145*, 77–86. [CrossRef] [PubMed]

34. Colaprico, A.; Silva, T.C.; Olsen, C.; Garofano, L.; Cava, C.; Garolini, D.; Sabedot, T.S.; Malta, T.M.; Pagnotta, S.M.; Castiglioni, I.; et al. Tcgabiolinks: An r/bioconductor package for integrative analysis of tcga data. *Nucleic Acids Res.* **2016**, *44*, e71. [CrossRef] [PubMed]

35. Kassambara, A.; Kosinski, M.; Biecek, P.; Fabian, S. Survminer: Drawing survival curves using "ggplot2". Available online: https://rpkgs.datanovia.com/survminer/index.html (accessed on 10 June 2019).

36. Chen, Y.; Lun, A.; McCarthy, D.; Robinson, M.; Phipson, B.; Hu, Y.; Zhou, X.; Robinson, M.D.; Smyth, G.K. Edger: Empirical analysis of digital gene expression data in r. Available online: https://bioconductor.org/packages/release/bioc/html/edgeR.html (accessed on 10 June 2019).

37. Kanehisa, M.; Goto, S. Kegg: Kyoto encyclopedia of genes and genomes. *Nucleic Acids Res.* **2000**, *28*, 27–30. [CrossRef] [PubMed]

38. Luo, W.; Brouwer, C. Pathview: An r/bioconductor package for pathway-based data integration and visualization. *Bioinformatics* **2013**, *29*, 1830–1831. [CrossRef] [PubMed]

39. Mandard, A.M.; Dalibard, F.; Mandard, J.C.; Marnay, J.; Henry-Amar, M.; Petiot, J.F.; Roussel, A.; Jacob, J.H.; Segol, P.; Samama, G.; et al. Pathologic assessment of tumor regression after preoperative chemoradiotherapy of esophageal carcinoma. Clinicopathologic correlations. *Cancer* **1994**, *73*, 2680–2686. [CrossRef]

40. Jomrich, G.; Maroske, F.; Stieger, J.; Preusser, M.; Ilhan-Mutlu, A.; Winkler, D.; Kristo, I.; Paireder, M.; Schoppmann, S.F. Mk2 and etv1 are prognostic factors in esophageal adenocarcinomas. *J. Cancer* **2018**, *9*, 460–468. [CrossRef]

41. Ahmed, M.A.; Selzer, E.; Dorr, W.; Jomrich, G.; Harpain, F.; Silberhumer, G.R.; Mullauer, L.; Holzmann, K.; Grasl-Kraupp, B.; Grusch, M.; et al. Fibroblast growth factor receptor 4 induced resistance to radiation therapy in colorectal cancer. *Oncotarget* **2016**, *7*, 69976–69990. [CrossRef]

42. R Development Core Team. *R: A Language and Environment for Statistical Computing*; R Foundation for Statistical Computing: Vienna, Austria, 2018; Available online: https://www.gbif.org/tool/81287/r-a-language-and-environment-for-statistical-computing (accessed on 10 June 2019).

43. Therneau, T. A Package for Survival Analysis in S. Version 2.38. 2015. Available online: https://cran.r-project.org/web/packages/survival/citation.html (accessed on 10 June 2019).

44. Heinzle, C.; Gsur, A.; Hunjadi, M.; Erdem, Z.; Gauglhofer, C.; Stattner, S.; Karner, J.; Klimpfinger, M.; Wrba, F.; Reti, A.; et al. Differential effects of polymorphic alleles of fgf receptor 4 on colon cancer growth and metastasis. *Cancer Res.* **2012**, *72*, 5767–5777. [CrossRef]

45. Dorkin, T.J.; Robinson, M.C.; Marsh, C.; Bjartell, A.; Neal, D.E.; Leung, H.Y. Fgf8 over-expression in prostate cancer is associated with decreased patient survival and persists in androgen independent disease. *Oncogene* **1999**, *18*, 2755–2761. [CrossRef] [PubMed]

46. Zhang, J.; Zhou, Y.; Huang, T.; Wu, F.; Pan, Y.; Dong, Y.; Wang, Y.; Chan, A.K.Y.; Liu, L.; Kwan, J.S.H.; et al. Fgf18, a prominent player in fgf signaling, promotes gastric tumorigenesis through autocrine manner and is negatively regulated by mir-590-5p. *Oncogene* **2019**, *38*, 33–46. [CrossRef] [PubMed]

47. Shim, H.J.; Shin, M.H.; Kim, H.N.; Kim, J.H.; Hwang, J.E.; Bae, W.K.; Chung, I.J.; Cho, S.H. The prognostic significance of fgfr4 gly388 polymorphism in esophageal squamous cell carcinoma after concurrent chemoradiotherapy. *Cancer Res. Treat.* **2016**, *48*, 71–79. [CrossRef] [PubMed]

48. Meyerholz, D.K.; Beck, A.P. Principles and approaches for reproducible scoring of tissue stains in research. *Lab. Invest.* **2018**, *98*, 844–855. [CrossRef]

49. Meyerholz, D.K.; Beck, A.P. Fundamental concepts for semiquantitative tissue scoring in translational research. *ILAR J.* **2018**, *59*, 13–17. [CrossRef] [PubMed]

50. Lin, F.; Prichard, J. (Eds.) *Handbook of Practical Immunohistochemistry: Frequently Asked Questions*, 2nd ed.; Springer: New York, NY, USA, 2015; p. xv. 764p.

Permissions

List of Contributors

Anuja Neve, Jessica Migliavacca, Charles Capdeville, Marc Thomas Schönholzer, Alexandre Gries, Karthiga Santhana Kumar, Michael Grotzer and Martin Baumgartner
Department of Oncology, University Children's Hospital Zürich, CH-8032 Zürich, Switzerland

Min Ma
Faculty of Biology and Medicine, University of Lausanne, Biochemistry, CH-1066 Epalinges, Switzerland

Jessica Guerra, Paola Chiodelli and Chiara Tobia
Department of Molecular and Translational Medicine, University of Brescia, 25123 Brescia, Italy

Claudia Gerri
Department of Molecular and Translational Medicine, University of Brescia, 25123 Brescia, Italy
Francis Crick Institute, London NW1 1AT, UK

Marco Presta
Department of Molecular and Translational Medicine, University of Brescia, 25123 Brescia, Italy
Italian Consortium for Biotechnology (CIB), 25123 Brescia, Italy

Monica Nanni, Danilo Ranieri, Flavia Persechino and Francesca Belleudi
Laboratory affiliated to Istituto Pasteur Italia—Fondazione Cenci Bolognetti, Department of Clinical and Molecular Medicine, Sapienza University of Rome, 00185 Rome, Italy

Maria Rosaria Torrisi
Laboratory affiliated to Istituto Pasteur Italia—Fondazione Cenci Bolognetti, Department of Clinical and Molecular Medicine, Sapienza University of Rome, 00185 Rome, Italy
S. Andrea University Hospital, 00189 Rome, Italy

Marta Latko, Aleksandra Czyrek, Natalia Porębska, Marika Kucińska, Jacek Otlewski, Małgorzata Zakrzewska and Łukasz Opaliński
Department of Protein Engineering, Faculty of Biotechnology, University of Wroclaw, Joliot-Curie 14a, 50-383 Wroclaw, Poland

Burcu Emine Celik-Selvi, Astrid Stütz, Christoph-Erik Mayer, Jihen Salhi, Gerald Siegwart and Hedwig Sutterlüty
Institute of Cancer Research, Department of Medicine I, Comprehensive Cancer Center, Medica University of Vienna, A-1090 Vienna, Austria

Shuyan Dai, Zhan Zhou, Zhuchu Chen and Yongheng Chen
NHC Key Laboratory of Cancer Proteomics & Laboratory of Structural Biology, Xiangya Hospital, Central South University, Changsha 410008, Hunan, China

Guangyu Xu
Key Laboratory of Chemical Biology and Traditional Chinese Medicine Research (Ministry of Education), College of Chemistry and Chemical Engineering, Hunan Normal University, Changsha 410081, Hunan, China

Patrick M. K. Tang
Department of Anatomical and Cellular Pathology, State Key Laboratory of Translational Oncology, Prince of Wales Hospital, The Chinese University of Hong Kong, Hong Kong, China

Jun Yu
Institute of Digestive Disease, State Key Laboratory of Digestive Disease, The Chinese University of Hong Kong, Hong Kong, China
Department of Medicine and Therapeutics, The Chinese University of Hong Kong, Hong Kong, China

Alfred S. L. Cheng
School of Biomedical Sciences, The Chinese University of Hong Kong, Hong Kong, China

Jinglin Zhang, Yuhang Zhou, Wei Kang and Ka Fai To
Department of Anatomical and Cellular Pathology, State Key Laboratory of Translational Oncology, Prince of Wales Hospital, The Chinese University of Hong Kong, Hong Kong, China
Institute of Digestive Disease, State Key Laboratory of Digestive Disease, The Chinese University of Hong Kong, Hong Kong, China
Li Ka Shing Institute of Health Science, Sir Y.K. Pao Cancer Center, The Chinese University of Hong Kong, Hong Kong, China

Katalin Csanaky and Lars Klimaschewski
Division of Neuroanatomy, Medical University of Innsbruck, 6020 Innsbruck, Austria

Michael W. Hess
Division of Histology and Embryology, Medical University of Innsbruck, 6020 Innsbruck, Austria

Ana Jimenez-Pascual and Florian A. Siebzehnrubl
European Cancer Stem Cell Research Institute, Cardiff University School of Biosciences, Cardiff CF24 4HQ, UK

Yong Teng
Department of Oral Biology and Diagnostic Sciences, Dental College of Georgia, Augusta University, Augusta, GA 30912, USA
Georgia Cancer Center, Department of Biochemistry and Molecular Biology, Medical College of Georgia, Augusta University, Augusta, GA 30912, USA
Department of Medical Laboratory, Imaging and Radiologic Sciences, College of Allied Health, Augusta University, Augusta, GA 30912, USA

Liwei Lang
Department of Oral Biology and Diagnostic Sciences, Dental College of Georgia, Augusta University, Augusta, GA 30912, USA

Else Munthe, Iwona Grad and Eva Wessel Stratford
Department of Tumor Biology, Institute of Cancer Research, the Norwegian Radium Hospital, Oslo University Hospital, 0379 Oslo, Norway

Robert Hanes
Department of Tumor Biology, Institute of Cancer Research, the Norwegian Radium Hospital, Oslo University Hospital, 0379 Oslo, Norway
Norwegian Cancer Genomics Consortium, 0379 Oslo, Norway

Jianhua Han
Centre for Cancer Biomarkers (CCBIO), Department of Clinical Sciences, University of Bergen, 5021 Bergen, Norway

Ida Karlsen
Centre for Cancer Biomarkers (CCBIO), Department of Clinical Sciences, University of Bergen, 5021 Bergen, Norway
KinN Therapeutics AS, 5021 Bergen, Norway

Emmet McCormack
Centre for Cancer Biomarkers (CCBIO), Department of Clinical Sciences, University of Bergen, 5021 Bergen, Norway

Department of Internal Medicine, Hematology Section, Haukeland University Hospital, 5021 Bergen, Norway

Leonardo A. Meza-Zepeda
Department of Tumor Biology, Institute of Cancer Research, the Norwegian Radium Hospital, Oslo University Hospital, 0379 Oslo, Norway
Norwegian Cancer Genomics Consortium, 0379 Oslo, Norway
Genomics Core Facility, Department of Core Facilities, Institute of Cancer Research, the Norwegian Radium Hospital, Oslo University Hospital, 0379 Oslo, Norway

Ola Myklebost
Department of Tumor Biology, Institute of Cancer Research, the Norwegian Radium Hospital, Oslo University Hospital, 0379 Oslo, Norway
Norwegian Cancer Genomics Consortium, 0379 Oslo, Norway
Department of Clinical Science, University of Bergen, 5020 Bergen, Norway

Aroosha Raja and Farhan Haq
Department of Biosciences, Comsats University, Islamabad 45550, Pakistan

Inkeun Park
Division of Medical Oncology, Department of Internal Medicine, Gachon University Gil Medical Center, Incheon 21565, Korea

Sung-Min Ahn
Division of Medical Oncology, Department of Internal Medicine, Gachon University Gil Medical Center, Incheon 21565, Korea
Department of Genome Medicine and Science, College of Medicine, Gachon University, Incheon 21565, Korea

Patrycja Szybowska
Department of Molecular Cell Biology, Institute for Cancer Research, The Norwegian Radium Hospital, Oslo University Hospital, Montebello, 0379 Oslo, Norway
Centre for Cancer Cell Reprogramming, Institute of Clinical Medicine, Faculty of Medicine, University of Oslo, Montebello, 0379 Oslo, Norway

Antoni Wiedlocha
Department of Molecular Cell Biology, Institute for Cancer Research, The Norwegian Radium Hospital, Oslo University Hospital, Montebello, 0379 Oslo, Norway
Centre for Cancer Cell Reprogramming, Institute of Clinical Medicine, Faculty of Medicine, University of Oslo, Montebello, 0379 Oslo, Norway
Military Institute of Hygiene and Epidemiology, 01-163Warsaw, Poland

Ellen Margrethe Haugsten, Michal Kostas and Jørgen Wesche
Centre for Cancer Cell Reprogramming, Institute of Clinical Medicine, Faculty of Medicine, University of Oslo, Montebello, 0379 Oslo, Norway
Department of Tumor Biology, Institute for Cancer Research, The Norwegian Radium Hospital, Oslo University Hospital, Montebello, 0379 Oslo, Norway

Gerd Jomrich and Felix Harpain
Department of Surgery, Medical University of Vienna and Gastroesophageal Tumor Unit, Comprehensive Cancer Center (CCC), Spitalgasse 23, 1090 Vienna, Austria
Department of Medicine I, Institute of Cancer Research, Medical University of Vienna, Borschkegasse 8a, 1090 Vienna, Austria

Sebastian F. Schoppmann
Department of Surgery, Medical University of Vienna and Gastroesophageal Tumor Unit, Comprehensive Cancer Center (CCC), Spitalgasse 23, 1090 Vienna, Austria

Xenia Hudec, Gerald Timelthaler, Thomas Mohr and Brigitte Marian
Department of Medicine I, Institute of Cancer Research, Medical University of Vienna, Borschkegasse 8a, 1090 Vienna, Austria

Daniel Winkler
Department of Statistics and Operations Research, University of Vienna, Oskar Morgenstern Platz 1, 1090 Vienna, Austria

Index

Printed in the USA
CPSIA information can be obtained
at www.ICGtesting.com
JSHW051407091023
49903JS00006B/310